From Decoding Turbulence to Unveiling the Fingerprint of Climate Change

Hans von Storch

From Decoding Turbulence to Unveiling the Fingerprint of Climate Change

Klaus Hasselmann—Nobel Prize Winner in Physics 2021

 Springer

Hans von Storch
Max Planck Institute for Meteorology
Hamburg, Germany

ISBN 978-3-030-91718-0   ISBN 978-3-030-91716-6   (eBook)
https://doi.org/10.1007/978-3-030-91716-6

Preface

This book brings together how we, friends, colleagues, and co-workers, see the scientist Klaus Hasselmann. We have tried to assess the legacy of his work, oceanography, climate science, economy, and physics.

The original momentum for this project came from Ola M. Johannessen. Hans von Storch designed the concept of the book, supervised the composition, and holds the overall-reponsibility of the book.

One group of authors has contributed to various aspects of the book, in particular to Chaps. 1 and 3; others also to the interviews in Chap. 4, and the library of MPI to the hopefully eventually complete publication list in Chap. 5: Ola M. Johannessen, Susanne Hasselmann, Gerbrand Komen, Peter Lemke, Dirk Olbers, Carola Kauhs, Martin Heimann, Hans von Storch, Dmitry Kovalevsky, and Lennart Bengtsson.

Another group has provided "personal accounts" in Chap. 3, additional to the first group, these are (in random order) Hans Graf, Jin-Song von Storch, Jürgen Willebrand, Achim Stössel, Mojib Latif, Jörg Wolff, Christoph Heinze, Ulrich Cubasch, Ben Santer, Patrick Heimbach, Robert Sausen, Gabriele Hegerl, Luigi Cavaleri, Kristina Katsaros, Peter Janssen, Jürgen Sündermann, Klaus Fraedrich, Hartmut Graßl, and Udo Simonis. These accounts were provided "upon invitation", and there are certainly many more companions of Klaus, whose account would have been interesting and even entertaining. But obviously, the book must have a limited size; thus, we owe an apology to those, whom we had not invited.

There was some funding needed for this project; a little for technical support, but in particular for having the book printed by Springer Publisher.

These costs were generously covered by the Max Planck Institute of Meteorology in Hamburg.

Cover-drawing: Victor Ocana. Two significant work contributions need to be recognized—the final check of the various references by Carola Kauhs, and the final reading and detecting of glitches of all kinds by Dirk Olbers; quite possibly Hans von Storch added new glitches when trying to do the final corrections. After completion, we were informed that Klaus Hasselmann was awarded the Nobel prize in Physics of 2021.

Hamburg, Germany   Hans von Storch

Contents

List of Facsimiles

1

Klaus Hasselmann—His Scientific Footprints and Achievements

1.1 Overview

Klaus Hasselmann was born in Hamburg in 1931. His family fled to England in 1934 because of the Nazis, so he grew up in an English-speaking environment, and returned to Hamburg after the war, where he studied physics, started a family, and became an innovative researcher. Later, he spent several years in the United States of America, but always returned to Hamburg, where he became the founding director of the Max-Planck-Institut für Meteorologie in 1975. His Institute soon became one of the world's leading research facilities in the field of climate science. He retired in 2000, but continued his work in climate science as a "grey eminence" in the background, whilst his heart and mind turned to particle physics. He recently turned 90, and we—a group of former co-workers, scientific friends and colleagues—decided that we had to tell the story of this remarkable man.

One of the challenges we were faced with was the sheer breadth of interest and commitment of Klaus's career. In this section, we shall attempt to evaluate his achievements as a Naturwissenschaftler[1] (Sect. 1.2), an enabler (Sect. 1.3), and as a public figure (Sect. 1.4). Klaus himself will have his say in Chap. 2 in which we reproduce an interview from 2007 as well as a recent

---

[1] We have chosen to use the German word "Naturwissenschaft" (and its derivatives) rather than "science", because of an important difference in meaning—the latter refers to "a branch of knowledge or study dealing with a body of facts or truths systematically arranged and showing the operation of general laws", and therefore to the product of scientific endeavor, whilst the former describes the process of creating knowledge about the character and dynamics of natural systems, i.e., the endeavor itself.

© The Author(s) 2022
H. von Storch, *From Decoding Turbulence to Unveiling the Fingerprint of Climate Change*,
https://doi.org/10.1007/978-3-030-91716-6_1

brief addition to it, and a chat between two Hasselmann acolytes, Dirk Olbers and Hans von Storch in May 2021. His key scientific achievements, which are summarised in Sect. 1.2, are discussed and evaluated in some detail in Chap. 3 in which a number of his original texts, some in German, are reproduced. In Chap. 4, some of Klaus' former co-workers and colleagues present their personal accounts and memoirs of what it was like to work with him: the reader will see a certain amount of overlap between the different accounts—recurrent themes include his personal friendliness but also his sometimes-rude insistence on scientific rigour. But the various accounts differ in terms of content, and the scientific issues addressed. Reading these accounts will give the reader an insight into the variety of endeavours, interests and successes of this man. The final Chap. 5 includes an overview of his major publications, awards and a CV.

1.2　The Naturwissenschaftler

When considering how to survey Klaus Hasselmann's scientific achievements, we were reminded of the old Indian parable[2] of the wise blind people, who want to understand what an elephant is. One examines a leg, another the trunk, the third an ear and so forth. They all understand exactly what they have in front of then, but none of them sees the whole animal. The authors of this book are the blind people, and Klaus Hasselmann is the elephant.

Illustration: Hans Møller · mollers.dk

---

[2] E.g., https://en.wikipedia.org/wiki/Blind_men_and_an_elephant.

We have identified seven major fields, to which Klaus made significant contributions; there may well be more, which we, the blind, have not yet classified and fully evaluated, one of which could be "internal' waves"—a subject with which Klaus probably became involved during his stay at La Jolla and his collaboration with Walter Munk, and which has much in common with the surface wave problem (generation, wave-wave interactions and dissipation), to which he devoted a great deal of his time. At that time, Walter Munk was working with Chris Garrett to establish the scientific basics of internal wave research in the form of a unified wave spectrum of the three-dimensional oceanic wave field. Klaus himself never published any substantial work on the subject (except for [27]) but he did encourage several PhD students (Kern Kenyon at La Jolla and Peter Müller and Dirk Olbers later at WHOI and Hamburg) to work on the question and provided a number of far-reaching new concepts.

Klaus has always been interested in ocean waves, which gave him his entry point into mainstream science (see Sect. 3.1). His PhD thesis from 1957 was titled "Über eine Methode zur Bestimmung der Reflexion und Brechung von Stoßfronten und von beliebigen Wellen kleiner Wellenlänge an der Trennungsfläche zweier Medien "(A method for determining the reflection and refraction of shock fronts and of arbitrary short wavelength at the interface between two media.) [191]. The first seminal paper he published on the subject was "Grundgleichungen der Seegangsvorhersage" (Basic sea state prediction equations) in 1960 ([3], see facsimile in Sect. 3.1). This paper provided a basic foundation for a reliable, generally applicable method of sea state prediction based on the basic energy balance equation of the ocean wave spectrum. He later published 3 papers entitled "On the nonlinear energy transfer in gravity-wave spectrum" in 1962–1963 [6, 8, 9], and "Propagation of ocean swell across the Pacific" [18] with Walter Munk in 1966.[3]

Klaus' work on remote sensing and the satellite ERS-1 (see Sect. 3.2) was also related to the field of ocean wave dynamics. He and Manfred Schieler published a paper on "Radar backscatter from the sea surface" [26] in 1970. Later papers, beginning with [45] from 1978, addressed aeroplane or satellite-based ocean wave spectra measuring methods. The MARSEN experiment (see below) led to another breakthrough when Klaus and his co-authors published the "Theory of SAR ocean wave imaging: A MARSEN view" [75] based on an imaging model, which was fundamental for SAR imaging of the ocean

---

[3] There was also a movie produced, mostly with Walter Munk, but with Klaus showing up every now and then–https://youtu.be/MX5cKoOm6Pk.

surface from future satellites SARs. Another fundamental paper, which he co-authored with his wife Susanne, was "On the nonlinear mapping of an ocean wave spectrum into an SAR image spectrum and its inversion" [102], a reproduction of which is included in Sect. 4.2. This paper serves as an example of one of the major contributions the Hasselmann couple made to the future of the retrieval of the ocean wave spectrum from the ERS-1 C band SAR on the global scale. Klaus' latest, and perhaps final contribution to this topic was the extensive review entitled "The ERS SAR wave mode: A breakthrough in global wave observations" [176].

When Klaus Hasselmann took on the responsibility and challenge of running the Max Planck Institute for Meteorology (MPI-M) in Hamburg, most of his time and attention was taken up by the subject of climate change. His initial thoughts on the subject were set out in his seminal "stochastic climate model", which was published in 1976 ([38], see facsimile in Sect. 3.3), which provided an insight into the formation of long-term internal variations excited by short term random fluctuations. Although this approach was not particularly surprising for a theoretical physicist, the concept did change the way climate scientists thought about the problem. The stochastic climate model firmly established the concept of a stochastic climate system, which included the separation of externally provoked variations ("signal") and unprovoked internal" variability ("noise"). This led to the emergence of the general concept of "Principal Interactions Patterns" ([86], see facsimile in Sect. 3.4), which included the key idea that the full infinite state space may be split into a low-dimensional "signal" space in which deterministic dynamics hold sway, and an infinite higher-dimensional "noise"-space, which is well approximated by stochastic dynamics. The "Principal Oscillation Patterns" [89] represented a special case. But the most important aspect of this approach was the question of detection and attribution [54, 110], i.e., of detecting the footprint of anthropogenic climate change in the empirical record of climate variation. This approach emerged as a key argument in the Intergovernmental Panel on Climate Change's (IPCC) assessment that anthropogenic climate change is real and will intensify if greenhouse gas emissions continue unabated.

When the science of the mechanisms of climate variability and change had matured [118] in the early 1990s, Klaus Hasselmann became interested in the interaction between climate and society and how mankind could deal with human-induced climate change. He understood quite early that the anthropogenic climate change problem goes well beyond the domain of climate science. It is not nearly enough to frame climate change research solely within the limits of what is sometimes referred to as "curiosity-driven science". Instead, research into anthropogenic climate change should support policymaking and coordinated climate action. These ideas prompted Klaus to create the Potsdam Institute for Climate Impact Research as a matter of urgency to deal with the economic dimension of climate change. In collaboration with Dmitry Kovalevsky and others, such as Michael Weber and Volker Barth, Klaus attempted to construct optimal policies to balance the expected costs of damages with the expected costs of mitigation. Section 3.5 provides more information about this field.

Following his retirement, Klaus Hasselmann became less interested in climate science, probably because he thought that he had already contributed everything that he could to the field and that the remaining challenges, such as the economic dimension, would be taken care of by others. Instead, he returned to a topic, which he had been thinking about in his spare time throughout his career. On his 60th birthday, he surprised his guests with the announcement that he would present something new—which would also explain to his family that he had really been thinking about particle physics (Sect. 3.6), when they falsely believed that he was just trying to get out of mowing the lawn. His talk took about two or three hours: "You can ask me, but you cannot stop me!" he said, and most of the audience did not understand a thing but enjoyed the show. Of course, Klaus was serious about the topic and his Metron concept. His wife, Susanne, volunteered to manage his schedule and he began to present his ideas to the physics community, the majority of whom were unfortunately not inclined to listen to him. The full concept has now been documented in a series of articles and in an unfinished book whose introduction we have reproduced in Sect. 3.6. All we can do at present is to wait to see if Klaus ideas' will eventually rule the waves, as they often did in the past.

1.3 The Enabler

When we talk about Klaus Hasselmann as an "enabler", we are referring to his ability to set things in motion, to create a scientific environment, which enables individuals to realise their full potential. Enabling activities may not leave a scientific footprint, but they do make an indirect contribution to the scientific process whether by creating a well-functioning working environment or by providing access to crucial empirical evidence. Klaus was an enabler in multiple ways.

His most important achievement was certainly the Max-Planck-Institut für Meteorologie, which provided many (at that time) young scientists with an environment in which to develop their skills. The various personal accounts included in Chap. 4 illustrate this breeding ground convincingly. "Do something that you consider interesting"—this Klausian request sounds like a recipe for disaster, an invitation to a hoard of intelligent young scholars to develop without coordination, without reference to a programme, and in various directions. But that did not happen. The young researchers all converged on the same problems but from their own unique angles. Of course, an Ernst Maier-Reimer would not take orders from anybody, but he would always have a suitable FORTRAN code for most problems in one of his desk drawers. In short, the Institute was a marvellous incubator, which was sometimes compared to an aquatic ecosystem populated with bottom feeders, primary producers, and gracious predatory fish. At the top of the chain there was just the one big fish, Klaus himself, who somehow managed to steer the dynamics within the incubator, with unconditional scientific rigour, personal friendliness and an endless reservoir of ideas.

One may well ask whether the MPI was organised in any way? There were a number of "Zwischenkapazitäten" (something like "lieutenants"), such as Dirk Olbers, Jürgen Willebrand, Mojib Latif, Martin Heimann, Peter Lemke, Ernst MaIer-Reimer and Hans von Storch, who acted as mentors to the younger researchers, but ultimately it was Klaus who guided and managed ideas, often by rejecting, replacing, or rectifying them. Even the management functioned smoothly. Klaus served as Managing Director throughout most of 1975–2000; only once did he wish to take a short break, and someone else took the helm and attempted to install a certain amount of administrative order, counting pens and the like. One of the then unhappy Zwischenkapazitäten went to the elder statesman, Reimar Lüst, at the end of the hallway to ask if we had misunderstood something? Nope, said Lüst, you've understood perfectly well; don't give in, it'll soon be over. And, really, Klaus was back after just a few days, and there was no more counting pens. Scientific paradise was re-established. Some of his co-workers, who later became Institute directors themselves, tried to copy his approach with some success.

As Institute Director, Klaus was also responsible for the financial side of things and had a special way of looking at this challenge, possibly guided by his colleague Hans Hinzpeter's famous dictum "a number is not a number", which played upon the empirical fact that any assertion about the financial situation would be preliminary, and prone to significant changes at short notice. Klaus spoke of "flying in fog", to suggest that knowing whether money would be available for hiring someone or making a capital purchase was largely based on a *feeling*. In a sense it was also a signal-to-noise problem. It worked out well and broke the potential spell of financial details involved in running a scientific Institution.

As outlined in his interview (Sect. 2.1) Klaus spent most of the first ten years trying to clarify the basic dynamic aspects of climate variability and change. Although the Institute has been founded to conduct research into climate change, most people there failed to notice this link. He also decided against purchasing a big computer, as most people had expected him to do but he worked instead with a relatively small group and a modest suite of hardware. But when the need for quasi-realistic climate models had become obvious by the early 1980s, Klaus established a close link to Günter Fischer's group at the Meteorological department of the University of Hamburg, which was located in the so-called in the Geomatikum some 10 m below the MPI. One member of that group, Erich Roeckner, was the then leading expert on atmospheric modelling. The first step was to replace the dynamical core that had been developed in-house with the model of the European Center for Medium Range Forecast, ECMWF, with parametrizations of the Hamburg model. The model was dubbed ECHAM—EC + HAMburg.

At about that time, Klaus had decided to set up a separate large computing centre, the Deutsches Klimarechenzentrum (DKRZ), which was to be led by Wolfgang Sell. This move made it possible to carry out gigantic simulations using a climate model based on Erich Roeckner's atmospheric ECHAM and Ernst Maier-Reimer's ocean LSG models. The computer system was updated regularly, with generous funding from the German Federal Ministry of Education and Research, whilst the costs for running the DKRZ were shared between a the MPI (the majority shareholder), the University of Hamburg, the Alfred Wegener Institute in Bremerhaven, and the GKSS in nearby Geesthacht. This system continues to run smoothly in 2021.

This process was eventually completed when Lennart Bengtsson was persuaded to become co-director of the MPI, where he would focus on atmospheric modelling (see Sect. 3.7). The ECHAM has become one of just a few leading quasi-realistic climate models, which are used in various institutions all over the world.

The incorporation of the DKRZ and the Bengtsson-department within the MPI were completed in the 1990s when the Potsdam Institute of Climate Impact Research, headed by Hans-Joachim Schellnhuber, was set up. This completed Klaus' original vision. He went on to think about metrons (Sect. 3.6).

Gathering data

Although Klaus Hasselmann was really a theoretical scientist, he often got involved in the practical challenge of gathering the relevant data, whether in preparation for satellite missions, or for setting-up and managing various campaigns (in atmospheric and oceanographic science jargon: experiments).

The first experiment he ran as the lead scientist was the JONSWAP project in 1969 (see Sect. 3.1). Wind stress, atmospheric turbulence and swell attenuation were monitored in the German Bight, and eventually the JONSWAP spectrum was derived.

Following the success of JONSWAP, Klaus apparently felt secure enough to initiate the IWEX (Internal Wave Experiment) mooring campaign during his stay at the Woods Hole Oceanographic Institution (WHOI) which he visited alongside Mel Briscoe, Terry Joyce, Claude Frankignoul, Peter Müller and Dirk Olbers. To our knowledge, the IWEX tripod was the first mooring capable of measuring current cross-spectra with sensors at horizontal and slanted separations. The experiment was carried out in the Sargasso Sea in 1973.

This culminated in the international Marine Remote Sensing Experiment, MARSEN which Klaus coordinated. It was carried out in the North Sea between the 16th of July and the 15th of October 1979 and was designed to achieve the following two objectives: (1) to investigate the use of remote sensing technology for oceanographic applications and (2) to utilise remote sensing technology in concert with in-situ oceanographic measurements to investigate oceanic processes in finite-depth water in the near-shore zone. MARSEN made use of 6 remote sensing aircraft including the NASA CV-990 with the JPL SAR. 60 scientists from 6 countries took part in the experiment, which spawned a plethora of papers, 14 of which were published in a special issue of the Journal of Geophysical Research in 1983.

In his later career, Klaus no longer participated in long-term empirical observational campaigns, perhaps because his attention was increasingly focused on the climate issue. Experimental work in ocean science is more about understanding and parametrizing various processes in ocean models—and the JONSWAP spectrum is an excellent example of this—whilst climate science mostly depends upon ongoing monitoring efforts. So, Klaus became involved in remote sea state monitoring technologies, which also, of course, had to do with his interest in the predictive potential of ocean waves. By the 1980s he had already become a key member of the ESA High Level Advisory Committee (EOAS), which had been set up by the ESA DG (Sect. 3.2). His commitment to the preparation of the ERS-1 satellite among other things was honoured by ESA, who invited him to join the launch of ERS-1 on the 17th of July at the Guiana Space Centre in French Guiana.

When Klaus became aware of the societal urgency of the climate problem in the 1990s, he founded the European Climate Forum (ECF), which was later expanded to cover a broader geographical range and renamed as the Global Climate Forum (GCF).[4] Klaus made an active contribution to various ECF/GCF operations, where was vice-chairman and a member of the board for many years.

The WAM group

In the spring of 1984 Klaus invited a number of ocean wave researchers to a meeting in Hamburg. There he proposed to jointly work on the development of an advanced numerical ocean wave prediction model. He wrote a new acronym (WAM, for Wave-Modelling group) on the blackboard and opened the discussion. The meeting supported the idea, because previous collaborations had created a shared feeling of urgency: model improvement was needed. Klaus recruited the participating Gerbrand Komen as chairman of the new group. He and Klaus would collaborate closely to "create a scientific environment, which enables individuals to realise their full potential", but which also enabled Klaus to achieve his goals.

One of the first challenges was to develop a scientific strategy. Some members focused on model development, but others simply wanted to collaborate on ocean wave research. This led to a number of subprojects, one of which was aimed at the development and implementation of global and regional versions of the model. Other subprojects focused on growth curve reanalysis, directional effects, shallow water effects and data assimilation.

Annual meetings were hosted at participating institutes on a rotational basis. They all opened with a review of the group objectives and what had been achieved in the past year. This was then followed by one or more formal presentations by guest speakers and a *tour de table* where each participant could say what he or she had done and was planning to do. After discussions, the tasks and commitments for the coming year would be listed. Klaus usually played a rather passive but very inspiring role in these meeting. Of course, some stricter co-ordination was later agreed upon in smaller groups, especially when the model was implemented at ECMWF.

Initially, there was no special funding. Most people contributed because the objectives of the WAM group aligned with the objectives of their home institutes. Once the research got underway the group became successful

---

[4] https://globalclimateforum.org/.

in acquiring additional funding. Outreach and enlargement of the group played an important role in generating support. Outreach took the form of many (invited) lectures at participating institutes and specialist conferences, sometimes as a showcase of a successful collaboration. The group was made truly international when it merged with a newly established working group (Working group 83, 'Wave Modelling') of the Scientific Committee on Oceanic Research. This brought in participants from China, the Soviet Union and elsewhere, to everyone's mutual benefit. The group ultimately included about 70 people from 15 countries.

One characteristic of the group was the spirit of collaboration. Group identity was reinforced by a newsletter in addition to informal contacts during working visits and the annual meetings. One of the participants wrote some alternative lyrics to the well-known Beatle song "Those were the days, my friend", which became "Those were the waves, my friend", which included the unforgettable line "Comparisons have shown // The physics are unknown". It was sung loudly after dinner during a meeting in Canada, which created a warm feeling of solidarity. In hindsight one may well wonder whether there was too much of a warm feeling, as it may have blocked some healthy dissidence. But so it goes.

Understandably, given his many other activities, Klaus was strongly focused on model development, which culminated in the implementation of the WAM model at the ECMWF. The other subprojects were also successful and resulted in several publications. One of the outreach highlights was a five-week course on ocean waves and tides at the International Centre for Theoretical Physics with support from the World Meteorological Organization and about 100 participants, mainly from developing countries. Another one was the successful completion of a jointly written monograph (Dynamics and Modelling of Ocean Waves [244]) which was published by Cambridge University Press in 1994. After that the group was dissolved. Mission completed.

1.4   The Public Figure

Klaus Hasselmann's work on ocean waves, which included remote monitoring and predictions was relevant and represents elegant science, which, however, hardly attracted public interest. The climate issue was completely different:

his two major achievements, namely the detection of the signal of anthropogenic change against the background noise of internal variability, and the attribution of this signal to human greenhouse gas emissions, as well as the provision of a scientifically first-class climate simulation platform, had an enormous impact on the relevant public discourse, not only in Germany but around the world. Nevertheless, he remained mostly unknown to the general public. This was not a matter of bad luck but of his own deliberate intent—he was simply not "interested in informing the public, [he was] always interested in basic research", as his wife Susanne put it in Sect. 2.3. Thus, he was happy when others offered to do the job for him. These others were Hartmut Graßl and Mojib Latif, who both became very well known to the German public. Both excelled in explaining complex dynamics and perspectives in a way that lay persons could readily understand. Interestingly, they did so without coordination between themselves or with Klaus, which obviously caused no problem.

Asked for his opinion on the public perception that the most important climate researchers in Germany were Graßl and Latif, whilst he himself was barely known, he responded (Sect. 3.3): "I was very pleased about that."

Thus, not surprisingly, Klaus left few traces in the media. Among the few examples he did leave were:

- Klaus Hasselmann: Die Launen der Medien. ZEIT 32/1997
- Johann Grolle: Wieviel ist der Wald wert?–Interview mit Klaus Hasselmann. Spiegel, 41/1992, 271–274
- Johann Grolle: Nobelpreis? Nee, daran hab' ich nie gedacht–Spiegel-Gespräch. Spiegel, 41/2021, 110–111
- Pieter Sartorius: In Sandalen die Welt von morgen suchen. Süddeutsche Zeitung, 31.10.1997s

The last piece is reprinted in the next Sect. 1.5. It has not been translated, as it is a wonderful example of living German whose special charm would hardly survive translation.

Klaus and two of his Zwischenkapazitäten, Ernst-Maier Reimer (left) and Mojib Latif (right).

(c) Günther Menn, Lea La Greca; mit freundlicher Genehmigung

1.5   In 1997, A Visitor Told His Perceptions When Visiting the MPI

Freitag/ Samstag/Sonntag, 31 Oktober/1./2. November 1997.

Süddeutsche Zeitung Nr 251 / Seite 3.

© *Süddeutsche Zeitung GmbH, München. Mit freundlicher Genehmigung von Süddeutsche Zeitung Content (* www.sz-content.de*).*

Das Foto auf der vorhergehenden Seite wurde mit dem Text zusammen veröffentlicht. Urheber: Günther Menn; Genehmigung des Nachdrucks durch die Nachlassverwalterin Lea De Greca.

Peter Sartorius

In Sandalen die Welt von morgen suchen

Klimaforschung: Welche Auswirkung hat die vom Menschen beschleunigte Aufheizung der Erde?Im Hamburger Max-Planck-Institut arbeiten 150 Wissenschaftler an Formeln, mit denen sich der Zustand des Planeten in 100 Jahren errechnen läßt

Hamburg,im Oktober

Bakan, der Bayer, studiert Wolken, manchmal von unten mittels eines Laser-Geräts aus der Abteilung des Kollegen Bösenberg, manchmal an Bord eines Spezialflugzeugs von oben oder, was noch spannender ist von innen, etwa über Spitzbergen. Meistens aber macht Bakan Beobachtungen von seinem Hamburger Arbeitsplatz aus, wenn sich die Wolken in physikalische und chemische Formeln aufgelöst haben. Dann stellt er Fragen an sich selbst - die Teilchengröße im Wasserdampf betreffend oder den Umstand, daß sich Wolken über dem Nordatlantik in Hunderte von Kilometern langen, parallel laufenden Wolkenbändern organisieren. Welche Bedeutung könnte dies für den Wärmeaustausch zwischen Atmosphäre und Ozean haben? Eminent wichtig, sagt Bakan, ein Arbeitsschwerpunkt

Aber was eigentlich ist nicht wichtig? Bakans Kollege Graf, der Sachse, verfolgt bei allem Mitgefühl für die Betroffenen begierig Vulkanausbrüche, egal in welcher Weltecke. Je mächtiger die Eruption, desto eindrucksvoller die Auswirkung auf Wolken, Winde, Wirbel und vieles andere, was das Klima der Erde bestimmt. Stundenlang kann Graf da erzählen, beginnend bei den vulkanischen Aerosolen, Rußpartikeln, die zur Wolkenbildung fuhren, was die Sonneneinstrahlung beeinträchtigt und die *Mittelatmosphäre* abkühlen läßt, auch wenn, andererseits, die Wolken die Wärmeabstrahlung der Erde absorbieren und eine Erwärmung bewirken. Jedenfalls verstärktt sich ein polarer Wirbel, der wiederum die planetaren Wellen beeinflusst, die als Führungsschienen der Tiefdruckgebiete fungieren Graf ist erst am Anfang seines spannenden Kollegs, aber Wesentliches ist bereits gesagt, nämlich daß alles, was in der Natur geschieht, zu Kettenreaktionen führt.

Zerfranstes Biotop

Gleiches gilt, wenn der Mensch lai die Prozesse der Natur eingreift und als Emission des Industriezeitalters seinen rasselnden Atem in die Atmosphäre stößt, anthropogene Treibhausgase, vor allem Kohlendioxid, wovon noch die Rede sein wird. Aber zu berichten ist zunächst von einer anderen Kettenreaktion, die ganz persönlich Hasselmann betrifft, einen in Ehren und mit silbernem Bart ergrauten Mann. Als Physikstudent hatte er vor fast einem halben Jahrhundert nach einer Systematik im Zusammenwirken von kurzen und langen Wellen im Meer gesucht, woraus sich folgerichtig ergab, daß er Ozeanologe, dann Klimaforscher, dann Professor und am Ende - indes auch schon vor mehr als 20 Jahren - Begründer und Direktor des Hamburger Max-Planck-Instituts für Meteorologie wurde.

Und dort haben wir uns eingefunden, um zu studieren, was das eigentlich ist, Klima. Und wie es erforscht wird in der doppelstöckigen Baracke, in der das Institut zwischen einem Uni-Hochhaus namens Geomatikum und einem zerfransten Biotop mit dazugehörendem Tümpel ein Hinterhofdasein führt, freilich eines, dem immer größere Bedeutung zukommt, jetzt, da eine vom Menschen mitverursachte allmähliche Klimaaufheizung registriert und die erhöhte Temperatur als ein ernsthaftes Krankheitssymptom der Welt erkannt worden ist. Wer sich professionell damit beschäftigt, muß nicht unbedingt Meteorologe sein. In der Mehrheit sind die 150 wissenschaftlichen Mitarbeiter des Instituts promovierte und habilitierte Physiker oder Chemiker, die indes mit ihren Titeln so lässig umgehen wie mit ihrem Outfit, zu dem kurze Hosen und Birkenstock-Sandalen gehören. Mit der mangelhaften Klimatisierung der Baracke könnte das zusammenhängen. Aber reizvoller ist der Gedanke, es könne sich um einen dezenten Hinweis der Fachwelt handeln, die weiß, was auf die Welt zukommt Zwar steigt deren Mitteltemperatur, kurz Klima genannt, scheinbar minimal und liegt heute lediglich ein halbes Grad über jener vor hundert Jahren, als das Abgaszeitalter begann. Aber in nochmals hundert Jahren wird nach allen Berechnungen die Zunahme eben nicht nur weitere 0,5 Grad, sondern zwei volle Grad ausmachen, was ersichtlich macht, daß da ein Schwungrad in Gang ist, das die Aufheizung beängstigend beschleunigt. Und was das zur Folge haben kann, wird jedem schlagartig klar, der Bengtsson zuhört, dem Schweden und Co-Direktor im Institut, der bei Tee und Keksen darauf zu sprechen kommt, daß das eigentlich Dramatische der Umstand sei, daß die Erwärmung, soweit der Mensch sie bewirkt, vermutlich nicht mehr rückgängig zu machen ist. Die Kettenreaktion, wie bei der Atomkraft. Aber Bengtsson hält sich bei der Analogie nicht auf, referiert jetzt über El Niño, das pazifische Klimaphänomen, das die Passatwinde erlahmen läßt und dadurch weltweit das Wetter durcheinanderbringt. Dürre in Australien, Waldbrände in Indonesien, Fischsterben vor Peru, winterliche

Kältewellen in Regionen, wo es erst Herbst sein sollte. Vielleicht kommt zum Jahresende noch Schlimmeres, wenn El Niño seinen Kulminationspunkt erreicht haben wird. Man wird dann, soviel ist jetzt schon sicher, den stärksten El Niño aller Zeiten messen. Nicht, daß Bengtsson dies unbedingt der allgemeinen Erwärmung der Welt und menschlichen Einfluß zuschreiben würde. Dafür gibt es noch zu wenig Erkenntnisse, außerdem ist es, ein paar Türen weiter in der Baracke. der Forschungsgegenstand des Kollegen Latif Aber Gedanken macht sich Bengtsson doch darüber, ob sich da nicht im Zusammenwirken von langfristiger Klimaerwärmung und kurzfristiger Wetteränderung Tiefdruckgebiete verschieben und dann eines Tages, zum Beispiel, in China der Gelbe Fluß überläuft, der bereits jetzt nur mühsam in seinen Dämmen gehalten werden kann. Das Oder-Hochwasser in fürchterlicher Potenz Millionen Opfer ist denkbar, sagt Bengtsson leidenschaftslos, Todesopfer.

Aber es ist nicht die Stunde von Horror-Visionen. Ums Prinzipielle geht es, darum, dass sich der Planet eben nur eben in beschleunigtem Tempo erwärmt. Schon vor Jahren hat die Wissenschaft in eine Art Bibel mit dem Titel *Climate Change 1995* darauf hingewiesen dass nicht nur die Natur mit ihren Zyklen sondern auch der Mensch mit seinen Emissionen schuld ist. Großen Eindruck hat dies nicht gemacht, und auch auf der bevorstehenden Weltklimakonferenz im japanischen Kyoto wird eine wirksame Reduzierung des Kohlendioxidausstoßes der Industriestaaten kaum zu erreichen sein.

Wer im Treibhaus sitzt, wirft ungern mit Steinen auf die eigene Industrie.

Angst vor dem Irrtum

Und ist es nicht tatsächlich beruhigend fürs Gewissen, dass man so ganz genau nun doch nicht weiß in welchem Ausmaß der Mensch der Übeltäter ist? Hat sich nicht die Wissenschaft schon einmal geirrt? Hat sie nicht den Kühleffekt durch vulkanische und industrielle Aerosole unterschätzt und deshalb vor einem Jahrzehnt eine zu starke Erwärmung vorausgesagt, woraufhin ein Hamburger Nachrichtenmagazin bereits aufgeregt die Ozeane wegen des Abschmelzens der Polkappen überschwappen und den Kölner Dom untergehen sah? Jetzt also ist die Forschung zurückhaltender geworden, und von der Steinkohleindustrie bis zur Automobilbranche wird dies triumphierend als ein Gegenbeweis für die Aufheizthese ins Gefecht geführt.

Ein zäher Kleinkrieg ist da zwischen Klimaforschung und Industrielobby im Gang, wobei die Wissenschaft den Vorteil hat, mit den ganz schweren Waffen anrücken zu können, mit Großrechnern, blaugrau, raumfüllend wie Banktresore. Auch die Hamburger haben sie, wenn sie auch nicht eigentlich im Besitz des Max-Planck-Instituts sind. Sie gehören dem im Geomatikum untergebrachten Deutschen Klimarechenzentrum, einer Gemeinschaftseinrichtung deutscher Wissenschaftsinstitutionen, die sich mit Meteorologie beschäftigen. Aber weil das Max-Planck-Institut der hauptsächliche Nutzer ist, amtiert Hasselmann im

Nebenjob auch als Co-Direktor des Rechenzentrums und kann alle paar Jahre für rund 20 Millionen Mark einen dieser Rechenkolosse anschaffen, der dann mit Kränen von außen bis hinauf ins 15. Stockwerk des Geomatikums gewuchtet werden muß - vermutlich, weil ein Gehirn immer irgendwie oben zu sitzen hat.

Praktischer wäre, es befände sich auf ebener Erde. Aber Hauptsache, es liefert Ergebnisse, und zwar schnell. Böttinger, der wissenschaftliche Betreuer des Computers, spricht zungeschnalzend von mehreren Milliarden Gleitkomma-Operationen in der Sekunde, was Hasselmann indes nicht zu beeindrucken vermag. Mit schnelleren Rechnern, sagt er, könnte man ganz anders auflösen – die Welt ebenso wie deren Rätseln. Denn auflösen heißt, dass man ein Raster ein Gitter um den Globus legt, indem erfasst wird, was die Natur an Zuständen, Aktionen, Reaktionen und sonstigem zu bieten hat. Je größer die Auflösung, desto schärfer das Bild – ähnlich wie am Fernsehschirm. 250 km beträgt die gebräuchliche Maschenweite des globalen Gitters und man operiert in der Atmosphäre in 20 Schichten nach oben und im Ozean mit weiteren 20 Schichten nach unten. Aber mit mehr power sagt Hasselmann, könnten mehr Gitterpunkte – Stop. Haben wir da nicht noch ein Wort von Graf im Ohr. Graf hatte bis zur politischen Klimawende 1989 an der Humboldt-Universität in Berlin/DDR geforscht. Jedesmal, wenn er rechnen lassen wollte, mußte er mit seinen Unterlagen unterm Arm Tagesreisen um Westberlin herum zum DDR-Wetterdienst unternehmen. Man habe, hatte Graf

gesagt, angesichts der Verhältnisse den eigenen Kopf zu gebrauchen gelernt. Tatsächlich ist die Unbestechlichkeit elektronischer Schaltkreise eine Sache, die Kompetenz des menschlichen Geistes eine andere. Bei Graf, bei Bakan, bei Bengtsson, bei Latif, bei Hasselmann und all den anderen geht es darum, gedanklich die Voraussetzung für eine Hochrechnung zu schaffen, für ein Modell, eine Weltformel, in der alle Erscheinungen der Natur enthalten und in Relationen gebracht worden sind. Man kann das als eine Denksportaufgabe nennen, bei der ist eine komplette Lösung nie geben wird, sondern immer nur verbesserte Annäherungen. Man kann es auch die Hersteller von Software nennen, deren Computer-Ausdruck Meter hohe Papierberge ergibt. Wie immer man es nennt: es bleibt eine geistige Anstrengung, die nur im Team und im internationalen Austausch zu bewältigen ist.

Als sichtbares Produkt fördern Sie zunächst einmal kunstvolle Bilder zu Tage, Schautafel mit roten, blauen, honiggelben Schlieren, die sich über Kontinente ziehen: Simulationen klimatische Zustände in einer existierenden oder nur vermuteten Welt, angefertigt in großer und geringer Auflösung, unter Verwendung aller verfügbarer oder auch ausgewählter Faktoren. Testbilder sind es oft nur, dazu dienlich, herauszufinden, wie fehlerhaft möglicherweise einzelne Vorgehensweisen beim Lösen der großen Aufgabe gewesen sind Gleichwohl, wie exotische Pflanzen wachsen die Bilder aus Aktenschränken, setzen sich an Bürowänden fest, wuchern hinaus auf Flure und von dort in andere Büros hinein - von Arpe zu Roeckner, von

Roeckner zu Latif. Von Latif zu Maier-Reimer.

Aber dort, bei Maier-Reimer, dem Schwaben, ist die Invasion zum Stehen gekommen. Ein anderes Kunstwerk, Botticellis Geburt der Venus, bedeckt die Wand, und man denkt sich: Wenn die Klimaforschung dort nur so bildhaft darzustellen wäre. Der Wind aus Engels Mund. Die Wolken als wehende Teppich. Der Ozean als gewellte Muschel. Maier-Reimers Welt ist profaner. Die Lösbarkeit von Kohlendioxid im Ozean. Maritime Biologie im Strömungssysteme. Die Konzentration von Phosphat, Nitrat, Silikat, Sauerstoff in unterschiedlichen Tiefen. Wie wird es sich aus auf das ozeanische Kohlendioxid aus, wenn sich der atmosphärische Kohlendioxidgehalt erhöht? Welchen Einfluss hat es auf Algen, wenn sie mehr Kohlendioxid aufnehmen? Algen verändern die Farbe des Wassers und damit die Absorption der Sonnenstrahlung - was sind demnach die Folgen für die Erdtemperaturen, falls sich die Algenkulturen vermehren? Experimentell schwer zugänglich, sagt Maler-Reimer mehr zu sich selbst als zum Besucher.

Das Gehirn der Natur

Antworten führen zu immer neuen Fragen in diesem tiefen, rätselvollen Forschungsgebiet namens Ozean. Und wieder muß man von Hasselmann reden, den das Meer nie losgelassen hat Das Phänomen der kurzen und langen Wellen in an anderer, größerer Dimension. Konzepte der theoretischen Physik hat Hasselmann auf die Wetterforschung

übertragen und für eine Formel verwandt, die den trägen reagierenden Ozean und die rasch veränderbaren Verhältnisse der Atmosphäre miteinander verrechenbar macht. Latif - ein waschechter Hanseat trotz des Vornamens Mojib - kann in einer schönen Metapher ausdrücken, was da dahintersteckt. Die kurzfristigen Witterungsschwankungen in der Atmosphäre, so seine Erklärung, müsse man als Tennisballe sehen, die unentwegt gegen einen Medizinball geschleudert werden: gegen den Ozean, der zunächst nicht reagiert, am Ende aber ins Rollen kommt, nein, etwas ins Rollen bringt - eben: diese langfristige Klimaänderung.

Will sagen: Der Ozean ist das Gedächtnis des Klimas. Wenden wir uns also an Roeckner, der in einem Zahlenwerk zu verrechnen versucht, was das Gehirn der Natur für die Zukunft gespeichert haben könnte. Umgehend wirft Roeckner eine Formel aufs Papier, dabei erläuternd, es handele sich um eine partielle Differentialgleichung mit lokaler Ableitung von Modellgröße nach Zeit, und er wolle jetzt Delta hinschreiben und F. Nicht, daß wir Roeckner gern unterbrechen, aber unsere Wißbegierde geht eher ins Grundsätzliche. Welche Komponenten müssen im institutseigenen Klimamodell enthalten sein, von dem immer wieder die Rede ist, in dieser großen Gleichung, die die eigentliche Rechtfertigung des Max-Planck-Instituts darstellt und mit der sich jede Art von Klima simulieren läßt - das vergangene, das derzeitige, das künftige, mit und ohne Treibhausgas des Menschen? Jahrelang feilte man an dem Modell. Und heute ist es der Stolz des Instituts. MP/ECHam heißt es. Bei einem internationalen Leistungsvergleich hatte es sich gegenüber anderen Modellen unlängst vorzüglich behauptet.

Also was alles ist in der Formel enthalten? Die Frage ist doch kaum ausgesprochen - schon ergießt sich eine Sturzflut von Fakten und Faktoren auf den Besucher. Atmosphären-Temperaturen, Wasserdampfgehalt, Winde in alle Himmelsrichtungen und in der Vertikalen, Luftdruck an der Erdoberfläche, Wolkenbedeckung mit Niederschlägen in flüssiger und gefrorener Form. Bakan sehen wir über Spitzbergen kreisen. Druck in allen Luftschichten. Prozesse im Erdboden. Wärme und Wassergehalt dort. Schneehöhen. Ozeantemperaturen an der Oberfläche und in allen Tiefen. Eisdicke. Eisverteilung. Salzgehalt. Dreidimensionale Strömung. Maier-Reimer kennen wir hinter rutschenden Bergen von Papier, nachdenklich dem Bart streichend.

Nicht enthalten, sagt Roeckner, seien Gletscher und Plattentektonik. Aber fertig ist er noch lange nicht. Die chemischen Komponenten. Grafs Vulkane und Aerosole Das troposphärische Ozon. Autoabgase. Kohlenstoffkreislauf in der Atmosphäre und im Ozean. Dann, natürlich, die Biologie. Waldgebiete absorbieren die Sonnenstrahlung, im Gegensatz zu Wüsten. Aber plausibles dynamisches Vegetationsmodell, sagt Roeckner, als ob er sich entschuldigen müsse, ein solches Modell gebe es noch nicht, also vom Klima keine Rückkoppelung zur Vegeta-tionsänderung und von dort zurück zur Klimaänderung.

Gleichwohl, eine schwindlig machende Fülle von Faktoren steckt in dem Modell,

geordnet in einem vierdimensionalen Koordinatensystem, in dem Höhe, Breite und Länge das Gitter um die Erde betreffen, aber die vierte Koordinate die wichtigste ist: die Zeit, die Achse, auf der in die Vergangenheit zurückgegangen und in der die Zukunft hochgerechnet wird. Und alles wird in Hasselmanns Großrechnern mit Milliarden Gleitkommaoperationen pro Sekunde verarbeitet zu diesen vo Computer ausgespuckten Kunstwerken mit ihren schillernden Schlieren, ein errechneter Zustandsbericht der Welt, wie geklaut, nein, geklont von der Wirklichkeit der Natur mit ihren über den Globus verstreuten Trocken- und Regengebieten?

Arpe hat die bunten Bilder vor sich ausgebreitet. Wahrheiten, sagt er, lägen in ihnen, wenngleich Wahrheiten unterschiedlicher Qualität - die eine geliefert von der Natur, die andere vom Computer. Identisch müßten beide sein, wenn die Denksportaufgabe richtig gelöst worden ist. Tatsächlich haben beide Versionen die gleichen Farben, die gleichen Schlieren. Aber Arpes geübte Augen entdecken feine Unterschiede. Ein pedantischer, unnachsichtiger Prüfer ist er, der festzustellen hat, wo die Theorie mit der Wirklichkeit kollidiert. Eine Stichprobe hat er gemacht für eine Qualitätskontrolle des hauseigenen Klimamodells. Aus Atmosphären- und Ozeanwerten ließ er vom Computer weltweit die Bodendruckverhältnisse im Winter und damit die Verteilung von Hochs und Tiefs errechnen. Nun hat er das Resultat vor sich, und sein Blick bleibt an Spanien hängen, dem winzigen Appendix eins kleinen Europas. Es ist das Gelb dort, das ihn stört. Ein winterliches Dauerhoch zeigt es an. Aber dieses Hoch kann nicht sein, darf nicht sein, weder im Interesse des Hamburger Max- Planck-Instituts noch in dem der Iberer.

Es grünt nämlich nur deshalb so grün, wenn Spaniens Blüten blühen, weil es im Winter in Spanien regnet. Und die Simulation besagt, daß genau dies nicht der Fall ist, sondern daß ein Dauerhoch herrscht und das Land mithin eine Wüste ist wie weiter südlich die Sahara. Keine Ahnung, sagt Arpe, was da schiefgelaufen ist, warum sich das Bild verschoben hat. Liegt die Fehlerquelle im Atmosphärischen oder eher im Ozeanischen? Wäre das Ergebnis besser ausgefallen, wenn man die Welt größer aufgelöst, das Gitter enger gezogen hätte? Hatte man es ursprünglich mit zwei Fehlern zu tun, die unbemerkt blieben, weil sie sich gegenseitig aufhoben, und von denen dann einer ausgemerzt wurde, so daß der andere in Erscheinung treten konnte?

Ein Joker bleibt

Es kann lange dauern, bis man Antworten finden wird. Aber keine Frage, daß die Annäherung an die Wirklichkeit immer genauer wird. Die großen klimatischen Zyklen, denen die Erde im Rhythmus von 20 000 und 40 000 Jahr en ausgesetzt ist und die auf Variationen der Erdbahn um die Sonne zurückzuführen sind, lassen sich längst durch exakte astronomische Messungen berechnen. Und was die kürzerfristig, in Jahrhunderten auftretenden Klimaperioden betrifft, so hat Hasselmann mit seiner Formel, die das Meer mit der Atmosphäre verrechenbar

macht, wichtige Voraussetzungen für ein Verständnis geschaffen. Die Tennisbälle, wie gesagt, die gegen den Medizinball prallen. Klimaschwankungen, präzisiert nun Hasselmann, entstünden in Analogie zu den langsamen Zufallsbewegungen schwerer Moleküle, die von leichten angestoßen werden. Das definiert einen Mechanismus. Aber erklären läßt sich damit natürlich nicht alles. Der variierende und sich in Sonnenflecken äußernde Energieausstoß der Sonne, der in seinen Auswirkungen nicht richtig abschätzbar ist. Ein Joker sei dies, sagt Hasselmann, einsetzbar sowohl für Polemik gegen die Schulwissenschaft wie auch für Erklärung von Unerklärlichen. Gleichwohl, der Einsatz von Computern wird die Klimaforschung ebenso revolutionieren wie dies der Fall bei der kurzfristigen Wetterforschung war, der Meteorologie, die sich schon lange nicht mehr der Meteorolüge bezichtigen lassen muss. Die Klimaforschung, notieren wir als Credo aus der Baracke, stehen erst am Anfang – dort wo sich die Meteorologie wo vielleicht 20 Jahren befand. Der große Entwicklungssprung also steht erst noch bevor.

Immerhin kann Hasselmann die globale Erwärmung fürs nächste Jahrhundert schon heute zuverlässig prognostizieren und immerhin kann Latif schon ein halbes Jahr und länger vor dem Auftauchen der El Niños die betroffenen Länder warnen und immerhin kann Roeckner kanadischen Fischern die freilich nicht sehr tröstliche Auskunft erteilen, ihre Lachsfischerei habe vermutlich keine Zukunft, weil das Wasser an der Pazifikküste im nächsten Jahrhundert vielleicht zu lau sein wird.

Den schon ist alles verpackt den Formeln,
im Rechner – auch die Antwort auf die
Frage wo auf der Welt die
Klimaerwärmung tatsächlich höherer
Temperatur erzeugen wird und wo nicht
und wo sogar die Quecksilbersäule fallen
könnte, zum Beispiel in Mittelindien, wenn
dort durch Industrialisierung vermehrt
schwefliger Ruß in die Luft geblasen wird,
sich daraufhin Wolken bilden und die
Sonnenstrahlung reduziert wird.

Wie im Fall der vulkanischen Aerosole.
Aber nicht bei Graf sind wir, sondern noch
immer bei Arpe, dem Wahrheitssucher.
Spanien, sagt Arpe, denken Sie an
Spanien! Und so begreifen wir am Ende
Spanien als Synonym für verbleibende
Unwägbarkeiten und als Metapher für den
zusätzlichen Faktor, der sich zu den
unzähligen physikalischen, chemischen
und sonstigen festen und variablen
Größen der großen Klimagleichung gesellt,
den Unsicherheitsfaktor.

A critical inspection of the older sister Almut. Hamburg, shortly before leaving for England in 1934 (left), and In Welwyn Garden City, England, shortly before leaving for Hamburg, 1949 (right).

With Karl Wieghardt, diplom thesis advisor and later post-doc employer in Institute for Naval Architecture, at inauguration ceremony, 1975.

With Susanne and two oldest children, Meike and Knut, in La Jolla, 1963.

At Woods Hole Oceanographic Institution, in front Research Vessel Knorr, 1972.

With Bob Stewart, Brian Tucker and Australian sheep during break of the Joint Organizing Committee meeting of the Global Atmospheric Research Programme in Melbourne, 1974.

With Reimar Lüst, President of the Max Planck Society, at inauguration ceremony of the Max Planck Institute, 1975.

With Peter Fischer Appelt, President of the University of Hamburg, Senator Biallas of the City of Hamburg and Reimar Lüst during the inauguration ceremony, 1975.

Explaining ocean wave prediction, 1982.

In the new prefab building ("pavillon") behind the Geomatikum, after creation of the DKRZ, 1989.

Making a point, 1988.

Robertson Memorial Lecture Award, US National Academy of Sciences, 1990 (proposed by Carl Wunsch, second row, first left).

With Hartmut Graßl, 1996.

Explaining the multi-agent aspects of a coupled climate-economy model, 2002.

Explaining the detection of an anthropogenic climate signal at 95% statistical confidence level, with the Federal Minister of Research and Technology, Jürgen Rüttgers, 1992.

60th birthday, Rissen 1991

With Walter Munk, during Hasselmann's 60'th birthday symposium, 1991.

With Wolfgang Sell, Lennart Bengtsson and wife Susanne during emeritus dinner, November 1999.

Ola M. Johannessen, Walter Munk and Klaus Hasselmann sailing the fjords near Bergen. Ola comments: "Walter and I are discussing my CO2-Ice paper from 2008, too simple for Klaus, who took a nap"

Rissen, 2011.

with Hans von Storch

Susanne, Klaus and Dirk Olbers 2012 in Fischerhude, discussing ocean physics

2

Klaus Hasselmann—His Own Account

In 2006, Hans von Storch and Dirk Olbers ran an interview with Klaus Hasselmann. This interview is here reprinted without alterations. The numbered references in the interview refer to the publication list at the end of this book. The interview was published as von Storch H., and D. Olbers, 2007: Interview with Klaus Hasselmann, *GKSS Report 2007/5*; 67 pp.[1]

2.1 The 2006 Interview

Question: How did you become interested in physics?

Hasselmann: One of my early experiences which kindled my interest in physics was buying a crystal detector from a school friend for two shillings and six pence—half a crown—or about the price of a movie ticket. I must have been about 13 years old. I was quite impressed that even without plugging the device into a socket, I could listen to wonderful music through the earphones. I wanted to better understand the puzzling phenomenon that you could get something from nothing. I went to the town library in order to find out in books on physics for beginners how electricity and radios work. That was my introduction to physics. At that time, it was an exciting experience for

---

[1] See   https://www.hereon.de/imperia/md/content/hzg/zentrale_einrichtungen/bibliothek/berichte/gkss_berichte_2007/gkss_2007/_5.pdf.

In 2013 it was also published online by the Niels-Bohr Library and Archives of the Center for History of Physics, https://www.aip.org/history-programs/niels-bohr-library/oral-histories/33645

The original interview featured the photos shown before plus a few more, and additional to a foreword an introduction by Reimar Lüst and a concluding comment by Walter Munk.

© The Author(s) 2022
H. von Storch, *From Decoding Turbulence to Unveiling the Fingerprint of Climate Change*,
https://doi.org/10.1007/978-3-030-91716-6_2

me, completely independent of the fact that I was taught physics in school. I did not see any connection between our physics lessons in school and my personal learning from the books in the library—I think this experience of personal learning and discovery was very important for me.

We have just heard that the detector had cost half a crown—so you did not attend school in Germany but in England. How did that come about?

Hasselmann: When I was close to three years old my family—my parents and older sister—emigrated to England. My father was a social democrat and did not want to stay in Germany in 1934. Our family moved into a so-called community, consisting mostly of Jewish emigrants from Germany. The English Quakers helped us a lot in those days. Until we returned to Hamburg in 1949, we lived in a very nice small town, Welwyn Garden City, 30 km north of London. I passed my A-levels there (then called Higher School Certificate). I felt very happy in England. So, English is in effect my first language.

Nevertheless you studied in Germany.

Hasselmann: I studied in Hamburg. I did a half year practical training in a machine factory first, because I was not sure whether I wanted to study engineering or physics. In addition, I was not yet at home living in Germany—neither were my parents, in fact, because Germany had changed. So I had to find my feet first. When I started studying, the idea of having to work hard for my studies was also a new experience. So I fell back a little during the first year. I had doubts whether I really was talented enough to continue with my studies, so—as a test—I took a study exam (Fleißprüfung), which I passed, and so I continued. I did not regret that period of adaptation, but it was a drastic change between my English school days spent in a healthy, suburban garden town north of London and living in Hamburg, where everything was bombed to ruins. However, I had always wanted to go back to Germany to explore my roots. My parents were always patriotic, in a natural, pre-nazi sense. But I was always very happy in England and did not really experience any difficulties due to my German origin, not even during the war. Still, I wanted to find out where I belonged. In spite of the difficult period of adaptation during the first one or two years, I did not regret returning to Germany.

Did you study only in Hamburg?

Hasselmann: I studied in Hamburg for eleven semesters until I obtained my diploma in physics, in the summer of 1955, with mathematics as a second subject. Then I obtained my Ph.D. at the Max Planck Insitute of Fluid Dynamics and Göttingen University from 1955 until 1957. Afterwards, I returned to Hamburg, where I spent three years as a post-doc working with my former diploma supervisor, Prof. Karl Wieghardt, at the Institute for Naval Architecture, before going to America in 1961.

Would you like to recount the theme of your diploma thesis?

Hasselmann: In my diploma thesis I worked on isotropic turbulence and found an—in my opinion—slightly more elegant derivation for the basic dynamic equations for isotropic turbulence [1]. For my doctoral thesis I changed subject to study the propagation of so-called von Schmidt head waves, elastic waves at the boundaries between two solid objects. In Hamburg I returned again to fluid dynamics research, mostly to experimental work on turbulence in ship wakes, using hot-wire instruments in a wind tunnel and a towing tank. But I also continued working on turbulence theory.

This did not correspond to the mainstream of education in physics. Were not atomic theory and nuclear research considered the normal case in physics already in those days?

Hasselmann: Yes, that was the mainstream, but I wanted to work in an area in which I thought I would be able to contribute something. I always had a practical bent, I wanted to work on problems which I thought I would be able to solve. I did not want to work on abstract, theoretical problems, and I did not have enough self-confidence to think I could make significant contributions to such difficult fields as general relativity or quantum field theory. So I went into fluid dynamics. I was always interested in the way planes and rockets worked. I liked my field of work, and I only gradually drifted into oceanography, meteorology and climate research. Later, I did then become interested in quantum field theory, elementary particle physics and general relativity, through my work on nonlinear interactions in geophysical wave fields, starting from ocean waves. I pursued these investigations for many years in parallel to my regular research, so to speak as a private hobby. However, all this developed in the course of the years. First I had wanted to work on a practical, solvable task as a physicist.

Then there actually was a practical task resolved by you?

Hasselmann: This is an embarrassing question.

The turbulence theory has surely not been resolved.

Hasselmann: Exactly, but then I was young and naive, and I hoped to make some progress in this problem, despite the fact that several generations before had failed. Nevertheless, my struggles with turbulence theory taught me a lot on stochastic processes and interactions in nonlinear systems. This enabled me to solve other problems later on. The first problem I solved theoretically was the question of the nonlinear coupling of ocean wave components. I would not have been able to solve this problem if I had not worked on turbulence before.

Which mark did you get in your doctorate thesis? This question may provide moral support for millions of others.

Hasselmann: Another embarrassing question. I received a 2 (corresponds to B). The reason was presumably that I solved the problem I was posed (propagation of von Schmidt head waves) in a different way than suggested by Prof. Tollmien's assistant. I found out quite early, after a few months, that the way suggested by my supervisor would not work. So I chose another path, which led to the goal, but my supervisor was not enthusiastic. Nevertheless he accepted my thesis and gave me a 2, because I had produced some very nice computational results obtained with Germany's first electronic computer, the G1, which had been developed in Göttingen. It is now in the German Science Museum in Munich. It had a total memory of–believe it or not—25. It was quite a challenge to use it to solve a system of several equations with many different parameters. I had access to the machine at night, and played table tennis with another student until the alarm bell of the G1 informed me that there was an error, which I would fix by cutting out and replacing part of a holerith paper tape, which was glued together in a closed loop. Different computational loops were realized by different holerith paper tape loops on different readers. One could follow the course of the computation as different readers were switched on and off. I presented my results very nicely in numerous graphs, which apparently impressed my supervisor. So I obtained my Ph.D. in less than two years [2, 191], in spite of the forbidden approach I had used to solve the problem.

Your family did not discuss physics at breakfast. How did you head towards science?

Hasselmann: I was always interested in understanding physical processes. As I already said, one trigger was the crystal detector. But I also constructed electrical motors and such things, and was continually producing short circuits at home. I got good grades in physics in my final school examinations, but without any relation to what I was taught in school. My physics teacher did not inspire me at all; for him I was an unruly trouble maker whom he often kept in after school. „Hasselmann, detention at four!" is still ringing in my ears.

Later at the university I was strongly motivated by my fellow students, particularly Wolfgang Kundt, Gerd Wibberenz and Ewald Richter. with whom I solved exercises together and had many discussions. That was a very intense period, forming lifelong friendships. Wolfgang Kundt and Gerd Wibberenz became Professors of physics in Bonn and Kiel, and we worked together occasionally also later. Ewald Richter became a professor of philosophy in Hamburg, and we had many interesting discussions with him too. I was also inspired as a student by Pascual Jordan, who taught theoretical physics in Hamburg. I was not in personal contact with him, but I really enjoyed his lectures. After the diploma I mainly instructed myself. I read interesting books and familiarized myself with the literature related to my research—as I suppose all young scientists do. But I never really had a proper mentor, neither at school, nor during my studies. In 1961, when I was already 29, I got to know Walter Munk,[2] who invited me to his institute in La Jolla. I have had a close relationship with him ever since. His open, generous personality as well as his enthusiastic approach to science have always impressed me. Nonetheless, although I wrote one or two joint publications with him, I regard Walter more as a personal than a scientific role model.

Would you say that you had a factual supervisor?

Hasselmann: For my Ph.D.? No, I did not have a real supervisor. Prof. Tollmien, then Director of the Max Planck Institute for Fluid Dynamics, was no longer active. As I explained, his assistant had a different idea on how I had to solve the problem posed for my thesis. I could not really discuss the

---

[2] After recording this interview, Klaus Hasselmann and Hans von Storch prepared an interview with Walter Munk, see: von Storch, H., and K. Hasselmann, 2010: *Seventy Years of Exploration in Oceanography. A prolonged weekend discussion with Walter Munk*. Springer Publisher, 137 pp, https://doi.org/10.1007/978-3-642-12087-9 (http://www.hvonstorch.de/klima/books/munk-springer-final.pdf).

problem with him. I worked and learnt independently and read the necessary literature. In the following three years in Hamburg I had very good relations with my former diploma supervisor, Prof. Wiegandt, but scientifically, we did not interact very strongly, as he was oriented more towards experimental work. Although I was also involved in experimental turbulence measurements at that time, using hot-wire instruments, I worked more or less on my own—with limited success experimentally, I have to admit. But it was still fun finding out how to build the equipment, learning about feedback systems and the havoc that they can create in trying to construct high level amplifiers to measure weak turbulence signals.

Then you went to America.

Hasselmann: Yes, this was through Prof. Roll, the former president of the German Hydrographical Institute, today called BSH. Parallel to the development of hot-wire measuring instruments, I had become interested in ocean waves. At the Institute for Naval Architecture there was considerable interest in the wave resistance of ships and ship motions in waves, motivated by the director of the institute, Prof. Georg Weinblum, a very kind and supporting person, who was an international expert in the field. The behaviour of vessels in rough seas in particular was a central topic at the institute. In this context, I read some very interesting papers by Owen Phillips and John Miles on the wind generation of ocean waves, which further stimulated my interest in the subject. My own first contribution to the subject was simply the introduction of the spectral energy balance equation for the prediction of ocean wave spectra, which, strangely, nobody had used before. Then it became clear to me that to understand the spectral energy balance of ocean waves, one had to solve the problem of the nonlinear interactions between wave components. I realized that the problem could be solved by the methods I had learnt in struggling with turbulence theory. Although the relevant closure methods were inadequate to solve the strongly nonlinear turbulence problem, they were directly applicable to the problem of weak interactions between ocean wave components. So I was able to derive a closed expression for the nonlinear energy transfer between ocean waves. It was represented by a relatively complicated five-dimensional so-called Boltzmann integral. Basically, I solved this problem to relieve my frustration at not being able to solve the turbulence problem.

I presented my results on the spectral energy balance and the nonlinear energy transfer in a seminar at the Institute for Naval Architecture [4]. Although most of the naval architects were somewhat confused by the mathematics, Prof. Weinblum was enthusiastic and encouraged me to continue with theoretical research. Prof. Wieghardt also concluded that I was probably more effective working theoretically than making painstaking experiments with hot-wire instruments, that had a troubling inclination to oscillate. Prof. Roll, who had been working in air-sea interaction for many years, was also there and was apparently favourably impressed. He proposed that I should attend the coming Ocean Wave Conference in Easton/USA in April 1961, to which he had been invited, but could not go. That is how I came to America, where I again presented my results. At that time—although I had not known this–the problem of the nonlinear interaction between ocean waves was seen as one of the central problems of ocean waves. I immediately received invitations to the Ocean Research Institutions in La Jolla, California, and Woods Hole, Cape Cod, as well as to the University of Illinois. I accepted the position of Assistant Professor in La Jolla offered by Walter Munk, whom I met for the first time at the Easton Conference. I found the atmosphere at the Institute for Geophysics and Planetary Physics that he had just founded at Scripps Institution of Oceanography very stimulating. So half a year later, at the end of 1961, I went to La Jolla, and enjoyed more than three very fruitful and stimulating years there.

Did you already have the complete resonant interaction theory on surface waves when you were invited to give a talk in the USA? It is known through your publications that the triple interaction of surface waves does not function and that, one must extend interaction theory to higher perturbation order to get reasonable results.

Hasselmann: Actually, independently of my papers [5, 6, 8–10], Owen Phillips had already shown that the necessary conditions for the resonant energy transfer between different wave components could not be satisfied by three wave components, but only by four. However, Phillips had not derived the Boltzmann equation. Before Phillips published his paper, I had already independently derived the complete Boltzmann equation for the lowest-order triple-wave coupling. When I wanted to calculate the integral, however, I found to my dismay that the resonance condition could not be satisfied. That was a shock. I had calculated the complete theory up to the third order, and understood all the details about the energy transfer through resonant interactions in a continuous ocean wave spectrum, only to discover that the third-order resonance conditions could not be satisfied due to the special dispersion relation of ocean waves. That meant that the calculations had to be extended to fifth order.

I went for a three-hour long walk in the town park in Hamburg and debated within myself whether I could muster the energy to carry through two further orders of these quite complicated calculations. I decided to go through with it and spent another two or three months working on the algebra. It proved not as bad as I had first feared, although I had to derive formulas extending over one or two pages. By the time I received the invitation to present my results at the Easton Conference, I had already found a very talented young student of applied mathematics, Herr Krause (students in those days were addressed rather formally in Germany), who programmed the numerical calculation of the Boltzmann integral for me. He used the highest possible resolution available on the computer of the University, which by now was more than the G1, but still quite limited. I was very impressed that within two or three months he came up with the first numerical results. Although we later obtained more accurate results with improved computers, his results were qualitatively correct. However, they did not agree in all aspects with what I had anticipated intuitively, and so when I gave my talk in Easton [10], I pointed out that they were probably incorrect in some details. Later it became clear, however, that his calculations had in fact been qualitatively quite correct. He had even correctly computed the most important process—which I had questioned intuitively–namely the transfer of energy from waves near the peak of the spectrum to still longer waves. Ten years later we were able to show–through the JONSWAP experiment–that this is the dominant process responsible for the continual growth of wind generated waves from shorter to longer and longer waves. I am still grateful for this impressive contribution by Herr Krause. It enabled me to present not only the theory, but also first numerical results in Easton.

Was it customary these days that you did not program yourself? I am slightly astonished that as a relatively young man, as a postdoc, you got someone to program for you. Were there special technical obstacles to be overcome?

Hasselmann: No, you only had to have some experience in programming. Of course, I cooperated with the student. I explained to him which numerical algorithms should be applied, but he implemented that knowledge into the program, carried out the computations, made the usual tests and searched for errors, etc.. He fully understood what he was doing. I simply hired him as a student assistant.

We are talking about 1960/61. Did FORTRAN already exist?

Hasselmann: I can't actually remember. FORTRAN may already have existed, but I cannot recall in which language Krause wrote the program. I know that the first programs I wrote for my Dr. thesis were in machine code, and my later programs were all in FORTRAN, but I am not sure whether Krause was alredy using FORTRAN.

Starting from 1960, can you please tell us when which persons entered your life?

Hasselmann: During the first period in Germany it was Professors Karl Wieghardt, Georg Weinblum and Hans Roll, and Pascual Jordan as a physics teacher and the usual mathemathics professors, but I was not in personal contact with them. In America, as I said, Walter Munk left—and still leaves— a lasting impression on me. I had already known his name from the first classic publication by Sverdrup & Munk (1947) on the prediction of ocean waves, from which I had concluded, however, that his knowledge of physics was rather limited. At first, I underestimated him as a scientist, but when I got to know him personally, I was very impressed not only by his clear scientific thinking but also by his open-minded, positive and supportive generosity. He had a Viennese charm. He was an Austrian, who had emigrated to America already in the twenties, but still spoke with a strong Austrian accent. I gladly accepted his invitation to his new IGPP in La Jolla. I had an office in the beautiful new redwood building of his institute, that his wife Judy had designed, overlooking the Pacific on a cliff. I felt very happy in La Jolla from the beginning, especially with the open American way of welcoming new visitors. Coming from the somewhat, well, perhaps not stuffy, but not particularly creative atmosphere of German science in the fifties and early sixties, to America, where everyone was really enthusiastic, was a great experience for me.

Walter Munk was the central figure, but there were also other very stimulating people in La Jolla, such as Michael Longuet-Higgins, a well-known applied mathematician and fluid dynamicist from Cambridge, who had contributed many basic papers on ocean waves, microseisms and other geophysical phenomena. He had a guest professorship in La Jolla while I was there. Other guests were Norman Barber from New Zealand, a pioneer in ocean wave research who had studied the propagation of ocean swell, and David Cartwright, a co-developer of the pitch-and-roll buoy for measuring directional ocean wave spectra, and also a leading expert on tides. At Scripps there were also John Miles, who had developed an important theory on wind-wave generation, and Hugh Bradner, an interesting former high-energy physicist, who measured pressure variations in the deep ocean. I further enjoyed the interaction with George Backus and Freeman Gilbert, two young geophysicists of more or less my age, who had done some very nice work on inverse methods in geophysics and whose basic mathematical knowledge was very impressive. Klaus Wyrtki[3] who later became one of the leading figures in El Nino research, and Carl Eckart, who had written an impressive book on theoretical oceanography, were also two well known figures in Scripps at that time, although myself had little direct contact with them. Another person who came to Scripps while I was there was David Keeling (he signs his papers Charles Keeling), who was making measurements of CO_2 on Mauna Loa in Hawaii. He had just started the measurements four years earlier. I didn't know at the time that I would later be continually referring to the now famous Keeling curve as the most important observational basis of the climate change debate. Our main contact at that time was through the madrigal choir that a few of us started. It later blossomed into quite a large university choir led by David until he died last year.

So I was immersed in a highly stimulating scientific environment. The discussions continued also in the weekly wine and spaghetti parties in Walter Munk's home—a beautiful spacious redwood bungalow overlooking the Pacific, which his wife Judy had also designed.

There were also many stimulating students. The first student I supervised was Russ Snyder, who worked later also in ocean waves. I kept in contact with him, and several years later we wrote a joint paper, together with my wife and two other colleagues [114]. My wife and I also joined Russ's family on a two-week sail in the Eastern Mediterranean along the beautiful Turkish coast. It was on their way back to America after a three-year sail around the world in a ketch Russ had built himself. My second student was Kern Kenyon, who visited me later in Hamburg and is still at Scripps today. Then there was Brent Gallagher, who also was very talented and did some nice work on nonlinear barotropic waves. He is now somewhere in Hawaii. Finally, there was Tim Barnett, who in his Ph.D. thesis developed the first model for ocean wave prediction based on a realistic representation of the spectral energy balance, including the nonlinear energy transfer. Some years later we worked together in the JONSWAP experiment, and still later, after the Max Planck Institute was created, we cooperated in several papers on climate. Today he is a well-known climate researcher. So, these were my first students. I am glad they all did well.

---

[3] Klaus Wyrtki has been interviewed in English earlier, see von Storch, H., J. Sündermann, and L. Magaard, 2000: Interview mit Klaus Wyrtki. http://www.hvonstorch.de/klima/Media/interviews/Wyrtki.pdf (GKSS Report 1999/E/74, 41 pp.).

I know that you were not always seated at your desk, interpreting integrals. You also did experimental research, e.g. on Hawaii.

Hasselmann: This was the first large, ocean-wide wave experiment organised by Walter Munk and coordinated by Frank Snodgrass, a technician and Walter's right hand man in all experimental matters. Similar to Norman Barber, Walter Munk had carried out continuous measurements of the spectral properties of swell arriving at a single coastal station, in his case near La Jolla. He had inferred from the gradual change in the observed swell spectra–the arrival first of very long waves, followed by waves with gradually decreasing wavelengths—that the swell must have originated in storms very far away in the South Pacific and Antarctic. Munk now wanted to find out how the energy of the swell changed as it propagated from its source somewhere south of Australia, in the high-wind region of the "fighting fifties", across the entire Pacific up to Alaska, over a distance of about two thirds of the earth's circumference. Some waves even originated in the Indian Ocean, propagating into the Pacific along a great circle between New Zealand and Australia. So Munk set up a series of wave measuring stations along a great circle extending across the entire Pacific, starting in New Zealand and ending in Alaska. In between there were stations at Samoa, Palmyra, an uninhabited atoll between Samoa and Hawaii, Hawaii, and "Flip". Flip was a special ship anchored between Hawaii and Alaska that could be flipped so that it stood vertically like a float in the water, the bows up high and the stern down below. The idea was that this way the boat stayed almost still in the waves and could be used as a wave measuring station.

Walter Munk, with Judy and his two daughters, stayed in Samoa, a scientist, Gordon Groves, and radio operator were flown to Palmyra, Frank Snodgrass and I myself, with my wife Susanne and three children, were in Hawaii. Frank Snodgrasss took care of the logistic organisation, and I had to tend a wave instrument and check the data from the entire experiment, which was flown to the computer center in La Jolla and then back to Hawaii for a first analysis. The experiment ran for the three summer months of 1963.

We had a wonderful time in Hawaii. One of the first things Frank Snodgrass did was to install a telephone connection from the swell measurement station off Honolulu to our house in Kailua, which was situated on the other (northern) side of the island. My measurement task was to turn on the tape recorder for an hour at 06:00 a.m. and again for an hour at 06:00 p.m, check for a couple of minutes whether the data on the paper tape looked OK and airmail the tapes to Scripps for spectral analysis. And occasionally I would plot up the analyzed spectra from all the stations that were sent back to Hawaii from La Jolla.

Unfortunately, this wonderful time was occasionally interrupted by the electric generators on Palmyra breaking down. They had five generators, of World War II vintage, which one would have thought was sufficiently redundant, but four were usually broken down. I had to drive around Oahu to find replacement parts. Palmyra had served as an airbase during World War II, but was now deserted except for our scientist and the radio operator. Frank Snodgrass felt rather uneasy about leaving two people alone on a deserted island for three months. So he had arranged that if Gordon Groves should inform him via the radio operator that "the second amplifier had failed", this was code for "urgent problem, come immediately". After two weeks we received the message. I went there by plane to find out what was wrong. In the meantime, however, the two had already patched up. Two weeks later the radio went silent and we did not hear anything from the two. Then I received a radio message that Gordon Groves had hurt his hand, which was bleeding strongly. This was followed by another week of total silence. We became quite worried and decided to go there by plane.

The first time I flew there it was in an old B25, a twin-engined bomber from World War II, used by former marine aviators to spray fields. A short time earlier, they had already tried unsuccessfully to fly to Palmyra. They did not have any modern navigational aids. They flew by Dead reckoning, i.e. like a sailor without navigational marks. You fly in a certain direction at a certain speed for a certain time and calculate your position accordingly. In addition, you must know the winds. They arrived at the calculated position, but Palmyra was nowhere to be seen. So they flew on to Tahiti. But there a thunderstorm prevented their landing. So they flew back, again over Palmyra without finding the atoll. With their last drop of fuel they just managed to land in Honolulu. The whole airport had been closed down. No other plane was permitted to land before they had landed. Directly after landing, the two pilots were taken off by the police.

That was the crew I flew to Palmyra with. If my wife had seen those bearded and dirty characters, sparsely clad in shorts, with or without T-shirts, she never would have let me fly. They again had problems finding the atoll. I was seated behind the navigator who was busy with his square search, and I could see pearls of sweat developing on his neck. But suddenly he cried: „There's the island!"

After that first time, Frank Snodgrass decided not to repeat the experience. He was able to obtain a transport aircraft of the US Coastal Survey, a large four-engined machine with a crew of eight, modern navigational aids etc. When we arrived and wanted to rescue our assumedly seriously ill scientist we were met by our two friends, both extremely cheerful, and with Gordon Groves sporting a small band-aid around one finger.

It was a time full of fun and adventure. Walter Munk, however, was a little disappointed by the outcome of the experiment [18] because he had hoped to observe the attenuation of swell by interactions with the local windsea, when the swell crossed the trade wind areas. However, no significant loss of swell wave energy could be found over the entire distance travelled by the waves, from Antarctica to Alaska. This was nevertheless an important result, which was used in the wave prediction models that were developed later. We did infer some energy loss immediately after the wind-generated waves left the area of high winds and started on their long journey as swell, that is, as long waves that are no longer forced by the wind. We were able to explain this by the nonlinear energy transfer. This was perhaps the first observational evidence of the significance of this process for the energy balance of the wave spectrum.

The Pacific swell experiment supplied also the idea for JONSWAP, the Joint North Sea Wave Project, which we carried out in the summer months of 1968 and 1969. JONSWAP was complementary to the Pacific swell experiment. Instead of studying the propagation of swell after the waves had left the wind-generating area, we investigated the growth of wind-generated waves themselves within the wind generating area. To understand the dynamics of waves, this question was clearly fundamental. We used the same strategy as in the Pacific wave experiment, but on a much smaller spatial scale: we observed the change in the wave spectrum under off-shore wind conditions at ten wave stations spaced over a distance of 160 kms off the West coast of Germany, off the island of Sylt near the Danish border, in the North Sea.

Nevertheless, many things were still to happen before the JONSWAP experiment. Your time in the USA ended, and you returned to Germany. Why?

Hasselmann: As I explained, the scientific working conditions in the USA were excellent. However, my wife was less happy, although this improved after we made friends, sang in the San Diego chorale and in the madrigal group that we had founded with Dave Keeling. Susanne had also made friends with a very stimulating piano teacher. But our children were also not as happy as they had been in Germany, especially our oldest, Meike, who had always been a beaming sunshine. At that time California was going through a phase of laissez faire, in which children grew up without any restrictions. They never knew any rules, what was permitted or forbidden, and they always seemed ill-tempered. At least in the kindergardens we knew the children did not seem to be really happy. Meike had become rather unstable. She had a pseudo croup, and we nearly lost her. In the end, we finally decided to return to Germany and bring up the children there.

But the decision was difficult and we did not make it immediately. Before going back I first tried a joint appointment, with six months in Hamburg and then six months again in La Jolla. But then we finally decided to return to Hamburg. It was not an easy decision.

How did you go on? Assistant at the Institute for Shipbuilding. Returning to the much more authoritatively organised German university must have been quite a difference from the more liberal structures in California? And to be taken up only as an assistant.

Hasselmann: No, I really had no problems. I had to give relatively few lectures, and this suited me, because I always felt that I could not explain things better than they were explained already in good text books. I was never a motivated lecturer on basic courses. I liked talking about research in seminars, but I was not motivated to repeat the basics that people could better study in text books that had been prepared with much greater care than I ever devoted to my lectures. I myself also preferred learning from books, at a pace set by myself, rather than being told things by someone else. Presumably, this influenced my attitude. So I was left in relative peace regarding lecture activities. And I tended to choose subjects which attracted only a small number of students, so that contact could be more personal.

Also, although I was in an Institute for Naval Architecture, I was able to follow up on my ocean wave research, in which I was still interested, and prepare the next JONSWAP experiment, which I mentioned earlier. So I was not really hemmed in by Germany's relatively conservative system, because I was in a rather unconventional position.

Concerning this back and forth between Germany and America. The Center for Fluid Mechanics in that time was in England. Had you any time, opportunities or desire to go to England and work there?

Hasselmann: I was in fact invited as a Visiting Fellow for half a year, in 1967, and visited the Department of Applied Mathematics and Theoretical Physics. But I did not have a strong desire to visit Cambridge while I was working in La Jolla because I was more interested at that time in oceanography. There, in Scripps, were the leading scientists in oceanography, in ocean waves, currents and so forth. In England, in Cambridge, the effort was more on pure fluid dynamics and turbulence theory, and my interests had already switched from turbulence theory to wave dynamics in the ocean. I enjoyed my later visit to Cambridge and the relaxed style there, but La Jolla was more stimulating.

So, you came back to Hamburg and to the Institut für Schiffbau and then something interesting happened, something what could not happen nowadays, namely people took very swiftly decisions of what to do.

Hasselmann: I was gradually becoming an embarrassment for the Institute for Naval Architecture, because their main interest was in ship resistance, ship stability in waves—and, of course, in the design and construction of ships themselves–but not in the dynamics of ocean waves as such, or in oceanography in general. And I had started a large international experiment to measure the growth of waves under off-shore wind conditions in the North Sea. It evolved into quite an extensive affair, involving several institutions from different countries: Scripps from America, the National Institute of Oceanography from England, the Dutch Weather and Oceanographic Service KNMI, and the German Hydrographic Institute. There were four or five research vessels and other ships, a lot of activity installing wave measurement masts and wind measurement stations etc. All this created a lot of logistic overhead, and so I was tying up the secretaries, technical people, the workshop and so on in the institute for a project that had nothing to do with naval architecture.

So my former diploma thesis advisor, Prof. Wieghardt, in whose department I was working when I came back from America, came in one day and said quietly: Herr Hasselmann, don't you think you should find some other position somewhere, because it is actually not the main task of the Institute of Naval Architecture to measure waves in the North Sea. I wondered what to do, and so I asked Prof. Roll, President of the Deutsches Hydrographisches Institut, whether he could give me a job. He thought about it for a minute and probably decided that it would be a nuisance to have me in his institute as well. So he called the Federal Ministry for Science and Technology and inquired whether they could not provide a position for me in some form or another.

What then happened was that, at very short notice, the Ministry provided the funds to create a Department (Abteilung) of Theoretical Geophysics at the University of Hamburg, of which I was to become the director. An Abteilung had to be part of some institute, so Professor Menzel, the director of the Institute for Geophysics, was asked whether the new Department for Theoretical Geophysics could become part of the Institute of Geophysics.

Professor Menzel, a very kind man, agreed. And so I became a member of the Institute of Geophysics. I received some research funds from the Ministry for Science and Technology, as well as a secretary, and a small apartment, of about six rooms, I think, next to the Institute for Geophysics, in the Schlüterstraße. I worked there until the Max Planck Institute for Meteorology was founded in 1975—apart from a two year stay in America between 1970 and 1972. So the department was created, basically, through an informal discussion between the Ministry for Science and Technology and the director of the Deutsches Hydrographisches Institut, with the good-willing cooperation of everyone involved.

"Short notice"—how short was that notice?

Hasselmann: I cannot remember exactly how short it was, but it was really fast, because I was in the Abteilung when JONSWAP started, already in 1968, and I had just come back from Cambridge in 1967 and was already strongly involved in the planning of JONSWAP when this development began. It must have been less than half a year or so.

This would not be possible nowadays.

Hasselmann: Well, that was in a period of rapid scientific expansion everywhere. The same atmosphere prevailed in America, where a position was offered to me more or less spontaneously and was formalized within a few months. That was a time when one was looking for good young people everywhere, trying to build up a good research environment in response to the challenge of sputnik. Everyone was trying to be in the forefront of science. This was particularly true in Germany, where in the wake of the Wirtschaftswunder one wanted to catch up also in science.

Other people known to work with you entered the stage at that time.

Hasselmann: That's right. When the Department of Theoretical Geophysics was created I took on some Ph.D. physics students who were interested in working in geophysics, in particular in ocean wave theory and in the general theory of nonlinear interactions in geophysical wave fields, such as internal waves. At that time I had a number of good young students, for example, Dirk Olbers, Peter Müller and Jörn Kunstmann.

Kunstmann did not do any oceanography, he was working on plasma physics.

Hasselmann: That's true, I remember. At that time I was interested also in plasma physics. I had written a couple of papers with my former student friend Gerd Wibberenz on the scattering of protons in the solar wind by irregularities of the solar wind magnetic field. As lecturer in physics in Kiel, Wibberenz was working on problems of interplanetary space. I found the problem intriguing because it could be treated by exactly the same formalism that I had applied to determine the nonlinear energy transfer in an ocean wave spectrum. I also found working on this problem was useful because I gained some practice in the notation of relativistic electrodynamics, which was helpful for my recent excursions into particle physics—another of my interests that we can discuss later. Actually, the solar wind community was also not used to the relativistic notation, so that they had some problems reinterpreting our results in their language, but our papers were well received nonetheless [23, 28, 29].

Anyway, to better understand plasma physics, I decided to hold a seminar course on plasma physics together with Gerd Wibberenz and my other student friend Wolfgang Kundt, who at that time was a physics lecturer at Hamburg University. That's how Jörn Kunstmann came to me. His Ph.D. thesis was on interactions in the solar wind.

You said, you took some students. What you really did was to ensnare a whole seminar group from your friend Wolfgang Kundt. You gave a half of them new topics to work on their diploma, because we did not know what to do at that time.

Hasselmann: Yes, I seem to have hijacked Peter Müller and Dirk Olbers and maybe some others. Arne Richter and Hajo Leschke were also in that group, I think, but they did their diploma and Ph.D.s with someone else, probably with Wolfgang Kundt. The people that came to me seemed to be quite content just learning methods, physics and mathematics, but had no clear idea of what they should do for their diploma or Ph.D. thesis. So they were quite happy when I suggested some topics to them.

There was an IUGG Conference in Bern in 1966. There you suddenly became the coordinator of the JONSWAP effort.

Hasselmann: I became coordinator to my big surprise, by default, probably because I initiated the idea that we should do a joint experiment. I invited some colleagues I knew—David Cartwright from the National Institute of

Oceanography in England, Tim Barnett from Scripps, Karl Richter from the Deutsches Hydrographisches Institut, and some colleagues from the Netherlands, to discuss the idea of a joint experiment on wave growth in the North Sea. We met at the IUGG in Bern. We wanted to measure wave growth under off-shore wind conditions. I remember I had the crazy idea—as a physicist and theoretician—that in case of an east wind, we could measure the waves off the west coast of Germany, and when we had a west wind, we could measure waves off the east coast of England. But then some experimental colleague pointed out that it would be impracticable to install wave measurement stations on both sides of the North Sea, and that ships can not steam fast enough to go from one place to the other when the wind changes. So we decided to have the experiment on the east side of the North Sea, off the island of Sylt.

All this was agreed upon in principle, and then we went off home again. And then we suddenly realized that we have not discussed at all how to organize the experiment, and who should be the coordinator. Everybody assumed that because I had proposed the experiment, I should be the coordinator. I thought this was not a very good idea at all, as I had absolutely no experience in seagoing oceanography, and my past experience with experimental work with hot-wire turbulence measurements had convinced me that I was better employed doing theoretical work. But anyway, I was landed with this task and had to organize it.

The experiment was planned for the three summer months of 1968. A few months before the experiment was due to start, and everybody was geared up to install their equipment, I received a telephone call from the German Ministry of Defence saying that we would have to cancel our experiment. NATO was planning a large sea-to-air missile test in the North Sea at the same time. They would be testing radar methods of tracking missiles, and the ships and wave masts that we were planning to deploy would interfere with their radar signals. I said that it is impossible to cancel our experiment at this late hour, as we had already spent at least two million Deutsch Mark preparing for the experiment. The Ministry of Defense said that this might be true, but that they already spent fifty million on their exercise, so we have to cancel ours. I said, well, we cannot cancel it this way. The only solution I can suggest is that we reduce our experiment this year, without the wave masts and some of the ships, on the condition that you fund us to carry out the full experiment as originally planned next year. The Ministry of Defence agreed, and so we carried out two experiments, a reduced trial experiment in 1968 and the full experiment in 1969.

In retrospect, we were very fortunate that this happened, because it turned out that, from the point of view of logistics, the first experiment was a complete disaster. I had worked out precisely when every wave-measurement station should start recording, and for how long and how often, based on the wind conditions and the speed of propagation of the waves from one measurement station to the next. So on one particular day a particular station, a wave mast, say, should start recording at 7:30, measuring for half an hour every three hours. Further out a ship, say, should start recording at 11:45, and so on. But the communication system we had installed turned out to be completely inadequate to transmit this information reliably. This was not helped by the Russians jamming our radio stations everytime we went on the air because they thought we were part of the NATO exercise. We did get some nice data in the end, more or less by chance, but much less than we had hoped for. The coordination of the experiment was a continual stream of improvisations.

But we gained a lot of experience, and the next year, when we carried out the full-blown experiment, everything went very smoothly. We had a functioning communication system, a reliable predetermined schedule of measurements, and well organized logistics. All the equipment worked fine, and we obtained a very good dataset. The analysis of the data laid the foundation for the modern wave models that we later developed. So we were very fortunate that the Ministry of Defence interfered with our original plans and gave us a free trial experiment, so that we could carry out a good experiment one year later.

Would you mind assessing the impact of this experiment on your personal career, standing and satisfaction?

Hasselmann: JONSWAP was certainly the most successful experiment I have been involved in. We were extremely lucky, not only because of the free trial experiment, but—still more important—because we were able to explain the principal results of the experiment by the one single process governing the dynamics of wave growth that we were also able to compute theoretically from first principles, without any empirical parameters—namely the nonlinear energy transfer I had derived earlier.

The idea of the experiment was that we would determine the processes governing the dynamics of ocean waves by measuring the change in the wave spectrum as the waves develop under an off-shore wind from small, short waves close to shore, to longer, higher waves further off-shore, out to still larger distances off shore where the waves had reached a fully-developed equilibrium state—assuming such a state exists. The spectral energy balance of the waves is controlled by three main processes: the generation of waves by the wind, the dissipation of wave energy by white capping, and the redistribution of energy across the wave spectrum by the nonlinear energy transfer. Prior to JONSWAP, we had assumed that the nonlinear transfer had only a minor impact on the evolution of the spectrum. This was based on the results I had presented at Easton, which were computed for a fully developed spectrum. But we discovered in JONSWAP that the spectrum of a growing wind sea has a much higher, sharper peak. This greatly enhances the strength of the nonlinear transfer. And it is this feature, the sharply peaked spectral shape, that is the origin of the transfer of energy from the peak to still longer waves—that is, for the continual increase in the wavelengths of a growing windsea. I still remember the excitement when we repeated the nonlinear energy transfer computations for the new JONSWAP spectra and the points came out, one by one, directly on top of the observed spectral growth.

Based on these results the wave community was then able—several years later—to develop the wave model WAM that is used today by more than 200 centres world wide, including operational global weather forecasting centers such as ECMWF, the European Centre for Medium Range Weather Forecasting, that produces daily global forecasts of the two dimensional ocean wave spectrum. The forecasts are supported today by wind and wave data from modern satellites, that the wave community also helped to develop in follow-up experiments of JONSWAP, and for which they developed the necessary retrieval algorithms and assimilation methods. But ultimately, the success of much of this development really hinged on luck: the fact that the one process that we could really compute rigorously, the nonlinear energy transfer, turned out to be the dominant process governing the form and rate of growth of the ocean wave spectrum.

Regarding my own personal career, I was recognized as the lucky person who happened to have developed the relevant theory, initiated the experiment and coordinated the analysis. We carried out the initial analysis first in our various home institutes and completed the analysis in a workshop at the Woods Hole Oceanographic Institution—which I was visiting at that time—in the spring of 1971. The results [35] were presented the same year at the IUGG Conference in Moscow.

For me it was also a great experience that you can carry out an experiment which was a complete fiasco in 1968 and still be respected by your colleagues. In the business world I would have been fired. But the scientific community is extremely tolerant and understanding. I had the same experience later with other experiments, some of which also turned out to be a flop. I was always encouraged by my colleagues, who stood by me and accepted the fact that not everything that you try to do in science works. I personally very much enjoyed the experience of JONSWAP and the follow-up experiments JONSWAP2—although this was a flop—and MARSEN—this time a full success—in which we tested various remote sensing techniques relevant for the new wave-measuring satellites SEASAT and ERS-1. I also enjoyed the work later in the WAM group, in which we jointly developed the global wave model WAM that I mentioned [90].

All in all, JONSWAP clearly had a positive influence on the way my life developed. Probably, the fact that I was able to combine a field experiment with theory, both of which I had been involved in, also helped when I was later asked to become the director of the Max Planck Institute. It was presumably assumed that this indicated that I had enough flexibility to develop a new research program in climate. But that is only my guess. Anyway, JONSWAP was a lot of fun. It was a period in which we generated many lasting friendships. We had many parties and get-togethers with everybody involved, from the technicians to the radio operators to the ship people to the scientists. There was a great team spirit.

Could you speak about the role of Wolfgang Sell?

Hasselmann: The success of the experiment was due to the team work of many people, but two people in particular deserve mention. One was Addi Hederich, a technician from the Deutsche Hydrographische Institut. He coordinated the entire logistics, the ship schedules, the installation of the wave masts and wave buoys, including the main tower PISA for meteorological and wave measurements, as well as the complex operations for servicing the equipment at sea. He worked tirelessly in 1968–1969 to bring everything together.

The other person was Wolfgang Sell. We had collected an enormous amount of data—for those days—nowadays it would be peanuts. But, for that time, we were immersed in an intimidating array of data from instruments of many different types, with different data formats, obtained at different times and different places. Nobody had really thought seriously about how to bring all these data together into a coherent dataset. Nowadays this is routine. But for us it was quite new. I personally did not think about it at all and simply assumed that we would muddle through somehow. Fortunately, there was Wolfgang Sell in the team who realized that we had a problem. So he immediately sat down and worked out a data analysis scheme of how to store the data, how to process them, bring them together and manipulate them with a single data processing software. Without that input from him we would never have been able to complete the analysis of the JONSWAP data within only two months in Woods Hole—in time to present the results at the IUGG conference later that year in Moscow. Wolfgang Sell and a few other stalwarts, Peter Müller and Dirk Olbers, stayed on after the main workshop and helped clean up the results for the IUGG meeting.

At that time also a number of new persons came on the stage. One was Elsa Radmann.

Hasselmann: That was my secretary, a very reliable person. She came in 1968 when the Department of Theoretical Geophysics was founded and stayed with me until her retirement some thirty years later. She helped first in the organization of JONSWAP. When I went to Woods Hole for two years, in the autumn of 1970, she took care of the institute while I was away, kept up the communication, and so forth. She was an extremely reliable, conscientious person that I owe very much to. If I had to travel somewhere, I never checked where I was staying until I arrived, she had always arranged everything perfectly. She also had various likes und dislikes. If you were unfortunate enough to belong to her few dislikes you had a hard time, but for all others she was very helpful and friendly.

You mentioned the data analysis. I remember that you were doing the energy transfer calculations on many different computers. We were in DESY, in Darmstadt, we were here in Hamburg, on the Hamburg computing center and we were also in Woods Hole. Why did you go to Woods Hole? As far as I can see, Woods Hole is not a classical research centre for surface waves, for ocean waves.

Hasselmann: That was basically independent of JONSWAP. I received the offer of a professorship in the Woods Hole Oceanographic institution, on a chair that had just been donated by the Doherty foundation, to develop a joint program on oceanography between Woods Hole and MIT. I said that I would be happy to accept the professorship for two years, but could not decide yet whether I would to stay longer or go back to Germany. However, one of the reasons I accepted was that Ferris Webster, who had made the invitation, said that Woods Hole had just obtained a new computer that would be ideal for the JONSWAP analysis. So when I arrived, I talked to Art Maxwell, the director responsible for research at WHOI, and explained that we had this experiment, and that we somehow had to get together to analyze the data. He immediately offered not only the use of the computer, but also all other needed facilities, as well as some funds so that we could carry out the workshop there. That is the reason we had the JONSWAP workshop in Woods Hole.

There must have been a little bridge nearby.

Hasselmann: I believe you are referring to my memorable encounter on a bridge with Peter Müller. Peter Müller was one of the members of the JONSWAP working group. We had exactly two months to complete the analysis, because then everybody had to go back home. We had a tremendous amount of work to do, a lot of computations, reorganizing and reanalyzing the data from different aspects, and so forth. I was running back and forth under enormous stress to get all this done, between the computer center and the operations room, where we were all working together. And while I was running back and forth and completely out of breath and stressed, I saw one of the members of the group, namely Peter Müller, leaning over this bridge looking calmly down onto the water. I said: "Hello Peter". And he answered dreamily, after a long pause: "Yes, life is good ... but one needs time for contemplation."

Peter Müller and Dirk Olbers were responsible for designing the particular parameter representation of the JONSWAP spectrum.

Hasselmann: Yes, that's right. Peter and Dirk were the creators of the so called JONSWAP spectrum, which has since been widely used. They proposed a very simple three-parameter representation which reproduced the spectral shape very well for the different stages of wind-wave growth.

From your publication list I can see that there were other issues you were interested in, besides the solar wind problem that you mentioned, for example sound waves in the ocean with Hans-Hermann Essen.

Hasselmann: Yes, I wrote a set of papers, mostly with other colleagues or Ph.D. students—although usually the Ph.D. students would carry out the work and publish on their own—looking at different interactions between different types of wave fields in the ocean, the atmosphere and the solid earth. One paper was with Heinz-Hermann Essen [25], on the generation and scattering of sound waves in the ocean by surface waves, one was on surface gravity waves scattering off the ocean bottom, one or two papers were on interactions between internal gravity waves in the ocean and atmosphere, although this subject was mostly well covered by several nice papers by Dirk Olbers and Peter Mueller. One of my early papers was on microseisms [12], the generation of random seismic waves through resonant interactions between surface gravity waves, and between surface gravity waves and the ocean bottom.

In most of these papers we applied the interaction-diagram formalism that Feynman had developed to summarize the interactions between particles. I had slightly modified the Feynman diagram rules in a 1966 paper [16] to adapt the formalism to classical random wave fields.

This brings me to a rather interesting comment on the communication between different scientific disciplines. My standing in the ocean science community was originally founded on my papers on nonlinear interactions between ocean waves. Shortly after coming to America I gave a talk on this work at the Californian Institute of Technology. After the talk my colleague Gerry Whitham came to me and said "That is an interesting talk you gave, but did you ever notice that the plasma physicists appear to be doing similar things to what you are doing?". I replied, no, this was new to me, could he give me some references? So I looked up the references and discovered that the plasma physicists had indeed been doing exactly the same things that I had been doing, except that they were looking at plasma waves instead of ocean waves. This was a bit easier because they did not have to go to fifth order, the resonances occurring already at third order. But to my surprise they never actually presented the nonlinear computations. They simply took the analysis for granted. Sometimes they quoted a paper by Peierls back in 1929, in which he showed that the diffusion of heat in solids could be explained by the nonlinear interactions between phonons. I looked up the paper and discovered that Peierls had carried out exactly the same analysis as I had, using a different notation, but based on exactly the same approach. At that point I realized that my reputation in oceanography was based on very old results in physics that were simply not known in oceanography. I then started reading other physics papers and discovered that exactly the same formalism was used everywhere in quantum field theory, in describing the interactions between different particles, which are represented in quantum field theory by wave fields. Feynman had developed a well-known set of diagrams and rules summarizing the algebra involved. So I wrote my 1966 paper in which I showed how Feynman diagrams could be applied to geophysical wave fields, with a few simplifications appropriate for classical rather than quantum theoretical fields. We applied this formalism subsequently to the various wave interaction problems we investigated.

It was really an eye-opener to realize how specialized we are in our fields, and that we need to know much more about what was going on in other fields. Through this experience I became interested in particle physics and quantum field theory. So I entered quantum field theory through the back door, through working with real wave fields rather than with particles. From this other vantage point I became convinced—and remain convinced today—that Einstein was right in his criticism of the conceptual foundations of quantum theory, and that there was more to the concept of a particle than can be captured by wave dynamics. So since 1966 I have been exploring other approaches to elementary particle physics, parallel to my official research work. But I did not publish my first results, on the metron theory, until thirty years later [121, 122, 131, 132].

You mentioned already that you carried out the JONSWAP workshop in Woods Hole. And after the workshop we all became engaged in internal waves and a large internal wave experiment, IWEX. WHOI was an institute of oceanography. They did completely different things. What was this about? Did they ask you to do this?

Hasselmann: No, I was already interested on internal waves before I came to Woods Hole. Not experimentally, but with respect to wave dynamics. At Woods Hole they were more interested in ocean currents and water masses in the ocean than in surface waves or internal waves. But they had also developed current meters and thermistor instruments, and had considerable experience in deploying current-meter and thermistor-chain moorings. So I thought that WHOI would find it a challenge to deploy a large triangular array of current meters and thermistors to measure the internal wave spectrum in the main thermocline. This they did, very enthusiastically and professionally. Dirk Olbers and Peter Müller, together with Mel Briscoe, analyzed the data and wrote up the results in some very nice papers.

You finally came back to get a professorship for theoretical geophysics in Hamburg in 1972.

Hasselmann: Yes, Professor Brocks, the director of the Meteorological Institute of the University and the Fraunhofer Institute of Maritime Meteorology and Radio Meteorology, had succeeded, with the support of other colleagues, to create a new chair for me in Theoretical Geophysics, which I accepted.

Also, at that time you became a member of the Joint Organizing Committee of the Global Atmospheric Research Program GARP. You were one of the two oceanographers in that committee. In this way you became acquainted with the issues of climate, climate variability, climate change and problems of that sort. How was that?

Hasselmann: I had become a member of the Joint Organization Committee of GARP already in 1971 or 72, before I returned to Hamburg. They were looking for some young scientist who could contribute to the strengthening of the Global Atmosphere Research Program with respect to climate, the second GARP objective. The first was improving weather prediction. They wanted an oceanographer, because of the importance of the oceans for climate, but also an oceanographer who had some experience in air-sea interaction. There was already one oceanographer with this background on the committee, Bob Stewart, and he probably proposed my name. The work in the JOC of GARP was quite fascinating, as we were laying the foundations of what was later to become the World Climate Research Program.

Then you participated in a number of historically important meetings, namely the first climate conference in Stockholm 1974, then another one which focused on ocean problems, in Helsinki. You did not present your own work there, but you were part of the overall brainstorming which took place at that time.

Hasselmann: That's right. The Stockholm Conference was on climate in general, with a number of different working groups looking at different aspects of climate. The working groups were introduced by a few general talks, but the purpose of the conference was to work out recommendations on which research should be done in which areas. I was chairing one of the working groups involved in oceans and climate. I had a similar coordinating role in the following Helsinki Conference on Oceans and Climate, which I convened together with Alan Robinson of Harvard University. The two conferences provided the basis for the creation of the World Climate Research Program a year or two later at a conference in Geneva.

There was something else in about 1971/1972, namely the formation of the Sonderforschungsbereich 94 in Hamburg, of which you became the speaker. That was then when you really became responsible for bigger organization of science, for coordinated and interdisciplinary science. How was that?

Hasselmann: The discussions for the Sonderforschungsbereich 94 began before I went to America—around 1968–69. The proposal was written and accepted in about 1971. The first speaker of the SFB 94 was Karl Brocks, who had been the driving person in the formulation of the proposal. I had very good relations with Brocks. His institutes participated in the meteorological measurements and telemetry in JONSWAP. And he gave me much fatherly advice on how to run big projects, of which he had considerable experience.

Unfortunately, he died in 1972 just before I returned from Woods Hole, and I was elected as his successor as speaker of the SFB 94.

That was a very interesting time, because the SFB 94 was the biggest Sonderforschungsbereich at that time—in fact, later, too. It was extremely broad in its ambitions, encompassing oceanography and meteorology, air-sea interaction, ocean chemistry and ocean biology, with many different participating institutions. The challenge was to bring all these research activities together into a joint program. Many of these groups had never cooperated before and had quite different research cultures.

My first task was to start a series of seminars to define the joint projects that we wanted to carry through. We had written down some general objectives in our proposal, but we really had no clear idea of how these objectives were to be achieved. In these seminars we first had to understand how the different groups thought, and had to learn to communicate between these different cultures. Out of these discussions then came some very interesting ideas, for example, the first Fladen Ground experiment FLEX. The experiment took place in 1976 in the so-called Fladen Ground area of the northern North Sea. It was designed to investigate the coupling between the thermocline and mixed layer and the biological productivity and phytoplankton distribution during the main phytoplankton bloom in the spring. It was carried out in corporation with British groups and I believe some Dutch groups. It was quite a successful experiment. I understand the data is still an important reference data set today.

This is just thirty years ago. Could you say something about how difficult you found it—this first time when you truly became interdisciplinary. So far you were just in the realm of physics and as a physicist you should feel confident. But now you suddenly met very different people, very different scientific cultures.

Hasselmann: That was indeed a very interesting period. I remember our first discussions with the biologists. As physicists, we would ask: what happens during a spring-time phytoplankton bloom in the mixed layer? The biologists would answer with a highly detailed description of the various interacting processes that produce the exponential growth and subsequent decay of the bloom. We would reply: that's great, you seem to understand what happens, so let's put that into a model and test the ideas against some measurements. They would reply: but that's impossible, its much too complicated. And we would say: but if its so complicated that you cannot express it in a model, you cannot say you understand it. And so we would talk around each other.

But once the biologists realized that they were not simply slaves making measurements to test the models of high-brow mathematical physicists, and the physicists realized they were not simply slaves producing computer models to test the ideas developed by better educated biologists, a fruitful cooperation developed. In fact, the phytoplankton model that came out of this cooperation with the biologists formed the core of the global carbon cycle model that later became part of the Max Planck climate model.

You mention the modelers. Maybe you can drop some names?

Hasselmann: The two main people involved in the biological modeling were Ernst Maier-Reimer and Günter Radach. Radach developed the details of the phytoplankton model, but Maier-Reimer was the driver. In fact, he was the driver in all areas of modeling. If you tell him any idea about any process, he immediately produces a model. Actually, I have the same mentality: I like to produce models. But I am not as efficient as Maier-Reimer. In one of our first SFB seminars we were listening to what the biologists were telling us about phytoplankton growth in the mixed layer, how the phytoplankton gets mixed down, and how its growth or decay depends on the depths of the mixed layer and the euphotic layer, the layer penetrated by light. I thought that this would be a nice example to demonstrate how such ideas can be expressed in a simple model. So I coded a simple conceptual model on our small computer in the Institute for Geophysics. At the next seminar I was just going to present my simple computations when Ernst Maier-Reimer produced the model he had developed independently. His model was much better than my simple model. It was a detailed one-dimensional mixed layer model including temperature, phytoplankton and the penetration of the light. And he had produced some very nice plots demonstrating how the phytoplankton distribution depended on the various mixed layer parameters. I was quite impressed, and so were the biologists.

The only thing I am surprised about is that Ernst Maier-Reimer came forward with his model.

Hasselmann: You are referring to the many drawers in which Maier-Reimer has stacked away models that he has not yet shown to others, let alone published. Anyway, in this case—and many others—Ernst had a strong positive influence on the cooperative programs we developed in the SFB 94.

So you became engaged in networking, in bringing large groups of different sorts of scientists together to tackle questions of a system—in this case the system of the

North Sea. You were also confronted with questions about climate and then, some day, Reimar Lüst[4] came into your office.

Hasselmann: I did not find out the background of why he came into my office until later. Apparently, the Max Planck Society had decided to accept the proposal of the Fraunhofer Society to take over the former Fraunhofer Institute for Maritime Meteorology and Radio Meteorology of Professor Brocks in exchange for an institute of the Max Planck Society. The Fraunhofer Society was dedicated to applied research, but Brocks' Fraunhofer Institute was engaged in basic research on air-sea interaction and radio meteorology. At that time the Max Planck Society had an institute in Würzburg that was engaged very strongly in applied research in solid-state physics. Thus the proposal was that the two societies should simply exchange institutes. It seems that the Max Planck Society had agreed. So the President of the Max Planck Society, Reimar Lüst, came into my office in 1974, apparently looking for a director of this new institute.

The concept was that the institute should not simply continue Brocks' work on air-sea interaction, but should focus primarily on climate research. The principal advisors of the Max Planck Society in this decision appear to have been Hermann Flohn in Bonn and Bert Bolin in Stockholm, the chairman of JOC. The Max Planck Society probably thought that, as a physicist, with experience in various areas of research in the past, I would have enough flexibility to develop an effective program in the new area of climate research. As member of the Joint Organization Committee of GARP, I had been involved in preparing what was later to become the World Climate Research Program, which was probably also one of the reasons they chose me.

The embarrassing thing was that when Lüst came into my office I had only met him once before—he was present at the most disastrous talk I had ever given in my life.

I was supposed to give a formal presentation about oceanography to a lot of high ranking people that were responsible for funding research in Germany. I had intended to work on my talk in the plane on my way over from Woods Hole, but I was tired and I could not concentrate. The next day I was still more tired with jet lag, and felt very uncomfortable when I entered the large lecture room full of people in suits and ties. So I thought that I would break the ice at the beginning by telling a little joke. But the microphone was not

---

[4] Reimar Lüst has been interviewed in German earlier, see von Storch, H., and K. Hasselmann, 2003: Interview mit Reimar Lüst. http://www.hvonstorch.de/klima/Media/interviews/luest.interview.pdf (GKSS Report 2003/16, 39 pp.).

working properly, and somebody in the front row said "could you please repeat what you said?" I did not see much point in repeating my feeble joke, and started off on my poorly prepared talk.

So I went off rambling about all sorts of vague things about ocean research in general. I finally tried to escape from this floundering by giving an example of research. I wanted to explain how the random spectrum of ocean waves is generated by superimposing many different sinusoidal waves. This part I had prepared back in Woods Hole with a set of transparencies which I superimposed one after another. The result was impressively realistic and quite convincing. This time, however, when I began overlaying the different transparencies, I noticed that the audience was getting uneasy, then it started tittering, and finally it broke down in uncontrolled laughing. So I looked back onto the screen and saw that it had become completely black. The projector was too weak to shine through more than one or two transparencies, and my harmonic superposition, instead of producing a random wave field, had gradually transformed my sinusoidal waves into pitch black darkness. I somehow stumbled through to the end of the talk, but it was the worst talk I have ever given in my life and long haunted my dreams.

This was in the hotel Atlantic in Hamburg. My colleagues were very mad at me because they thought that this was hardly the way to convince the people that held the purse strings that investment in ocean research was a good idea.

So I was very surprised that, despite having witnessed this disaster, Reimer Lüst was offering this position to me.

So you were suddenly confronted with this Max Planck Society. Have you met with people in that group before? There was no Max Planck Institute, there was just the Max-Planck Society President who came in your office offering the position of the director of a new institute. What were the constraints of this offer? Did he provide you up front with a generous budget?

Hasselmann: When he made this offer, I had of course a discussion with him over the level of support the institute would have. I said that I would need one director for the group from the former Fraunhofer Institute for air-sea interaction.[5] Lüst accepted. I added that I probably would need two more directors, one for climate data, one for the atmospheric part of the climate system. Lüst replied that that would be very difficult, because the

---

[5] This position was later taken over by Hans Hinzpeter, wo was also earlier interviewed in this series, see: von Storch, H. and K. Fraedrich 1996: Interview mit Prof. Hans Hinzpeter, Eigenverlag MPI für Meteorologie, Hamburg, 16 pp, http://www.hvonstorch.de/klima/Media/interviews/hinzpeter.pdf.

Max-Planck Society did not have the budget for this now. But if it turned out to be necessary later on, the Max-Planck Society would consider a third person, at least. This was a gentleman's agreement. We did not have it written down anywhere.

Reimar Lüst then asked whether we needed a computer. I said that I did not need a large computer straitaway, but would want one later. First, we would need to develop our research program. It was clear to me that we had to solve many fundamental issues first. Once they were clarified, we would come back to the issue of a large computer. That we would need a supercomputer sooner or later was clear to me from the beginning. Lüst accepted this too.

So, essentially, I started the institute on the commitment of one additional professor to take over the former group of Professor Brocks and the gentleman's agreement of a possible third director and a supercomputer at a later time. The staff for the climate group consisted of five scientists and some additional technical and administrative staff. The group was not large, but this complied with the general Max Planck Society policy of not assigning more than about five scientists to a director, otherwise the director would turn into a manager rather than remaining a creative scientist.

It took three or four years before I had gradually filled the five scientist positions and the climate research program began to take shape. So this was the starting basis of the institute. Later on, as the institute developed, the other elements of the gentleman's agreement with Reimar Lüst were also eventually realized.

The budget—I forgot what the actual value was—was more or less fixed. It was agreed that it would not be changed significantly from one year to the next. This is also general Max Planck policy. A constant, dependable funding level is clearly a necessary requirement for the development of a long-term research program. If we needed additional funds we could apply for these from third sources, which we did later when it became necessary. The Max Planck Society also had additional funds for special projects, but we normally received supplementary funds later through the climate programs of the Federal Ministry of Science and Technology (BMFT) and the European Commission. I was very grateful that the basic funding through the Max Planck Society was reliable and did not require a fight each year to become renewed.

Concerning models—here was a running atmospheric model in the group of Günter Fischer in Hamburg.

Hasselmann: Yes, the atmospheric model was not a problem. There was a good atmospheric general circulation model available already from Günter

Fischer at the Meteorological Institute of the university. And there was a still better operational model developed by the larger group at the European Center for Medium Range Weather Forecasting (ECMWF) in Reading.

Thus, these models were around and here you were with a new institute without a computer. You pushed for analytical approaches and indeed, the first publications and ideas were analytical.

Hasselmann: When the institute was created, I had two goals. One was understanding the origin of the natural variability of climate. This was not understood at all, but was clearly a key issue if we wished to distinguish between natural climate variability and human made climate change. I had just developed my stochastic model of climate variability [38], so I could build on that work as a starting point—we had a ready-made core program. Our first publications were, as you said, in this area. The other goal was developing a good ocean circulation model for climate studies. I knew from the Helsinki meeting that the biggest gap in the development of a climate model was the ocean model. We needed a good coupled atmosphere–ocean model, but we had no global ocean circulation model of comparable quality to the available global atmospheric circulation models.

Kirk Bryan had his model at the time?

Hasselmann: Yes, it was a start, but it was not generally regarded as adequate for climate studies. It was a highly diffusive model, with a thermocline that was much too deep.

Later Maier-Reimer's model was based on similar numerics, but maybe the idea was to go different.

Hasselmann: Our goal was to produce a better model. We developed the model concept in a series of mini-seminar meetings in my office. We first explored the idea of building a composite ocean model consisting of different components for different regions, with different resolutions and different physics. The idea was to distinguish between the fast barotropic and slow baroclinic components of the system and treat them separately, and to combine these with models of, say, the Gulf Stream, the equatorial-wave system and the surface layer, all within a complete coupled system. However, we ran into severe problems already through the coupling of the barotropic and baroclinic components via the bottom topography. In the end,

Maier-Reimer wisely dumped all these ideas and quietly produced a traditional gridded model, the Large Scale Geostrophic (LSG) Model, but with improved numerics. The LSG model used an implicit scheme that allowed much larger time steps, so it could be integrated over much longer times. The model was also no longer as diffusive as the Bryan model.

At the same time we were developing the global ocean circulation model, we were looking also at the carbon cycle. Maier-Reimer produced a first global carbon cycle model by incorporating the uptake and transport of CO_2 in the LSG ocean circulation model. This he successively extended in the following years by including various biological sources and sinks. The chemistry was also gradually generalized to include further constituents and tracers.

Thus we soon had a full climate model consisting of a coupled ocean–atmosphere general circulation model and the carbon cycle. The improvement of the global climate model, and its application to predictions of both natural and human made climate change, later became the main thrust of the institute's climate program.

Hans von Storch: I think it was one of your weaknesses that you have not been very good in telling the full picture. You had that vision, but you did not really share it with your coworkers—maybe you believed everybody would know, because it was so obvious to you. From my time at the Max Planck Institute we had not understood the grand strategy in the beginning.

Hasselmann: That surprises me. I hear this for the first time. So I suppose I was not clear in describing the goals that we were following. But as you say, I thought it was obvious.

Dirk Olbers: The SFB was going on all the time. I remember many, many meetings with the atmospheric modeling group of Günter Fischer, with Erich Roeckner and others. But our message was that we wanted to make progress with analytical means. All the Postdocs and the Ph.D. students in the first years were working on simpler subsystems like ice propagation, like mixed layer physics etc.

Hasselmann: I think you are confusing the two main branches of research I mentioned. One was looking at natural climate variability. This we could study using simple energy balance models, sea-ice models or mixed-layer models. That was what Klaus Herterich [83], Ernst Walter Trinkl [59], Peter Lemke, Claude Frankignoul [39], Dick Reynolds and others were doing.

That was one aspect. I was simply exploring what could be done with the stochastic climate concept that already existed, and a number of publications came out of this approach quite quickly. These efforts were independent of the parallel development of a realistic comprehensive climate model. This took longer, involved more discussions, and the publications came later. The strategy was to first demonstrate the basic principles of how long-time-scale climate variability can be driven by stochastic short-time-scale forcing by the atmosphere, using simple climate models. Once this was achieved, we could apply the concept later to the more sophisticated climate models that Meier-Reimer, Günter Fischer, Erich Roeckner and others were developing. This in fact happened. After Maier-Reimer had developed the LSG ocean model, he wrote an interesting paper with Uwe Mikolajewicz[6] on the natural long-term variability of the ocean circulation generated by short-term fluctuations in the atmospheric forcing. I had assumed that this strategy was obvious, but perhaps it wasn't.

Hans von Storch: I understood that much later, but now I see it and it makes very much sense. The relatively simple concept of a stochastic climate model was very useful for the overall debate because it helped overcoming the traditional concept that if climate is changing then there must be a driver. The role of internal dynamics was simply not seen. On the other hand, the nonlinear issues, chaos and so on, were coming up at that time, to which the stochastic climate model was a useful simple alternative.

If you now speak to students, also here at the Max Planck Institute, hardly anyone would know anything about the stochastic climate models. Even though you have brought it down to a form which is very easy to understand nowadays. In those days it was very complicated. How do you feel or observe that this aspect, at least in the present Max-Planck-Institute, is almost forgotten?

Hasselmann: I think it depends on your background training. If you are used to working with a high resolution general circulation model, looking at all the dynamics and interactions and so forth, you probably never think about Brownian motion or may not even have heard of the Langevin equation. These are simply not part of your basic research experience. If you are accustomed to only one way of thinking, you simply cannot see problems in another way. People are too specialized in the particular techniques they

---

[6] Mikolajewicz, U. and E. Maier-Reimer, 1990: Internal secular variability in an OGCM. Climate Dyn. 4, 145–156.

have learned. They are not able to cross their narrow boarders and see things from a different–often simpler and more elegant–perspective. But I don't see this as a basic problem. Sooner or later, ideas that are fruitful will always find acceptance.

In principle these ideas are now well known and this is why we quote it. Also people speak about this concept and your name is associated to it. Hardly anybody has read the 1976 Tellus paper but very many are quoting it.[7]

We should hear some more about the stochastic model. You mentioned that you came from turbulence theory, which you were then able to connect to the ocean wave problem. But you had learned all the techniques already. Was this the same situation with the stochastic model?

Hasselmann: Yes, but the stochastic model is on a much simpler level. It is just an application of the concept of Brownian motion as developed by Einstein in one of his famous 1905 papers. Like many of Einstein's concepts, the idea is elegant but basically very simple. The fact that the short-time-scale Brownian forcing is non-differentiable is a slight complication, but otherwise the basic diffusion process is quite elementary. I became acquainted with stochastic processes in various forms through my work both in turbulence theory and with hot-wire turbulence measurements. If you are trying to build a high-level amplifier which is continuously on the verge of oscillating because of feedback, you start reading about systems analysis and very soon come to stochastic processes. Brownian motion is one of the simplest stochastic processes. The idea that one could explain long-term climate variability very simply by the short-term fluctuations of the atmosphere in analogy with Brownian motion came to me while I was sitting in a plane somewhere, I believe on the way to the Helsinki conference. The idea is really rather obvious, and I thought I would write it up somewhere in a little note.

But it came as a very big surprise in the meteorological and oceanographic quarters.

Hasselmann: And it took a surprisingly long time until it sank in. For many years people did not really look at the paper. The interesting thing is that it was not even the first paper on the subject, as I discovered after I had written the paper, I believe through a reviewer. J.M. Mitchell had expressed the same

---

[7] In June 2006, scifinder was listing 513 quotations of this paper.

concept, on the generation of different frequency domains of climate variability by the successive forcing of longer time scales by shorter time scales, already in a very nice paper in 1966. Mitchell's analysis was more qualitative, but he had captured the main idea quite clearly.

How careful have you been reading the literature?

Hasselmann: I tend to read very diagonally. But when I find something interesting then I read it very thoroughly. When I read diagonally I try to grasp the basic idea.

Dirk Olbers: When you were going to Woods Hole, I was sitting in the Schlüterstraße in your room and, there was a huge pile of reprints which had not at all been touched by you. And I, of course, had time enough to look through all these reprints and I was amazed how many things one could pile up without reading. The papers were yellow and dirty from the sun and from the dust. It was clear that you had never read anything from that pile.

Hasselmann: Not all things we plan to do but fail to are so embarrassingly visible.

Dirk Olbers: You said, the first part of the Max Planck story were these more fundamental conceptual aspects of understanding climate dynamics, and the stochastic climate model was an important element to it. The second part was something like the technical challenge, namely to construct a reasonable ocean model which can be integrated over long times. These two efforts took your attention until about the early 80 s. The people engaged in these efforts were Peter Lemke, Jürgen Willebrand, Klaus Hererich, but also Claudia Johnson, Harald Kruse, Volker Jentzsch and Gerd Leipold.

There was a three-level hierarchy. At the top was Klaus, and at the bottom all the Ph.D. students, in the middle level, I think, Kruse had generated this word 'Zwischenkapazitäten' (middle experts). We, Peter Lemke, Jürgen Willebrand and myself were the ZK's. So we were running from one Ph.D. student to another and were engaged in trying to solve their problems with them.

In those times you would still know most developments in some detail that were taken place. So you were intellectually participating, while at later time your control, your participation became more distant.

Hasselmann: I was always looking for experienced people to whom I could transfer some of my responsibilities These either came new to the institute or, more often, evolved from the scientists already there as they gained more experience. Also, we later had a much broader range of activities, so that I could not keep up to date with all activities all the time. In those days of the ZK's—a new term for me, a typical Kruse creation!—we used to have seminars in my office to work out what the next steps should be in a particular program. It was a much more intimate style of research. It was an exciting period, but one which could not be maintained in the same way as the institute became larger.

We had this weekly seminar and Klaus was really very much engaged. We had created these two minutes seminar. Do you know what this means?

Hasselmann: Yes, I used to interrupt every two minutes.

*No, you were **allowed** to interrupt the speaker only **after** two minutes. This was really very lively.*

Hans von Storch: I think that we are now in the early 80s and I remember the Lütjenseer Wende-Parteitag. This was the first time I was confronted with Klaus. The Fischer group of the University of Hamburg, of which I was part, was invited to participate in building this climate model. You persuaded Erich Roeckner to do something very wise, namely to replace his own atmospheric model by the European Center's model. Could you elaborate a bit on that as it was a pretty important decision?

Hasselmann: It was clear at that time that we needed a good general atmospheric circulation model as part of the climate model. One needs a critically sized group to do this. The groups that had done this successfully were GFDL, NCAR in the US and—in particular—ECMWF in Europe. ECMWF was producing the world best-global medium range weather forecasts on an operational basis and had at that time the leading general circulation model of the atmosphere. It had a large group of experts working on the model. It was quite obvious that it was rather a waste of time to have excellent people like Günter Fischer and Erich Roeckner trying to compete with this large group, trying to do the same thing.

So the obvious thing was to take the ECMWF experience and to improve upon it using one's own expertise. Everybody agreed, also Günter Fischer and Erich Roeckner, although perhaps with less enthusiasm. Both are extremely competent modelers. After Günter Fischer's retirement, Erich Roeckner moved to the MPI, where he developed the original ECMWF model into the—in our view—world-best climate model, under the later directorship of Lennart Bengtsson. So I think the scientific reputations of both Günter Fischer and Erich Roeckner were enhanced by the decision. And it was, of course, essential for the development of the Hamburg climate model.

Then we are in 1982, you then had the Large Scale Geostrophic ocean model, you were to get the needed atmospheric model, you had a good conceptual framework, but you had no computer. What did you do then?

Hasselmann: In 1979, the World Climate Research Program was created, and one year later, in 1980, the German Climate Research Program. So there was obviously a need for the German climate research community, and not just the Max Planck institute, to have a good climate model.

But it was also clear that only the Max Planck Institute, together with the Meteorological Institute, would be able to provide the model. However, since there was a general community need for a state-of-the-art climate model, it was also logical that the super-computer needed to run the model should be provided for, and therefore be funded by, the community, in other words, by the Federal Ministry of Science and Technology. This is what ultimately happened, but the route there was not straightforward.

To spin up our modeling activities, we had first applied for a medium sized computer from the Max Planck Society—in accordance with my gentleman's agreement with Reimar Lüst. This we obtained in 1979, I believe a CDC Cyber 173, but only after lengthy battles with lobbyists in the computer committee of the Max Planck Society, who argued that we would be better served by a remote access to the large computer at the Max Planck Institute for Plasma Physics in Garching, near Munich. The next step was to upgrade the Cyber 173 to our first supercomputer, a Cyber 205. This occurred around 1982. The investment was funded already by the BMFT, but the running costs were taken still from the budget of the institute.

Our computer staff was not really sufficient to run a supercomputer, and the few additional people we had taken on were already straining the institute's budget. Wolfgang Sell headed the computer staff, Dirk Schriever, who had been responsible for data processing at the former Brocks institute, organized the data archive, and we had a few operators.

But we also had a problem with developing the comprehensive climate model. Günter Fischer, who had headed the atmospheric modeling group of the Meteorological Institute, had retired, and it was clear that his successor, whoever it would be, would not be a numerical modeler.

We found a good solution to both problems. I approached Reimar Lüst and reminded him of our second gentleman's agreement. I explained that the time had come when we really needed a third director to take care of the atmospheric modeling activities. His response was positive—in principle. I then approached Frau Tannhäuser, the administrator of the German Climate Research Program, and proposed that our supercomputer should be transferred from the Max Planck Institute to a new-to-be-created German Climate Computing Center (the DKRZ), and that the BMFT should carry also the associated staff costs. She also responded positively—in principle. There followed a period of negotiations between the parties involved regarding the distribution of costs, the distribution of computing time between the Max Planck Institute and other users from the general climate research community, legal formalities, etc.

The net result was that our computing staff was transferred from the Max Planck Institute to the DKRZ, which freed a number of positions that we could now offer to the new third director of the institute. The DKRZ was founded in 1985, with Wolfgang Sell as Technical Director and myself as Scientific Director. The third director of the Max Planck Institute, Lennart Bengtsson, came a few years later, at the end of 1990.

Who, among other appointments, then got Eric Roeckner to move from the Meteorological Institute of the University of Hamburg to the Max Planck Institute?

Hasselmann: This was a very good move. But Lennart also had a lot of experience in atmospheric modeling too, of course, as well as a great deal of organizational experience. He knew the Centre's model very well, and his arrival, together with Roeckner's expertise and hard work, gave us a big push.

He also hired Ulrich Cubasch at that time.

Hasselmann: That is right. Ulrich Cubasch used to be at the European Center. He was very effective in analyzing the results of our simulation experiments. Lennart Bengtsson also hired Lidia Dümenil, Klaus Arpe, and Bennert Machenhauer, who developed a nested regional atmospheric model. So he built up a very good group. The Hamburg version of the ECMWF atmospheric model, ECHAM was then coupled to our LSG ocean model, including the carbon cycle, to create the ECHAM-LSG coupled climate model. This was done in cooperation with a number of visitors, both to Lennart's group and to my group. Lennart had a continual stream of guests, many of whom had previously visited the European Centre, while we had stimulating visits, for example, from Wally Broecker from the Lamont Observatory and Bob Bacastow from Scripps, who both collaborated with Ernst Maier-Reimer in developing the carbon cycle model.

At the same time people like Dirk Olbers left. There was a change in the general direction. It was more towards the dynamical, quasi-realistic complex models, less dynamical conceptualization, more brute force implementation of experimental tools.

Hasselmann: That's true. We first had to demonstrate some basic concepts regarding natural climate variability using simple models. But once that had been achieved, there was obviously no point in pursuing the analysis further with simple models. We had to first construct more realistic models. So as soon as the LSG ocean circulation model had been created, Maier-Reimer and Mikolajewicz computed its response to stochastic forcing, as I mentioned. The next step would have been to apply these ideas to the full climate system, the coupled ocean–atmosphere general circulation model. But somehow we got side-tracked. I am glad to hear that Jin von Storch has started looking at this problem with one of her Ph.D. students. But there is much that still needs to be done. I think the distinction between the three possible sources of natural climate variability, namely stochastic forcing by short-time-scale atmospheric variability acting on the slow climate system, internal nonlinear interactions on comparable time scales within the slow climate system itself, and external forcing, for example by volcanic activity, or by variations in the sun's radiation or in the earth's orbit, has still not yet been properly clarified.

We were probably distracted from this straightforward goal by the many interesting new problems that came up in connection with the modeling effort. For example, we began looking at the feasibility of the prediction of natural short-term climate variability on time scales up to a year. I worked with Tim Barnet on this, applying purely statistical methods, based on linear multi-time-lag regression models [50, 61, 64]. Later we applied also a realistic GCM model to El Nino predictions, and a reduced-complexity coupled model of the type was used very effectively by Mojib Latif. Tim Barnett used another, still simpler linear feedback model, also in collaboration with Mojib, which worked quite well too. So we had opened another arena in which we could apply relatively simple dynamical concepts without a full-blown global climate model.

But we also became involved in improving the global climate model itself, by extending the biology and chemistry representation in the ocean subsystem, by improving the sea-ice model, by adding atmospheric chemistry, in collaboration with Paul Crutzen's group at the Max Planck Institute in Mainz, by including surface vegetation, and so forth. This is, of course, an endless task.

Another question I pursued relatively early as a side-line in our modeling activities was the projection of complex models onto simpler models using so-called Principal Interaction Patterns (PIPs) and Principal Oscillation Patterns (POPs) [86, 89]. A basic difficulty of complex models is that, as they become more realistic by incorporating more processes and degrees of freedom, they become just as difficult to understand as the real systems they simulate. I tried to devise methods for constructing simpler models that capture the dominant processes that govern the dynamics of the full complex system in terms of just a few basic interaction patterns—in the general nonlinear case, in terms of PIPs, in the special case of a linear system with stochastic forcing, in terms of POPs.

Finally, we also became more strongly engaged in later years in IPCC activities, in scenario computations of anthropogenic climate change over the next 100 years.

All these tasks were quite fascinating and distracted from our original goal of sorting out the different forms of natural climate variability. But now that the question of anthropogenic climate change has become much more center stage in the public awareness, I believe the distinction between anthropogenic climate change and natural climate variability will rise to high priority in the climate research agenda. We will have to look in earnest again at the structure of natural climate variability. The increased public interest this problem is apparent in the recent discussions over the possible impact of anthropogenic change on the frequency and intensity of extreme events such as hurricanes, flooding and droughts.

In that sense it had a revival or an important implication in the last years of your directorship. It would not have made sense to think about detection of anthropogenic climate change without a stochastic concept.

Hasselmann: I am not so sure that the stochastic concept as such is important for the detection and attribution problem. The main point is that you are trying to distinguish between the anthoprogenic climate signal— or some other externally forced climate change signal, for example, due to a volcanic eruption—and the internal natural climate variability. The origin of the natural climate variability, whether through stochastic forcing by the short-term climate variability or through nonlinear interactions within the climate system itself, is irrelevant. The central issue is to distinguish between an externally forced climate change signal and natural climate variability, on the basis of the frequency spectra of the two signals. This is another example of applying a ready-made theory from another field—in this case signal processing in communications—to a climate problem. I pointed this out in a 1979 paper [54], but the paper lay dormant until the detection problem became relevant in the mid 90's, when a spate of papers [110, 125, 129, 133, 135, 138] demonstrated that the anthropogenic climate change signal had now indeed become detectable above the natural climate variability noise.

In the 60s and 70s, people would not necessarily have agreed that there is variability for no specific reasons.

Hasselmann: I think there were already two schools of thought at that time. One school thought that climate variability must indeed be produced by some external forcing mechanism, such as volcanic eruptions or variations in solar radiation. But the second school recognized that you could explain natural climate variability simply by the fact that climate is a nonlinear system containing feedbacks. Such systems, for example, turbulence, are known to exhibit random variations. Both mechanisms can contribute to climate variability. The stochastic forcing model merely points out that there exists a particularly simple realization of the second mechanism, since the climate system contains a ready-made source of natural variability in the form of the turbulent atmosphere. All one has to do is separate the time scales, that is, distinguish between the fast atmosphere and the rest of the climate system, consisting of slow components such as the oceans, cryosphere and carbon cycle. But the idea that internally generated natural variability can be expected in a nonlinear system such as climate was already around at that time.

Hans von Storch: My understanding of stochastic variations is that we have very many chaotic components in the system, so that the overall behavior cannot be distinguished from the mathematical construct of noise. Therefore, we can describe the nonlinear dynamics very efficiently as noise. In the same way as a random number generator is also a deterministic algorithm on a computer.

Hasselmann: Well, I think, we find this in any nonlinear system.

But it would not necessarily look like noise if you have a few degrees in a system. So for the Lorenz' system you would not conceptualize the behaviour as noise.

Hasselmann: It depends on what you define as noise. If you define noise simply as a statistically stationary stochastic process, then the Lorenz system, in the appropriate parameter range, produces noise—although it is certainly not Gaussian, as assumed in many noise analyses. No, I think the essential point about the stochastic forcing concept is not that one has noise, or that the system has very many degrees of freedom, but that one can understand the origin and structure of the noise in the climate system very simply by separating the time scales. The origin of the noise is the short-time-scale turbulent atmosphere. This then generates variability on much longer time scales in the rest of the climate system. There is no need to understand the detailed dynamics of the atmosphere. It is sufficient to know that the turbulent atmosphere is characterized by a noise spectrum that is concentrated in frequencies corresponding to time scales of hours and days, but—because the system is nonlinear—also extends down to a finite level at very low frequencies. It is this low-frequency range, corresponding to time scales of months, years, decades and even longer—that can be treated as white, i.e. simply as constant—that generates variability in the rest of the climate system, the slow climate system.

In most of our initial applications of the stochastic climate model, we considered some simple component of the climate system–for example, the temperature of the mixed layer, or the sea ice extent–which we could linearize. So there was a popular misconception that the stochastic model could be used only to describe the response of a linear system to white noise forcing. But the concept is valid generally for any climate model, whether linear or nonlinear, as demonstrated by the application of Maier-Reimer and Mikolajewicz to the LSG ocean circulation model. This misunderstanding is perhaps related to the fact that some people may have had difficulties understanding my original stochastic climate model paper. To treat the general nonlinear case, I used the Fokker–Planck equation, the generalization of the Liouville equation of statistical mechanics to a system including diffusion, as required for Brownian motion. While most people can be assumed to have been familiar with the Liouville equation, the Fokker–Planck equation was perhaps less well-known.

You outlined this whole set up of the Max Planck Institute with the different models and couplings, ideas and so on. At the same time, we had a German climate science program. From outside it looked as though MPI ran this program. The MPI made many attempts to draw in people from outside, but other meteorological institutes were only marginally involved with respect to the global modeling efforts. Is that the same as you see it?

Hasselmann: Yes. I think the explanation is in human nature. We certainly tried to draw other groups into the program, but the problem was that to run or contribute to the development of a complex global climate model system, you have to be willing to get your hands dirty, you really have to become involved. You cannot just sit around and have some clever ideas. You cannot work on a complex model some 500 km away. The people we collaborated with came from India, Canada or somewhere else for a year or so. Most Germans—most of them had a family at home—were not willing to come for a longer visit. Another reason that our attempts were not very successful is that most scientists do not get excited at the idea of becoming involved in larger and somewhat anonymous activities.

So it was typical that in the German climate research program we had one global climate modeling group stationed in Hamburg, at the Max Planck Institute and the University Institute of Meteorology, and several smaller groups distributed everywhere else, at the GKSS in Geesthacht, in Jülich, in Karlsruhe, in Bonn and Cologne, all working on regional climate models, because they could do that on their own. I thought it was a waste of time and resources producing five or six different regional models, all of similar quality. We had a regional model in Hamburg, too, nested into the global model. This was a typical case of unnecessary parallelism because people simply had problems in getting involved in a joint program. I tried to overcome this, but I have to admit that I was not successful.

We were more successful with groups that were analyzing the outputs of our models, for example in Cologne, Munich or, later, in Potsdam. But there were rather few groups engaged in such activities. I believe the same problems are encountered everywhere by groups developing large models. One cannot yet effectively decentralize this type of work.

Concerning ocean models you see there was this division between LSG, which was large scale, and the rest of the oceanographers in Kiel and also in Bremerhaven who did eddy resolving models. But my impression was that you did not really value these.

Hasselmann: Well, yes, I was not convinced that the eddy-resolving models were really worth the effort.

They were or were not?

Hasselmann: I thought they were not. They burnt up a lot of computing time. Essentially, they showed that there were eddies, which we knew anyway. I was not convinced that the interaction between the eddies and the mean flow could not be parameterized sufficiently well for climate modeling purposes with a standard eddy transfer approach. Or, at least, the eddy-resolving simulations had not come up with a better parametrization. I am not convinced that we were discovering something basically new. What I have seen in talks to this day are beautiful pictures of the Gulf Stream and all these eddies floating around, but what have we actually learnt? If one can demonstrate that the impact of these eddies is radically different from what we have been putting into our coarser-resolution models, then I will admit that we have to start thinking of something radically different, or maybe even have to give up working with non-eddy-resolving models. But I have not seen this yet. What I have seen are mainly nice movie presentations that are good for public relations.

What do you think about visualization?

Hasselmann: I have mixed views. I think there has been an unnecessary polarization of viewpoints on this topic. The presentation of the results of a complex time-dependent simulation in a visualized form that the non-expert can quickly grasp can be very helpful. For somebody who has never seen satellite or other data on Gulf Stream eddies, the simulation with a good eddy-resolving model of the Gulf Stream can be very illuminating. On the other hand, my experience is that the active scientist doing quantitative data analysis seldom uses visualization. There can be a few cases in which it is useful. I remember one case in which watching a video sequence helped us discover an intermittent instability at a particular gridpoint that we had missed in the snapshot pictures. So I think, even it is not used routinely, it is certainly worthwhile to have a good visualization facility available.

Have you ever been in the caves, this three-dimensional visualization?

Hasselmann: I get sick in these things. I find them terrible. I experienced one in the Tyndall Centre in Norwich. Maybe I am too sensitive, but the three-dimensional projection did not seem to work properly, and I got giddy. After a certain time I got really sick. Perhaps I was not sitting in the right location. And maybe the techniques will improve with time. But I was not convinced that the additional information of seeing the data in three dimensions rather than two—in other words. with one eye closed—was terribly important for scientific purposes and justified the technical effort. But again, it may be OK for public relations, once the technique is sufficiently mature.

One climate component which has been tackled by the Max Planck Institute and others as well is the ice sheet. But I've never really seen ice sheets incorporated in climate models at MPI. Is that something which is too complicated?

Hasselmann: I don't think it is terribly complicated. There was probably just not enough push on my part. We had Klaus Herterich's ice sheet model. His model described very nicely how ice sheets grew and melted and when they start to surge.

I was interested in coupling an ice sheet model with an ice-shelf and a sea-ice model. A coupled model of this kind would be very useful to address the question of the stability of the Greenland or Antarctic ice sheet, whether the ice sheet can break down through ice surges. And if this model had been incorporated into our global climate model, we could have carried out simulations to investigate the origin of climate variations on century and millennium time scales, which still pose many open questions. The Milankowitch theory explains only part of the variability. I think that is a very important area of research, and it was probably my fault that I did not apply enough leadership to ensure that such studies, using an ice sheet model coupled with an ocean model and an atmospheric model, were pursued more seriously. It would have required a stronger group than just one person, Klaus Herterich, who later went on to a professorship in Bremen.

Was this overrun by the IPCC scenarios for the next hundred years?

Hasselmann: No, I don't really think so. This was carried out by other people, in particular, Ulrich Cubasch [106]. The IPCC scenarios were, of course, important for IPCC and the general international climate research effort, but they were also important for us. They demonstrated what the models could do. And they were important for the German Climate Research Program, which had to justify its program to policy makers and the public.

We participated also in the international climate model intercomparison project, which involved similar scenario computations. This was an important exercise to identify the strengths and weaknesses of different climate models.

From a scientific point of view, this work was not very exciting, but I don't think it was in the competition with the ice-sheet modeling. I was probably also distracted following up on other problems.

Hans von Storch: Perhaps it would be more honest to say we are now in a less focused period of the institute? After 1985, you let the reins loose more and more and at the end you became less and less interested in climate. That is my impression; I would not criticize you for that. Lots of things happened in the institute and this was one just one of these issues. There were many studies which were not related to this big modeling building and the IPCC.

Hasselmann: Yes, maybe that was the case, if you look at the many publications on different topics that were coming out the institute. We had also expanded the research on the carbon cycle and tracers using inverse modeling techniques, led by Martin Heimann, who came to us from Scripps in 1985. With highly competent scientists around like Martin Heimann, who is now director of the Max Planck Institute for Biogeochemical Cycles in Jena, I did indeed let the reigns a little loose and let group leaders take over in many areas—which I don't think was a bad thing.

Global warming was not a dominant issue at the institute in the late eighties. Lots of studies were done which had nothing to do with the overarching goal you just described. People were just entertaining, enjoying themselves.

Hasselmann: I would not put it that drastically. They were exploring many different interesting topics, and quite successfully. But we were also carrying out a good deal of work on global warming too, for example in the scenario computations you referred to. It is true that I myself did become involved in problems other than global warming at that time. However, I was still interested in ice sheets, although, admittedly, not aggressively enough. We had good contacts with Johannes Oerlemans, an international expert in ice sheet dynamics from Utrecht, who visited us several times, and with Bill Hibler from Canada, an expert in sea-ice modeling who stayed with us for a year. As a result, we did incorporate a good sea-ice model into the global climate model, but unfortunately not an ice-sheet model.

Perhaps I should honestly admit that I was also getting a little bored with always having to organize things and was quite happy that the so-called ZK had matured to a level of expertise and international recognition where I could happily let them take the lead in many areas.

I remember in the first period, when we were developing our work on stochastic models and so forth and also on the ocean modeling in the early eighties, Fritz Schott had visited us from Miami and talked to many people at the institute. He came to me afterwards and said that he had never been in an institute where the Ph.D.s and post-docs were so closely guided as in the Max Planck institute.

When did he say that?

Hasselmann: It must have been around the early eighties. I suppose that at that time I was indeed guiding people more strongly than in most institutes in the US, but I think that later on, I tended to let people loose to develop on their own—make their own mistakes rather than mine.

I heard stories that it was really tough for Ph.D. students in the late seventies to work with you.

Hasselmann: We had tough discussions. That is true. But it was never personal. I tried to support the students as well as I could. I can't remember any student actually failing, although one student did decide after a year to become a pastor. He thanked me later for motivating him indirectly to that decision. I'm not sure how. Perhaps I was a little tough.

On the other hand, you were also riding a lot of horses. The climate business was evolving and became useful—if we may call it this way—and this IPCC engagement also and our efforts to come up with prediction schemes for El Nino and things of that sort. This all went very smoothly and nicely and you were guiding all these things. But you did other things as well! We others did not really notice that but you were still engaged in wave aspects, still engaged in remote sensing with respect to wave activity. Can you tell us about that a bit?

Hasselmann: Well, I had decided more or less to stop my ocean wave research around the late 70s. But there were two developments that brought me back into the subject. One was that ESA was preparing to build ERS-1, the European follow-on of SEASAT, the US satellite that had operated for only 100 days in 1978, but had demonstrated the feasibility of measuring ocean waves from space. ESA asked me to serve on the ERS-1 advisory panel. The second development was that my wife Susanne—after a 15 year interruption bringing up children–had just completed her diploma in mathematics. We wanted to do work together. I did not want her to work in the climate area, because there she would have been in direct competition with other members of the institute. So I suggested finding some area where we could work together without overlap with the main work of the institute. Ocean waves was a natural choice.

This was also good timing, because we now understood ocean wave dynamics rather well, through JONSWAP, and we faced the challenge of translating this knowledge into a numerical ocean wave prediction model. Susanne, as mathematician, would be well able to do this. Also, we would need a good global ocean wave prediction model to assimilate the global wave height and two-dimensional wave spectral data that we hoped we would be obtaining continuously in a few years from the altimeter and SAR instruments aboard ERS-1.

So I renewed my activities in ocean wave research. Together with former JONSWAP colleagues we formed the WAM (Wave Model) group, with the goal of developing what was to be called the third generation wave model

3G-WAM. The 3G was dropped later as too cumbersome. We first carried out a comparative study of all existing ocean wave models [242], in which we concluded that the so-called first and second generation wave models were inadequate. First generation models, developed in the sixties, were based on our incorrect understanding of the wave spectral energy balance prior to JONSWAP. Second generation models included the nonlinear transfer in accordance with the JONSWAP picture, but the parametrization was too crude to reproduce the wave spectra for complex wind fields. We needed a third generation model with an improved representation of the nonlinear transfer. So Susanne and I first developed a more realistic approximation of the five-dimensional nonlinear transfer integral that could be implemented in a wave model [77, 78], and Susanne incorporated this in a first version of the WAM model. The model was then tested and further improved by other members of the WAM group [90]. Heinz Günter from GKSS cleaned up the numerics and documentation and ran the model at the European Centre, while others tested various other aspects of the model. It is now used world-wide in many operational forecasting centers and research institutes.

My work in the ERS-1 advisory committee also took a fair amount of time. I frequently had to travel to ESA headquarters in Paris or to the ESA Technical Centre ESTEC in Noordwijk in Holland. Through ERS-1 I met many interesting people involved in remote sensing, such as Ola Johan-nessen, director of the Nansen Center in Bergen, Norway. But ERS-1 also involved interesting scientific challenges. One was developing algorithms to retrieve the two-dimensional wave spectrum from the nonlinear ERS-1 SAR image spectra [102]. Another was assimilating the resulting wave spectra in the WAM model [120]. I worked on this together with Susanne. But there were so many other interesting problems, particularly when ERS-1 was launched in 1991 and began producing data, that I also took on some Ph.D. students, contrary to my original intentions. We had a small but very active ocean wave and remote sensing group consisting, in different periods, of Claus Brüning, Susanne Lehner, Patrick Heimbach, Eva Bauer and Georg Barzel. They worked independently of the climate groups, with relatively little interaction apart from seminars and other general institute activities.

What about Werner Alpers?

Hasselmann: Alpers was not a student of mine. He was a post-doc in the Sonderforschungsbereich. He worked with me on the remote sensing of ocean waves in my first 'ocean wave period', before the Max Planck Insti-tute was created. He then went to the University of Bremen as Professor for Remote Sensing, and later returned to Hamburg, again as Professor for

Remote Sensing. I worked together with him again after I revived my ocean wave and remote sensing interests. But I stopped working on ocean waves and remote sensing—this time, for real—after Susanne retired in 1996, and I turned to other interests.

You became interested in what some people say was a very naïve way of describing economics, dabbling in economics. What was that?

Hasselmann: It came through my involvement with the media and public audiences. In the late eighties and nineties, the media, general public and politicians began to become increasingly aware of the climate change problem and wanted to hear more from the climate experts themselves. So I was often invited to interviews on TV or the radio, and to give talks to the general public on climate. At the end of my talks I was always asked the same question: What should we do? And I would say: Well, I do not really know. I'm a climate scientist, not an economist or politician. But they would never let go, and kept persisting until I came up some off-the-cuff answer. So I decided I had better find some better answers and began looking into the problem of the impacts of climate change, and the possible economic and policy responses. I could find little reliable information on climate impacts, and was rather disappointed with the analyses of the economists, who were using—in my view—inappropriate outmoded economic equilibrium models. They were also distorting the critical issue of the proper discounting of future climate change costs. And the political stage, of course, was beset by lobbyists of all hues, which made it difficult to detect a signal in the noise.

So I began developing some simple coupled climate-economic models to determine the optimal CO_2 emission path that minimizes the net economic costs of anthropogenic climate change and climate change mitigation, with emphasis on the intertemporal discounting issue [143, 146]. At the same time Hans von Storch wrote some similar papers with Olli Tahvonen, an economist from Finland, whom Hans von Storch had interested in the problem.

I followed up this work with somewhat more realistic but still relatively simple economic models based on non-equilibrium multi-agent dynamics. Two nice Ph.D.s theses came out of this, by Volker Barth, Michael Weber and Georg Hooss. As a side product, we created a climate computer game based on our coupled climate-economic model that was implemented in a climate exhibition for a year or so at the German Science Museum in Munich. The game was quite popular.

Coupled climate-economic modeling is still a hobby of mine today. I believe there is an urgent need for the economic profession, in cooperation with physicists and social scientists, to develop realistic dynamical non-equilibrium socio-economic models that combine the climate change problem with the general societal issues of globalization, employment, limited resources, etc.

At the time I was becoming interested in these problems, in 1990, I was asked, together with my colleague Hans Hinzpeter, to become a member of an Evaluating Committee of the Academy Institutes of the former GDR. Our task was to recommend what should become of the Academy Institutes in the area of geophysics and the environment, now that the two German states had become unified. We came across a young group doing interesting interdisciplinary work on various climate-change impact problems. We recommended that they should be integrated into a new institute designated to study the societal and economic impacts of climate change and climate change policies. That was the origin of the Potsdam Institute for Climate Impact Research that was created two years later in 1992. PIK developed a good cooperation with the Max Planck Institute, analyzing many of our climate change simulations.

We tried to establish a similar activity on a smaller scale also in Hamburg. I suggested to the president of the University of Hamburg, Jürgen Lüthje, at a cocktail party given by Reimar Lüst in the Bobby Reich Restaurant next to the Alster, that the university should support a group to study the impact of climate change on the economy and society. This was becoming an increasingly important area of research and would be a good bridge between the climate activities at the Max Planck Institute and the strong economics department of the university. Lüthje straightaway talked to Michael Otto, the head of a large mail-order firm and a well known sponsor of environmental projects, and convinced him of the idea. Michael Otto offered to endow a professorship for environmental economics for five years and asked for proposals. The first time round the university proposal was not accepted, as the university had not committed itself to provide the necessary follow-on funds for the chair after the first five years had elapsed. But in a second round the university made the commitment, and the chair was created. Richard Tol, a very young scientist from the Vrije Universiteit Amsterdam who already had an impressive list of publications, was elected to the professorship.

Unfortunately, an intense cooperation did not emerge with Richard?

Hasselmann: It is the old problem of getting two disciplines to work together. Richard Tol turned out to be a rather traditional economist who looked rather skeptically on the attempts of physicists to get involved in economics. For this reason I think not everybody that he could have collaborated with—including myself—was enthusiastic. But Richard is very young and could develop. So perhaps there may be more collaboration in the future—unless Richard decides to accept positions he has been offered elsewhere, as has been rumoured.[8]

When you retired in 1999, you did something, which—I thought—was rather unexpected or unpredictable. You had already withdrawn to some extent from the climate field but you engaged in a new issue. The first time you spoke about that publicly was at your 60th birthday, when you gave a talk for something like two hours about your approach to particle theory. You withdrew from the climate field, which is quite something for a person with your authority and recognition in the field. You said I do not mind, I am going on to something else that I am more interested in.

So far you won all battles, you were the young attacker bringing down sclerotic old ideas and replacing them with more modern ideas. This was well done, you were successful in doing so and then you suddenly decided, no, I am doing something else now. I am really attacking something totally different and this would be an uphill battle. You would start as newcomer with all the difficulties; you could not really use your recognition in the field. How was that?

Hasselmann: Well, I realized that that would be the situation. I was not surprised. I was a bit surprised at the level of denial—in some cases, even antagonism—of the established particle physicists. Other physicists were more open to my ideas. Of course, they were skeptical, but they were willing to discuss, and in a few cases were even quite positive. But I was aware that for most physicists I would be regarded as slightly crazy, since I was seen as a climatologist who could clearly have no idea of particle physics. I was seen as a dreamer without really knowing what I was talking about. This is perfectly understandable. I have the same reaction to the strange people who sometimes drifted into my office without the slightest knowledge of climate and explained to me why we were or were not experiencing global warming. It did not bother me too much. In my career I have always found that the newer the idea, and the more distant the field it originates in, the more skepticism one encounters. Unfortunately, a skeptical reaction is no guarantee that you have a good idea. It can indeed be a crazy idea. The only way to find out is to press on regardless.

---

[8] Richard Tol has in the meantime moved from Hamburg to the Economic and Social Research Institute in Dublin.

I've been looking at particle physics ever since the mid-sixties when I wrote my Feynman diagram paper on wave-wave interactions in geophysical wave fields. I was convinced that something was basically wrong in quantum field theory. I did not know what it is, but I think many physicists would agree that Einstein had a point in his criticisms of the conceptual foundations of quantum theory. But, of course, everybody says that Einstein worked all his life to find another approach, so why should somebody like Hasselmann be able to solve the problem? Well, I thought it was worth trying. After all, we can't all be paralyzed for ever by Einstein. As you say, I have won most of my battles in the past, and what is the point of having some reputation capital if you cannot spend it on something that's fun?

I published a lengthy four-part paper [121, 122, 131, 132] on the basic ideas of my metron theory in 1996 and 1997, expanding on the first talk I gave on my 60th birthday in October 1992. This was in a journal on the basics of physics, which I discovered later, however, was not taken very seriously by most physicists. I have also published two other papers since then [139, 157] and am right now writing up two further papers on my recent results. Once the theory is published in accepted journals, it will become either accepted or rejected. This is as it should be. I am not really concerned about the outcome, which is beyond by control.

As I mentioned, besides this venture into a new field, I am also still working on coupled climate-economic models. I created the European Climate Forum, chaired by Carlo Jaeger, in which we are trying to bring the stakeholders in the climate change debate—business enterprises, energy companies, manufacturers, insurance companies, NGOs and so forth—together with climate scientists and economists to study the climate change problem, to analyze the various possible mitigation and adaptation policies options.

But your heart is with particle theory?

Hasselmann: Yes, my heart is with the particles.

Dirk Olbers: I had the pleasure to attend your 60th birthday meeting and to listen to your metron talk. I thought I understood most of what you said. My impression was that in just a few years and we would see a new Nobel Prize winner. Others thought the same, not only myself. Then I met you here and there, and you always said that you were almost there, you only have to solve these very complicated equations.

My problem with this answer was there was this equation and mathematicians, they know that there are existence theorems, and they do not bother at all how the solution looks. We have the Schrödinger equation and we know for any complex molecule whatever you can in principle say that the wave function must exist. What is the problem with this equation?

Hasselmann: The problem is that the basic metron equations, the Einstein vacuum equations in a higher—eight—dimensional space, are nonlinear equations without an external source term. The hypothesis is that besides the trivial zero solution, the equations have nonlinear eigenvalue solutions of a special soliton type, for which there exists no analogy that I am aware of in other branches of physics. It is not at all clear whether or not the equations have non-trivial solutions. In the Schrödinger equation for the linear eigenfunction of the hydrogen atom, in contrast, the electromagnetic field that traps the eigenmode is given, as the electromagnetic field of the hydrogen nucleus. In the metron model, the trapping field is not given, but is generated by the trapped eigenmodes themselves, by their nonlinear radiation stress. It is not at all obvious whether the two sets of interacting fields, the trapped eigenmodes and the trapping field, a distortion of the higher dimensional metric, are mutually consistent, as I had hypothesized. In my 60th birthday talk and published papers, I demonstrated that solutions of this type do indeed exist for a much simpler scalar analogue of the Einstein equations, but the problem was to show that they exist also for the much more complicated Einstein tensor equations in eight-dimensional space.

I believe that I can now indeed show that such solutions exist, by a numerical perturbation expansion, but only if one postulates that space is discretized at the smallest Planck scale. Or, alternatively, if one introduces an additional diffusion term into the Einstein equations that becomes effective only on the Planck scale.

Constructing the nonlinear eigenvalue solutions for the Einstein tensor equations in eight-dimensional space was a complex task that took several years. I did this together with Susanne, who wrote the complicated code for the algebraic tensor manipulations. But there is still a long way to go. I have to show that the metron solutions reproduce all the symmetries of the Standard Model of elementary particles, including the 23 or so empirical constants. And I have to show, too, that the metron model is able to explain the enormous amount of empirical data on atomic spectra, scattering cross-sections, superconductivity and so forth that quantum theory has been able to explain in the last eighty years. So the metron model is really more a program than a theory. But if the program is successful, it will automatically unify gravity and microphysics and resolve the many conceptual problems and formal shortcomings, such as divergences, of quantum field theory.

You are referring to numerical solutions. Could it be that there is a convergence problem? So that someone comes along and says this is a numerical solution, I do not believe you.

Hasselmann: That is always a problem with numerical perturbation solutions. But this is not my main concern. I have computed the solutions to nine'th order, and they have every appearance of a well converging series. Once I have written up my results and have them off my chest, I will be happy to discuss existence problems with mathematicians. As an applied mathematician, I tend to be more sanguine about such issues. I have given many talks on the metron model to physicists, and there was never a concern about the formal existence of a numerical series that appeared to be converging. The reactions always concerned the basic ideas, whether they were only odd or outrageous.

I should like to give some more talks to different audiences with a social scientist in attendance. He or she could analyze the different reactions of the audience and correlate them with the various fields of the people that were making comments. The closer the person was to elementary particle physics, the more aggressive were the comments—not the more critical, which I expected and would have understood, but the more aggressive.

I think one of the problems is that as physicists, we have all been brainwashed into believing that quantum theory is an admittedly unusual, but the only possible way of resolving the wave-particle duality paradox of microphysics. Philosophically, one has not been able to refute the fundamental quantum theoretical rejection of the existence of particles or waves as real objective entities in the classical sense. One can object only on aesthetic grounds. Einstein objected strenuously, but did not offer an alternative solution. He is generally seen as having failed. It has even be argued, such as in Bell's famous no-go theorem, that it is in principle impossible to explain quantum phenomena by classical theories. However, it has been shown—although this is widely ignored—that these arguments are all based on the existence of an arrow of time, which is not acceptable for microphysical phenomena. Nevertheless, anybody who tries to propose a classical theory is swimming against a mighty mainstream.

But, finally, must it be that one of the theories is correct and the other one is incorrect? Or could it be that, as in the case of a spectral model or a grid-point model, they are simply different ways of finding the same solution.

Hasselmann: I don't think so. The way I see it is that the problem with quantum field theory is that the theory captures only half the truth, the wave aspect of the wave-particle duality problem. In the metron picture, both particles and fields exist as real objects in the classical sense. Particles are the source of the fields, which therefore do not exist independently, but only together with their particle sources. The different types of fields—electromagnetic, weak and strong—are basically the same as in quantum field theory. And the interactions between the fields are also essentially the same. In addition, the metron model has gravitational fields, since it is a unified theory encompassing all fields. But apart from the additional gravitational field, the field content of the metron model is essentially the same as that of quantum field theory.

The difference is that quantum field theory doesn't have the concept of a particle as a real existing object. It is thus forced to negate also the existence of fields as real objects. Fields are interpreted only as abstract operators acting on a Hilbert space of states. From these states one can infer probabilities for the outcome of experiments—which must be described, nevertheless, in terms of the particles whose existence one has just negated. This is the strange construct that creates not only philosophical unease, but also the technical difficulties of quantum field theory, the divergences and difficulties in unification with gravity. So I don't see the two theories converging to simply two mathematically equivalent pictures of the same physics.

Hans von Storch: I would suggest that you read Ludwik Fleck's book "Die Entstehung einer wissenschaftlichen Tatsache", because I think you are just in the centre of the storm which this guy is describing.

Hasselmann: Maybe I should. I had not experienced such strong antagonism before. I had expected scepticism, but not antagonism. I presented a talk at a physical colloquium in Oldenburg, and a couple of people sprung up afterwards and shouted that it was a scandal that somebody should give such a talk in a physical colloquium. It was almost a religious reaction. I felt I was in one of those pre-election political talk shows that sometimes get out of hand.

I had not experienced such violent antagonism before. When I first presented the nonlinear wave interaction theory, people like Bill Pearson or Francis Bretherton emphatically said I was all wrong, but this was in the normal civilized framework of people being skeptical and arguing. And the established SAR experts were critical but not outright hostile when I trespassed in their area to develop a theory for the SAR imaging of ocean waves. Traditional economists also showed only mild irritation, or simply smiled condescendingly, when I came up with alternative economic models. I suppose there was never this feeling that I was attacking anybody's foundations. The Oldenburg hecklers were—I suspect somewhat frustrated—elementary particle physicists.

Hans von Storch: This is just demonstrating for me very clearly that science is a social process. We are a social group, physicists of whatever, and we have certain rituals or ways of defining authorities, who is right or wrong. You were confronted with a different band that has different rules and their authorities try to defend their status. So I find it very brave of you that you changed roads. You had been in one band one of the chiefs. Then you suddenly decided that you would be one of these silly unimportant footsoldiers in another band.

Hasselmann: I find it is a lot of fun. As I say, what is the point of having a reputation if you cannot use it to play.

Hans von Storch: This Fleck book analyses what happens when science is in a phase when people just try to repair their knowledge claims. They are inventing new rules and refining old ones and so forth, even though the whole system is already wrong. Then it takes a while until it breaks down.

Hasselmann: I personally am convinced that quantum common field theory as it now exists will break down. That it has basic problems nobody can seriously argue against.

I presume that you do not say that it is no good. It is good for a certain range of phenomena but then if you try to extend it as an explanatory tool to different phenomena, then it fails, it then needs to be re-written fundamentally.

Hasselmann: There is no doubt that quantum theory and quantum field theory work extremely well for a wide range of phenomena. But I think the problem is different from, say, Newtonian physics needing to be replaced by special relativity, or special relativity by general relativity. I believe that the problem of quantum field theory doesn't lie in the finite range of phenomena it can describe, characterized by some parameter range. It lies rather in the fundamental concepts as such, in the negation of the existence of real objects. Conceptualization in terms of real objects endowed with particular properties is, after all, the foundation not only of classical physics, but of all natural sciences since humankind has started to think scientifically.

But regarding the introduction of new ideas, I take solace in the famous physicist, I forget who it was, who observed that advances in physics are a natural phenomenon that takes care of itself. The old physicists die out and the young ones are not afraid of new ideas. I am encouraged that young physicists are much more open to my ideas.

I don't think that this is a problem of physicists, I think this is a problem of all scientists.

Hasselmann: Yes, of course, this is not limited to physicists or even scientists. People obviously build up their view of the world, everything, the interconnections, the values and so forth. And if that is being attacked, they feel threatened.

Another question. What are perspectives on bringing numerical mathematics into the field of climate sciences? Do we need that? Would you expect that we can come up with better algorithms which will help us in a significant way?

Hasselmann: Well, I am not a theoretical numerical mathematician, but an applied numerical mathematician. I simply apply whatever mathematics offers to solve problems. In the particular area in which I work, I find that the numerical techniques that people use have not been developed by mathematicians for their particular application, but are general off-the-shelf methods that have been adapted by meteorologists or physicists for their particular application. When they find them inadequate, they improve them themselves, such as in the question of whether to use Lagrangian or Eulerian propagation schemes in atmospheric models, or whether to use spectral or grid-point representations. The modifications normally evolve from actual practical applications. There have been very few, to my knowledge, really original new ideas that mathematicians have applied to particular problems in our area.

There had been some attempts to use multi-grid or adaptable grids and so forth, but these are again off-the-shelf mathematical methods that the scientists simply apply and adapt as the need arises. Often the theoretically more accurate methods turn out to be computationally less efficient when applied in vector or parallel supercomputers, so that in most of the larger climate models one tends to find rather conventional numerical methods. I know of no real examples where theoretical numerical mathematicians have been called in to upgrade the numerical performance of models. But perhaps I am no longer up to date.

Apart from Klaus Hasselmann, who relied on Herrn Krause in 1961.

Hasselmann: Well, that is in fact just an example that underlines my point. I chose the appropriate numerical algorithms, for example for the treatment of the resonant delta-function factors in the integrand, and the mathematics student implemented them on the computer. It was basically all off-the-shelf.

I have one more question about the relationship with the media or the way scientist should/can/should not/cannot speak to the public through the media. You started as a climate physicist because you were curious to try out certain things, then you found it interesting to construct a wave model and things of that sort. Suddenly you are in the midst of a great public concern and public interest and the public is asking all kinds of questions. Could you tell us about how you experienced that?

Hasselmann: Most scientists are not well prepared to do this job. But it is an obligation for scientists to present their results to the public, as I think we all agree. The only way to present the results effectively to a broader public is through the media. This is particularly true if the results, as in the case of climate change, affect the policies that a country or the society as a whole needs to pursue.

Few scientists have the talent to interact with the media effectively. Fortunately, at the Max Planck Institute we have had two people that could that very well, and also liked doing it. One was Mojib Latif, who was in my group and is now Professor at the Leibnitz Institute of Ocean Sciences in Kiel. He is probably the publicly best-known climate scientist in Germany today. Everybody has seen his clear expositions of the climate problem on TV. The other is Hartmut Graßl, a co-director of the Max Planck Institute who succeeded Hans Hinzpeter as head of the air-sea interaction and atmospheric remote sensing group. Graßl was not only an equally effective communicator with the media, but was also heavily involved in advising policy makers, as chairman or member of various high level Federal advisory committees. For these activities he received the prestigious German Medal of Merit. Through the excellent communication activities of Latif and Graßl, much of the pressure of interacting with the media, public and policy makers was taken off my shoulders, although I also had to carry my share.

This was sometimes a little frustrating, as the media like to report things that people like to read rather than what they should be reading, namely the facts. These can be rather boring, particularly if they are always the same, as they are for the slowly changing climate. So the media like to present extreme ideas that are not supported by the science community as a whole. The result is that the public tends to be rather confused regarding the climate change problem. But that is something that we have to live with.

Maybe one final question. It is quite personal. You sit on the beach in Sylt and you look out on the ocean, on the waves and on the climate and so on. You see the turbulence. You were in control of wave and climate studies in this early stage of the Max Planck Institute with all these small growing Ph.D. students and then this later stage. What do you think, what period was the most satisfying for you? Were all of the same kind or is there anything which you said I was really satisfied with this.

Hasselmann: I enjoyed all of these phases in different fashions. I was always very satisfied when I discovered some new insight, or when something finally worked.

For example, I was exhilarated when I carried out the computation of the nonlinear energy transfer for the JONSWAP spectrum and compared it with the growth data, and they agreed precisely. It took us ten years of work before we achieved this result.

I was absolutely elated when I watched the launch of ERS-1 in Kouru in 1991. It was incredible that after all those many meetings in ESA, discussing an abstract project in endless variations in innumerable committees, the satellite really existed and was roaring up there into space.

And I was enthusiastic when ERS-1 began providing ocean wave images with the SAR, from which we could retrieve two-dimensional wave spectra using the algorithm we had developed. When Patrick Heimbach compared the first three years of retrieved wave spectra in his thesis with the spectra produced with the operational WAM model at ECMWF, he found very good overall agreement [140]. But he also discovered a slight shortcoming of the model, in the propagation of swell, which needed to be brought into closer agreement with the old results of the Pacific swell experiment. All this was very pleasing.

I was also emotionally strongly moved on my 60th birthday surprise colloquium, when suddenly all the people I had worked with in different fields from different countries over many years turned up and gave talks. I had never realized until then how fortunate I had been in experiencing so many rich friendships in my career.

But I also had many satisfactory experiences that did not have this delta-function characteristic. For example, the strengthening and dissemination of the stochastic forcing concept through a number of very nice Ph.D. theses or post-doc papers, or the many influential detection and attribution papers that followed our first paper, in which we had come up with a quantitative estimate of the—very small—probability that the observed recent global warming could be attributed to natural variability. This led very soon to the general acceptance that anthropogenic global warming was real and had been detected.

In your list, you did not include the creation of the DKRZ.

Hasselmann: I did a lot of things that were simply my obligation as director of the Max Planck Institute, or as the member of some committee, but these were not things in which I was strongly involved emotionally. I pushed, for example, for ERS-1, in various committees—well, I guess I was emotionally involved there and did in fact battle with some lobbyists pushing other priorities. But one of the things that were simply necessary and didn't run into any opposition was the creation of the Climate Computing Center. This was, of course, a key component of the German, and later also the European, climate program, but not something for which I personally deserve particular credit.

You said, there were always two roles you played. One is the wage earner, just doing what you have to do; on the other hand you are the unruly scientist who is just following your curiosity. I guess the answers you gave just to those questions was the unruly part.

Hasselmann: Well, they were both parts. In fact, the successful parts were really the wage-earning parts. I believe most scientists, unless they are obviously geniuses, need to have a professional commitment to work in some field in which they can be reasonably sure to produce results that justify their salary. Climate, ocean waves and satellite remote sensing are three such typical fields. It is clear what needs to be done—within a spectrum of viable options—and if you work on the problems, you can expect to get useful results.

On the other hand, the things that really interested me, like turbulence theory or now quantum phenomena, were problems where it was not at all clear that one would ever be successful. If I were a young physicist today working officially in elementary particle theory, I would have great problems. It is quite clear that there is not an obvious road to a successful solution. But as a young scientist, you need to publish. So you have to jump on some bandwagon which the establishment has created, such as string theory, which joyfully leads everyone to nowhere.

So I think it is important—if you do not regard yourself as a genius—to have a serious obligation to society to do some useful research. This gives you the freedom to engage also in problems that cannot be solved from one day to the next, without the pressure of having to continually publish. But now that I am retired, of course, I am completely free to pursue these hobbies anyway.

2.2 Supplement 26 June 2021

When and why did you, or your family, return to Germany after the war. How did this happen?

My mother was suffering from MS and never really settled in England. She was quite unhappy there. My father was offered the position of CEO with the GEG (Großeinkaufs Gesellschaft Deutscher Konsumgenossenschaften) in Hamburg. He had already worked for the Cooperative Society in England for several years. This was an opportunity for the family to return to Germany in 1948. They lived on the top floor of the only non-destroyed apartment block known as the 'Beim Strohhause' block. All around there were only ruins. I stayed in England for another year to complete my Grammar School Certificate. Had I been a British national, I could have received a scholarship for Cambridge University but being German, this was not possible. I joined my family in Germany in 1949.

On my train ticket it said that I had to get off in Hamburg Altona, so I didn't get off the at Hamburg Hauptbahnhof and the next station was Flensburg, which is right up north near the Danish border. This was my first experience of Germany.

When did you know that you were destined to become a physicist? Did you consider any alternatives?

I've always been interested in Physics. My first really exciting experience was building a radio detector I couldn't believe that I could hear music through it. I pursued this interest on my own because my physics teacher didn't like me and was not very inspiring. However, I was also very interested in art, which I also pursued with a passion although my senior school exam results in the subject were not good. I took this as a sign that I should go into Physics. Having returned to Germany, I completed an internship at Menck and Hambrook, which was a mandatory requirement for a German Diploma in Physics at the time. I had some problems with my colleagues there: I heard them addressing each other with the familiar 'du'[9] and assumed that this was part of the corporate culture there … it wasn't—at least not for a trainee—and they let me know it in their own way.

Having seen Oxford students wandering around in the park, pondering and discussing things, I had gained the impression, that the life of a scientist was very laid back. So, I only rarely attended the lectures at the University of Hamburg and, by the end of the semester, found that I no longer understood a single word. So, for the first time in my life, I finally had to knuckle down and study zealously. I met some very lively and stimulating lifelong friends there including Wolfgang Kundt (Bonn), Gerd Wibberenz (Kiel), Ewald Richter (HH).

2.3   Conversation in 2021 About Climate Science Becoming a Political Actor

In June 2021, Klaus Hasselmann (KH) joined Susanne Hasselmann (SH), Dirk Olbers (DO) and Hans von Storch (HvS) to discuss how climate science had entered the social arena.[10]

HvS: Ola Johannessen once asked how the initially rather academic subject of climate change and climate dynamics became a dominant topic in political discourse. In your interview you said that your Institute was founded at the time explicitly with reference to the social importance of this topic and Reimar Lüst has made similar comments.

KH: Yes, that's right.

HvS: But in the early years it was rather abstract.

KH: Yes, that's the way it was. Everyone expected us to immediately buy a huge computer and start calculating. And it was clear to me that we had not yet understood many of the basic questions about climate change and human impact on the climate. I was particularly interested in clarifying the basics of human influence on the climate. How can we distinguish between natural and man-made climate fluctuations? Initially, that was my real motivation.

HvS: Dirk, you were an early member of the Max Planck Institute. Did you realize at the time that this was the overarching topic?

---

[9] There are three ways of saying "you" in modern German: du, Sie and ihr, whereby "du" is reserved for close friends and family as well as other extremely informal relationships.

[10] The discussion took place in German; translation by Dirk Olbers.

DO: No, not at all. We all had our own little niches in which we tried to solve various physics problems. Every now and then we would be requested to give a lecture on questions such as what climate is and what we can expect. In my opinion, this was initially explained in a relatively vague way. I gave lectures in which I really explained the spectrum of atmospheric radiation, although that probably didn't interest people, who just wanted to know whether it would rain or not at a given location. But we didn't know all that. We knew the basic physics, but the effects on humans and on the regional climate were unknown.

HvS: Klaus, who approached you about this topic? I know it was Reimar Lüst but there must have been others. Did certain politicians also approach you in the seventies asking for clarification of this issue?

KH: The climate problem had gradually entered public discourse and the political arena by the mid-seventies. The Institute was founded to address the climate problem. That was Lüst's idea, and it was he who convinced me to join him although I had nothing to do with the climate problem at that time. However, I was on an advisory committee that also dealt with human impact on the climate so, I was already familiar with the fact that climate was set to become the subject of political and scientific interest. So, whilst the topic was not new to me, I did wonder how I could best address it as a member of the Institute. The answer to that was developed at a later stage but it was certainly expected that I would take up this topic.

HvS: "It was expected"? By Reimar Lüst or were there other?

KH: By the public. By Lüst to a lesser degree: he always gave me a free rein and I never felt pressured by him.

SH: It was mainly the Swede Bert Bolin who recommended Klaus. The meteorologists were very much against it because he was better known as a physicist and oceanographer. He actually had to prove to them that he understood something about climate, because he didn't understand anything about it at the time and then quickly produced this simple stochastic climate model in order to have a basis on which to work.

KH: My first challenge in the climate field back in 1975 was to assess human impact on climate; to be able to distinguish man-made climate change from natural climate variability.

HvS: Who were the leading meteorologists at that time?

KH: Flohn was reasonable; but a meteorologist from Berlin was upset that a non-meteorologist had been given a job in climate research.

DO: What surprised us very much at the time was that the Institute was called the MPI for Meteorology rather than for Climate Research.

KH: I actually wanted to call it "for climate research" or something like that, but Lüst said it should be called "for meteorology" instead because that's very general and you never know in which direction things would go. It's a tradition at the Max Planck Society that people drift into something completely different from what they're supposed to be doing, which is why he wanted to leave it as open as possible.

HvS: So, in the first phase it was more academics like Bolin or Flohn or people who knew about it who approached you, but probably not heavyweight politicians?

KH: Not really.

HvS: When did politicians first take an interest?

SH: The first political interest in us was by Angela Merkel when she was minister for the environment in the early to mid 90s.

HvS: And this contact with Mrs. Merkel—how did that go?

SH: She had a perfect understanding about everything to do with the climate.

KH: I don't really remember that.

HvS: Initially, there were the general academic questions on climate: how does the climate work in the first place? and how can one analyse the climate? The effect or impact was not a relevant topic, which was also dealt with at one time. But then reunification of the two Germanys brought with it the opportunity to found the Potsdam Institute for Climate Impact Research (PIK) on your recommendation.

Politically speaking, it all happened very quickly back then. And it was a matter of transferring the strong aspects of GDR science into new forms. There was also a significant GDR weather service meteorology department at the Telegraphenberg in Potsdam. So, the idea was that something should be created on the Telegraphenberg, and I think it was then that you and others came up with the idea of founding an Institute for Climate Impact Research there, which would complement the work being carried out at your own Institute.

KH: Yes, the idea was that we should do the basic science and they should research the impact on society.

HvS: There was a paper by Nordhaus entitled "to slow or not to slow", which was quite important for me. And I think that was the moment when your commitment to the question of climate and society became much stronger.

KH: Yes, it grew. I don't know specifically when that was. Certainly, before the foundation of the PIK. After all, I had suggested the foundation of the PIK at that time to study the impact on society.

HvS: My perception is that at the end of the nineties or the beginning of the 2000s, the German public looked to the PIK, and no longer to the MPI for information on climate research.

KH: Yes, in terms of impact research. The impact of humans on the climate was the central question, not the climate itself as a scientific problem, as a physical problem, but as a human problem. That was also the intention, that the PIK should focus on a precise investigation into the anthropogenic impact.

SH: You weren't interested in informing the public, you were always interested in basic research, and you were glad that Mojib [Latif] took care of that on behalf of the MPI.

KH: I was glad that we had Mojib Latif and Hartmut Graßl. They took care of certain tasks that were important for the Institute. I had no desire to do that myself.

HvS: That led to a perception that the most important researchers in Germany were Graßl and Latif. Mr Hasselmann was hardly known.

KH: That suited me down to the ground.

DO: Initially, this division of labour between the MPI and the PIK didn't work properly because the PIK sometimes hired our doctoral students and colleagues, who then ran the same models as we did, and tried to answer the same questions. In this respect, the impression arose that rather than doing climate impact research, the Institute was conducting basic climate research. They were actually competing with us for a very long time. They were doing exactly the same ice sheet models—everything the same. That went on for ten years or so. Later on, research was actually conducted into such things as, let's say, the drought in Brandenburg in soil models and things like that. But today that's no longer the case.

HvS: Yes, they have a very strong economic dimension with Edenhofer. They also do that beautifully; I think the thing with the budget approach is downright ingenious. But, the tipping points, or the expectation that there will be a whole series of them are a dominant element in public discourse.

KH: Schellnhuber always zeroed in on a specific buzzword and that then had an effect in the discourse.

HvS: That is also correct from a communication perspective; it's how it should be done.

KH: Well, he overdid it a bit at times and sometimes introduced a bit of a weird aspect. I found the interaction between climate research, impact research and climate impact research exciting. I was very interested in this in general, but I was quite happy when the PIK really took it up and followed up on it. I thought it was an important topic and it was also my intention when I initiated the founding of the PIK.

DO: Were there any examples in other countries that already had similar institutes?

KH: To some extent, Smagorinski's group had done something similar. But, no, as far as I know, there was no institute doing exactly the same thing.

HvS: What is your idea of the ideal policy consultation?

KH: Well, it would involve people actually listening to what I have to say. That would be nice!

SH: For a long time, you were of the opinion that your role is to present the research, the facts, and that what is done with them is someone else's business. But certain things, such as the photo of Cologne Cathedral under water on the front cover of the *Spiegel* magazine etc., made you realise that that wasn't working. That was when you started to develop these socio-economic models and decided to make a contribution to the question of how to deal with the problem and thought that politicians would listen to you. But it didn't work out that way.

Why did a young girl sitting down in front of the parliament in Stockholm and going on a school strike start such a big movement? She came at it from an emotional standpoint; you came at it from a scientific base. But public demand is what influences the wider public and politics, and that goes via the emotions.

KH: I hadn't really considered the interaction between politics, the public and the media. The fact that it was important had always been clear to me.

The job of researchers is to clarify scientific connections. And I already knew that the ways in which the results are then put into practice and enter into the public domain to stimulate those mechanisms that are important in the process was an important field, but it was not my field.

My idea was always to set up a research institute to study the interaction between climate science and politics and society—the PIK.

HvS: Did the PIK succeed in this task?

KH: Yes, it has managed to do so in its own way and has come into the public eye—sometimes a bit distorted. But the question of how to mediate between scientific knowledge and policy implementation has already reached the public. That is still an issue.

HvS: At some point, we also started discussing the philosophy of science in Salzau. It turned out that none of us—except for Martin Heimann, who had heard a lecture on the subject in Switzerland—had ever really thought about it.

KH: I didn't know that at all; I must have forgotten. I always had a great respect for Heimann.

HvS: These questions did not figure in our curriculum at all.

KH: What questions?

HvS: History: the history of Science. Norms: what standards do we have? What does "good science" actually mean? You view that in different ways. But this whole philosophical dimension was non-existent at the MPI.

SH: That was already a question with Walter Munk, because American scientists feel a greater responsibility to the public than the Germans, so there was definitely a discussion with you on this topic.

KH: But I didn't discuss much with Walter Munk beyond science.

SH: Yes, yes, you had a lot of discussions about the fact that, as a scientist, one is also obliged to write popular science books. You said that your time was taken up with theoretical research and so on. But afterwards you got fully involved in these socio-economic models.

HvS: Your later turn towards the metrons then also completely took you over.

KH: Yes, yes.

HvS: One could see the situation like this: Hasselmann laid the foundations for understanding the climate problem and thus also for solving it. But when it came to actually translating this into a politically concrete situation, he then said "nah, others can do that better—I'm now focusing on the only topic that is really relevant, the metrons".

KH: Yes, the metrons had always interested me in the past in addition to my main work.

DO: You say the others can do it better—who can really do it better? I would say that they have failed just as much as we have. After us came the sociologists, such as Harald Welzer, who has written an incredible number of popular books, but they didn't listen to them either, just as they didn't listen to us.

SH: But that's not a question for the physicists, it's a question for the sociologists and psychologists.

HvS: Have you ever met a politician who internalised your arguments to such an extent that he wanted to implement something?

KH: Really from my ideas?

DO: Yes, about the climate problem, that he really wanted to do something? Politicians also came to the AWI. I wrote a brochure for Riesenhuber. The foreword, which he probably didn't write himself, is published under his name. My impression was that the essential thing for him was the photo opportunity when you handed this thing over. But what was written in it probably didn't interest him in the slightest.

KH: There are a number of different forces at work to which mankind is reacting. The climate problem is a long-term issue, but most people want to achieve short-term success. That's the main problem, that the long-term problems are always kicked into the long grass and then neglected. Whenever I'm supposed to do something for politics, I'm always a bit demotivated because I think, "but we already said that 40 years ago".

DO: Are the arguments of the sociologists, psychologists and economists who talk about climate today any more effective than ours? I look at it like this: we've done, let's say, 40 years of climate research. We understood everything and said everything we could say. Today, it's the turn of other sciences to talk about climate conditions on television. Most of the talk show guests are really sociologists and economists. Not much has actually happened.

KH: Humans are so trained in all the conflicts they have to resolve that this has to happen in the next 1, 2, or maybe 10 years. But here is a problem that requires planning on time scales of 30–40 years. Mankind is simply not used to that—with the exception of foreseers, who may be able to do it.

HvS: I wouldn't think so. So, if you think of the UN's 16 Millennium Development Goals, many of them are of this long-term character. Poverty for example: people in India who have to live in cardboard boxes. That's a long-term problem. But we no longer pay any attention to it; we are no longer interested in it. When it is said that the climate problem is the most important issue of all, then this corresponds to our local social perception. We no longer have that kind of poverty here—people living in cardboard boxes, for example. Other Millennium Development Goals do not play a role for us either. But in India they play a massive role. Since these are international global problems, there is a competition of concern that we cannot handle well.

KH: Yes, that's right.

HvS: There is an aspect of the North–South conflict, the consequences of colonialism. Climate is an important issue, I would say. But I doubt that it is really recognised globally as the most important and we can't judge or decide that either. We can judge the climate problem wonderfully, but we cannot judge the poverty problem.

KH: How to classify the climate problem and all the other issues facing humanity is something we don't really have a handle on. The climate problem is treated in the abstract and not embedded in all the other problems that global society is trying to solve at the same time. It's not just clear-cut problem solving. It's embedded in general politics.

3

The Strands of Klaus Hasselmann's Science

While Klaus Hasselmann began, as many theoretical physicists do, expecting to find a solution to the "turbulence problem" (whatever that is), he noticed that this would be a rather big challenge, and that it may also be good to tackle easier problems. And that is what he did before returning to the old dream when he retired, although he was not particularly successful in attracting praise and recognition. His thinking was informed by the dominance of a low-dimensional subspace, within which the dynamic lives and acts, while it influences, and is statistically influenced by myriads of factors in a high-dimensional space. This approach is sometimes obvious and sometimes under the surface, but it is ubiquitous.

The different strands of Hasselmann's interest and effort relate to:

1. ocean wave theory and prediction
2. remote sensing
3. stochastic climate model
4. reducing phase spaces
5. climate and society
6. building the modelling strategy of MPI
7. METRONs—particle theory.

For each of the above areas we shall first provide a brief overview of the topic, try to determine the significance of Hasselmann's work for the field, and provide a facsimile of a key publication, often his first on the subject. These first papers were often rather complicated and sometimes difficult to comprehend. In many cases, they were later followed by other versions

© The Author(s) 2022
H. von Storch, *From Decoding Turbulence to Unveiling the Fingerprint of Climate Change*,
https://doi.org/10.1007/978-3-030-91716-6_3

characterised by a remarkable clarity. The reason why, wherever possible, we present the first papers, is to provide the reader with a glimpse into Hasselmann's thought processes. As a rule, if something is truly great then the original ideas become simple. And indeed, Klaus was unwilling to update his original stochastic climate model-paper, "because it is too simple".

The numbered references below relate to the publication list in Sect. 5.

3.1   Ocean Wave Theory and Prediction: From Basic Physics to an Integrated Wind and Wave Data Assimilation System[1]

The challenge

There are two main challenges involved in ocean wave research:

1. One would like to better understand the basic physics, which is really quite complicated: even today many aspects are not fully understood.
2. There is a great need for practical applications: reliable forecasts and climatologies.

What was known in the 1950s?

Ocean wave research was booming in the 1950s,[2] with exciting progress being made along several lines. The semi-empirical forecasting methods of Sverdrup and Munk, based on wave height observations, came into wider use. Visual observations were complemented by instrumental observations, both in the laboratory and in the field. Bill Pierson introduced ocean wave spectra, applying results from studies on random noise, and he developed practical methods for ocean wave forecasting using wave spectra and statistics. Owen Philips and John Miles made significant contributions to the understanding of basic processes.

---

[1] By Gerbrand Komen after some discussions with Luigi Cavaleri.

[2] Details and references can be found in Hisashi Mitsuyasu's excellent Historical Note on the Study of Ocean Surface Waves (Journal of Oceanography, 58, pp. 109–120, 2002), and also in Klaus' own account [95].

Basic equations

Then, in 1960, Klaus Hasselmann published *Grundgleichungen der Seegangsvorhersage* (Basic equations for sea state predictions) in German in the journal *Schiffstechnik* (Maritime Engineering) ([3], see facsimile f1 below). The paper opens by noting that knowledge of the forces acting upon developing ocean waves ("wind sea") is insufficient, but also—more optimistically—that recent advances are encouraging in terms of attempts to develop a reliable, general method of sea state prediction and that this should be based on an equation that represents the energy balance that shapes the ocean wave spectrum. This is then followed by what is now known as the energy balance equation, aka the radiative transfer equation, which expresses the rate of change in energy of a spectral component as a result of advection, wind input, dissipation, and the exchange of energy between different wave components due to nonlinear resonant interactions.

In the paper, Hasselmann expresses his surprise that this equation had not been included in previous approaches. However, this is not quite correct, because, in fact, Gelci and his colleagues had formulated and used a similar equation in 1957 in a paper entitled *Prévision de la houle. La méthode des densités spectroangulaires*, which was published in the *Bulletin d'information du Comité central d'océanographie et d'études des cotes*. Obviously, this was not known to Klaus at that time. Anyway, his treatment contained an important new element, namely the inclusion of the wave-wave interaction term.

Some readers of *Grundgleichungen* (for example, Richard Dorrestein, director of Oceanography and Maritime Meteorology at the Royal Netherlands Meteorological Institute) were surprised that the paper did neither derive nor justify the correctness of the energy balance equation. In fact, a decent derivation was not provided until 1975, when Jürgen Willebrand published his 'Energy transport in a nonlinear and inhomogeneous random gravity wave field'.

Grundgleichungen not only includes the basic equations, but also discusses several applications in special situations, namely for fully developed wind sea and for the "development phase, in which the non-linear effects are still negligible". Later it would become clear that this second application was rather academic as nonlinear interactions were found to be strong for young wind sea. Finally, the paper includes a section on finite depth effects, with an application of generation in the Neusiedler See, a lake in Austria, south of Vienna.

Grundgleichungen has a modest citation record. Nevertheless, its impact has been enormous, as it not only provided a basis for further work, but also set out an agenda for ocean wave research by stating that:

- (...) more precise observations would be required (a theoretical calculation might fail for the time being due to the turbulence problem) to determine those terms used in the energy equation that are still uncertain with greater precision.
- the method could be expanded with the aid of a suitable computer programme for an electronic digital system, to calculate fast and accurate sea state and swell forecasts for any wind fields identified on the weather chart.

Klaus himself would actively pursue these objectives over the next few decades, with help of the global wave research community which he successfully mobilized. This is now history, with several well-written and well-documented accounts.[3] Here a short overview will be given.

Nonlinear interactions

Grundgleichungen contained an explicit expression for the exchange of energy between different wave components due to nonlinear resonant interaction, the so-called Boltzmann integral, a five-dimensional integral containing the products of wave spectra and a number of exchange functions. The exchange functions were not included in the 1960, but appeared in follow-up papers [6, 8, 9] in 1962 and 1963, and in a comprehensive and more general account which appeared in 1968 as "Weak-interaction theory of ocean waves" (21).

The Boltzmann integral is actually a 6-dimensional integral in wavenumber space, constrained by the resonance condition, namely that the frequency of the 'forced' component is equal to the sum of the frequencies of the 'forcing' components. Its numerical integration is challenging because wave spectra are typically sharply peaked. To obtain reliable results these peaks have to be represented with a high degree of accuracy in high resolution. Initial results were already available in 1961, indicating that energy from waves near the peak of the spectrum was transferred to still longer waves, but integrating the Boltzmann integral with sufficient accuracy and affordable computing costs remained a challenge for the next 25 years or so [198, 77, 78, 114]. An initial successful application emerged in 1972 when it was

---

[3] For example: Young and van Vledder 1993: A review of the central role of nonlinear interactions in wind-wave evolution; Janssen 2007 Progress in ocean wave forecasting; The WISE-group (Cavaleri et al.) 2007 Wave modelling—The state of the art. This in addition to Mitsuyasu's 2002 Historical Note mentioned in the previous footnote.

found that nonlinear resonant interactions were essential for understanding the spectral evolution observed during JONSWAP. Later applications from 1980 onwards were used in numerical wave prediction models.

JONSWAP

Klaus Hasselmann was involved in several large-scale field experiments. The first was the Pacific swell propagation programme [18] with Walter Munk and others. Another major campaign was MARSEN in 1979 in the North Sea. Perhaps best known is JONSWAP, which Hasselmann coordinated, which took place in the German Bight in 1969 following a pilot experiment in 1968. There were several objectives, such as measuring wave growth, wind stress, atmospheric turbulence, and swell attenuation. The development of sea states was studied by continuously measuring wave spectra along a line extending 160 kms into the North Sea westward from Sylt under (fairly) stationary offshore wind conditions.

One important result was the parametrisation of the observed spectra. The starting point was an earlier parametrisation by Pierson and Moskowitz for fully developed seas. The most remarkable difference was the strong enhancement of the energy level at the spectral peak during growth. Mitsuyasu, who had performed similar measurements at about the same time in Hakata Bay, proposed a somewhat different parametrisation, however the JONSWAP spectrum would be used more widely in later studies and applications.

A second important result of JONSWAP was the determination of the fetch dependence of the spectral parameters, where fetch is defined as the distance to shore. Ideally, one would like to perform these studies for a constant wind blowing perpendicular from a straight coastline. In reality this never occurs, which results in a lot of scatter in plots of measured wave parameters against fetch. This is usually somewhat hidden in log–log plots. Another problem relates to the choice of scaling variable. Quantities such as wave height and wavelength are usually presented in nondimensional form with the aid of either the wind speed at a given height or the frictional velocity. The choice is important when one extrapolates the JONSWAP results—which were obtained for fairly moderate wind speeds—to higher wind speeds, as the windspeed/friction velocity ratio is itself a function of wind speed.

Perhaps the most rewarding outcome was a better understanding of the mechanism of wave evolution. Using computations of the Boltzmann integral and simple parametrisations for wind input and dissipation it could be shown that wind input mainly occurs at medium and high frequencies and that the generation of low frequency waves—and the associated mean wavelength increase with fetch (and wave age)—is due to nonlinear interactions.

Models

The JONSWAP-results formed essential ingredients for the realisation of the second objective set out in *Grundgleichungen*: "to calculate fast and accurate wind sea and swell forecasts".

Numerical wave models represent the wave spectrum on grid points and simulate their evolution in small time steps. As numerical integration of the full Boltzmann integral was prohibitively expensive, several ocean wave models were developed in Hamburg and elsewhere in which the effect of the nonlinear transfer was modelled by prescribing the spectral shape and imposing the observed dependence of the spectral parameters on the wave age. These models had skill and were used for many applications, but then an international model intercomparison (SWAMP) found that different models produced very different results in particular situations. An important step forward was made by the development of EXACT-NL [76], a model that used an approximation developed by Klaus and Susanne Hasselmann. Results were presented in Miami in 1981 but were not published until 1985.

Hasselmann launched a new initiative in 1984 known as the WAM (Wave Modelling) group in which an international team of researchers would collaborate on the further development of a model based on the *Grundgleichungen*. This involved the further improvement of the source terms, a new more rapid approximation to the Boltzmann integral, and implementation in many different centres. At ECMWF much work was done by Susanne Hasselmann and others at ECMWF, in particular by Liana Zambresky, Peter Janssen and Heinz Günther, each of whom spent several years in Reading installing the model on the CRAY-1, coupling it to wind fields, performing test and validation runs, introducing the model into the operational forecast cycle, and setting up routine validation against observations. Visitors from the WAM group (such as Anne Guillaume, Vince Cardone and Luigi Cavaleri and their colleagues) also made significant contributions. The model became known as the WAM model. Results were published in 1988, and later, in 1994, in the monograph "Dynamics and Modelling of Ocean Waves" [244]. This was all done under the continuous guidance of and was inspired by Klaus Hasselmann.

There was a certain amount of consensus that models constructed on the basis of fundamental physics, such as that described in *Grundgleichungen*, and the WAM-model in particular, would be superior to more empirical models. However, reality is complex: the WAM-model had some shortcomings, in particular in the numerics, whilst some models that did not integrate the

Boltzmann equation were very well tuned and performed quite well. In practice, the quality of wind forcing was often a limiting factor. In fact, WAM was so reliable that it could detect errors in the atmospheric model used to generate surface winds.

Towards an integrated wind and wave data assimilation system

In 1985, when work on the WAM model was under way, and remote sensing from earth observing satellites became feasible, Hasselmann came up with a new and ambitious vision [74, 79, 95], namely, to run a coupled atmosphere/surface wave/ocean model, which could provide first-guess information for the retrieval of useful information from satellites, and which would assimilate all available observations in real time. This would then provide the best possible forecasts as well as an archive for climate and other research. This seemed like a pipedream in 1985, and some people were critical because of its inductive structure, as it would use model results to interpret measurements which were then used to validate the model. Nevertheless, it became a reality in the nineties, and was highly successful, helping to improve forecasting expertise and providing huge and useful datasets (ERA) for earth system research.

Heritage

After 1994, Hasselmann put his energy in other endeavours, while ocean wave research continued, building upon what he had already started. "Dynamics and Modelling of Ocean Waves" [244] became a standard reference book and remained so throughout the years, when many groups attempted to improve the representation of the various source terms. The WAM-model is still in use for both forecasting and wave climate studies.

Klaus' dream of an integrated wind and wave data assimilation system became reality in 1998 when ECMWF started running a coupled forecasting system, where the atmospheric component of the Integrated Forecasting System (IFS) communicated with the wave model through the exchange of the Charnock parameter, which determines the roughness of the sea surface.

New ocean wave models, such as SWAN and WAVEWATCH, were developed. They are still essentially based on the *Grundgleichungen* as described by Hasselmann in his 1960 paper.

Grundgleichungen der Seegangsvoraussage

K. Hasselmann

Institut für Schiffbau, Hamburg

Das Problem der Entwicklung der Windsee, die von einem zeitlich und räumlich veränderlichen Windfeld über einem Meer erzeugt wird, ist bisher vorwiegend von der empirischen Seite untersucht worden [1, 2, 4, 5, u. a.]. Zur Aufstellung einer exakten Theorie waren die Kenntnisse der auf den Seegang einwirkenden Kräfte noch nicht ausreichend. Es erscheint jedoch fraglich, ob im komplizierten allgemeinen Fall räumlich und zeitlich veränderlicher Windfelder eine befriedigende Seegangsvoraussage erreicht werden kann, ohne auf die wichtigsten Grundvorgänge der Seegangsentwicklung einzugehen. Im wesentlichen lassen sich drei Haupteinwirkungen auf den Seegang unterscheiden: die Anfachung durch Windkräfte, die Dissipation durch turbulente Reibung und die nichtlinearen Einflüsse, die durch die endliche Steilheit des Seegangs bestimmt sind. Einen wichtigen Beitrag zum Verständnis der Windkräfte lieferte O. M. Phillips [6] durch seine Erklärung der Seegangsentstehung als Folge der Einwirkung statistischer Druckschwankungen auf die Wasseroberfläche. Der Einfluß der Turbulenz auf den Seegang ist im einzelnen noch ungeklärt, jedoch wird allgemein angenommen, daß die erhöhte Dissipation infolge der Turbulenz in erster Näherung durch einen zusätzlichen konstanten Reibungskoeffizienten berücksichtigt werden kann. Das vielleicht größte Hindernis zu einer exakten Seegangstheorie waren jedoch die bisher ungeklärten nichtlinearen Effekte, welche vor allem bei der Ausbildung der ausgereiften Windsee eine entscheidende Rolle spielen. Es lassen sich zwei unabhängige nichtlineare Effekte unterscheiden:

1. der Vorgang der Wellenbrechung, bei dem die Wellenenergie unmittelbar in Turbulenz verwandelt und gleichzeitig infolge der erhöhten turbulenten Reibung schneller dissipiert wird, und

2. die nichtlinearen Wechselwirkungen zwischen verschiedenen Wellenkomponenten des Seegangs, die zu einer Energieumschichtung innerhalb des Seegangsspektrums führen.

Die Wellenbrechung ist vermutlich auf eine Instabilität zurückzuführen, die in den Wellenkappen auftritt, wenn die lokale, nach unten gerichtete Beschleunigung die Erdbeschleunigung übertrifft [8]. Die Berechnung der Häufigkeitsverteilung der brechenden Wellenkappen sowie zumindest eine Abschätzung der dabei verlorengehenden Wellenenergie dürfte mit üblichen statistischen Methoden durchzuführen sein. Eine Theorie, die den Energieaustausch im Seegangsspektrum infolge der nichtlinearen Wechselwirkungen beschreibt, ist bereits vollständig durchgeführt worden und wird in einer weiteren Veröffentlichung ausführlicher dargestellt werden[1].

Obwohl die Grundphänomene, die die Entwicklung des Seegangs beeinflussen, demnach noch nicht in allen Einzelheiten geklärt sind, ermutigen die jüngsten Fortschritte in diesen Fragen doch zu dem Versuch, an Hand der exakten Grundgleichungen des Seegangs eine zuverlässige, allgemeine Methode der Seegangsvoraussage zu entwickeln. Ausgangspunkt einer exakten Theorie muß zwangsläufig eine Gleichung bilden, die die Energiebilanz des Seegangsspektrums darstellt. Es überrascht etwas, daß diese Gleichung in den bisherigen empirischen Ansätzen durchweg nicht mit herangezogen worden ist. Auch ohne genaue Kenntnis der einzelnen Terme der Energiegleichung lassen sich aus der Struktur der Gleichung mehrere wichtige Aussagen ableiten. Insbesondere ergibt sich eine sehr einfache Antwort auf die vieldiskutierte Frage nach der relativen Bedeutung von „Fetch" und „Duration" bei der Ausbildung der Windsee.

Für die Entwicklungsphase, in der die nichtlinearen Effekte noch vernachlässigbar sind, läßt sich die Energiegleichung sofort integrieren. Einige charakteristische Lösungen werden diskutiert. Insbesondere ergibt sich, daß auch relativ kleine Gebiete großer Windstärke einen hohen Seegang erzeugen können, falls die Wanderungsgeschwindigkeit des Gebiets gerade halb so groß ist wie die Windgeschwindigkeit innerhalb des Gebiets. Der stationäre Endzustand des Spektrums für ein ausgedehntes Windfeld wird ebenfalls kurz diskutiert, die Lösung dieses Problems würde jedoch über den Rahmen dieser Arbeit hinausgehen.

Mit Hilfe eines geeigneten Rechenprogramms für eine elektronische Digitalanlage ließe sich die Methode ausbauen, um schnelle und genaue Windsee- und Dünungsvoraussagen für beliebige, durch die Wetterkarte bestimmten Windfelder zu berechnen. Hierzu müßten allerdings noch genauere Beobachtungen (eine theoretische Berechnung dürfte vorerst an dem Turbulenzproblem scheitern) zur näheren Bestimmung der noch unsicheren Terme in der Energiegleichung vorliegen.

1. Formulierung des Problems; Energiegleichung

Der von Windkräften angefachte unregelmäßige Seegang auf einem Meer läßt sich bekanntlich vollständig durch ein zweidimensionales Energiespektrum $F(\mathfrak{K}; \mathfrak{r}, t)$ charakterisieren[2], in dem der Wellenzahlvektor $\mathfrak{K}$ Fortschreitungsrichtung und Wellenlänge $\lambda = 2\pi/k$ der Wellen angibt und $\mathfrak{r}$ der zweidimensionale Ortsvektor an der Oberfläche ist. ($|\mathfrak{K}| = k$; Komponenten von $\mathfrak{K}$: k_x, k_y, k_z). Das Problem der Seegangsvoraussage formulieren wir dann folgendermaßen:

In einem Oberflächengebiet $\mathfrak{G}$ mit (allgemein zeitlich veränderlichem) Rand $\mathfrak{R}$ ist das Spektrum $F(\mathfrak{K}; \mathfrak{r}, t)$ zu bestimmen für $t \geq t_0$. In $\mathfrak{G}$ ist das Windfeld für $t \geq t_0$ und zur Zeit t_0 das Anfangsspektrum

$$F(\mathfrak{K}; \mathfrak{r}, t_0) = F_0(\mathfrak{K}; \mathfrak{r}) \qquad (1)$$

vorgegeben. Auf $\mathfrak{R}$ gelten stückweise eine der folgenden beiden Randbedingungen:

1. Randbedingung:

$$F(\mathfrak{K}; \mathfrak{r}_r, t) = F_1(\mathfrak{K}; \mathfrak{r}_r, t) \text{ für } (\mathfrak{v}(\mathfrak{K}) \cdot \mathfrak{n}) - U_r \lesseqgtr 0 \text{ und } t \geq 0 \qquad (2)$$

wo $\mathfrak{n} =$ äußere Normale des Randes,

$\mathfrak{v}(\mathfrak{K}) =$ Gruppengeschwindigkeit $= \dfrac{\mathfrak{K}}{2k} \sqrt{\dfrac{g}{k}}$

(g = Erdbeschleunigung)

$U_r =$ Normalgeschwindigkeit des Randes

und F_1 eine bekannte Funktion ist.

---

1) In einer kürzlich erschienenen Arbeit von O. M. Phillips [7] werden ebenfalls nichtlineare Wechselwirkungen von Schwerewellen betrachtet. Die Untersuchung beschränkt sich jedoch auf die Wechselwirkungen bis zur dritten Ordnung zwischen zwei diskreten Wellenzügen, während im Fall eines statistischen kontinuierlichen Spektrums der Energietransport durch Wechselwirkungen bis zur fünften Ordnung zwischen drei verschiedenen Spektralkomponenten hervorgerufen wird.

2) Die statistischen Eigenschaften des Seegangs sind streng genommen nur im Falle statistisch unabhängiger Fourierkomponenten vollständig durch das Spektrum beschrieben. Ein solches „Rausch"spektrum liegt sicherlich vor, solange die nichtlinearen Effekte vernachlässigbar sind. Es läßt sich jedoch auch im nichtlinearen Fall nachweisen, daß die Abweichungen des Seegangsspektrums von einem Rauschspektrum klein bleiben.

2. Randbedingung

$$F(\mathfrak{K}; \mathfrak{r}_r, t) = \lambda(\mathfrak{K}) F(\mathfrak{K}'; \mathfrak{r}_r, t)$$

für $(\mathfrak{K}\mathfrak{n}) < 0$ (3)

$$\text{wo} \quad \mathfrak{K}' = \mathfrak{K} - 2\mathfrak{n}(\mathfrak{K}\mathfrak{n})$$

der am Rand reflektierte Wellenzahlvektor ist.

Die erste Randbedingung gilt für Ränder auf dem offenen Meer. Sie gibt die Wellenenergie vor, die für $t \geq t_o$ von außen her in $\mathfrak{G}$ einströmt. Die zweite Randbedingung gilt für die Reflexion an Küsten. Um Fallunterscheidungen im folgenden zu vermeiden, betrachten wir nur den wichtigsten Fall, daß der Reflexionsfaktor $\lambda(\mathfrak{K})$ identisch verschwindet. Die 2. Randbedingung geht dann über in den Spezialfall

$$F_1 = 0, \quad U_r = 0 \quad \text{der 1. Randbedingung.}$$

Zu diesen Anfangs- und Randbedingungen ergibt sich nun aus der Energiebilanz des Spektrums eine partielle Differentialgleichung, durch die $F(\mathfrak{K})$ dann eindeutig bestimmt wird. Sie lautet:

$$\frac{\partial F(\mathfrak{K})}{\partial t} + \mathfrak{v}(\mathfrak{K}) \cdot \text{grad } F = \alpha(\mathfrak{K}; \mathfrak{r}, t) + \beta(\mathfrak{K}; \mathfrak{r}, t) F(\mathfrak{K}) -$$
$$- 4 \nu_t k^2 F(\mathfrak{K}) + N_1 + N_2. \quad (4)$$

Der Konvektionsterm auf der linken Seite der Gleichung ist die Divergenz des dem Spektrum $F(\mathfrak{K})$ zugeordneten Energiestroms $\mathfrak{v}(\mathfrak{K}) \cdot F(\mathfrak{K})$.

Der erste Term auf der rechten Seite stellt die Anfachung durch die Druckschwankungen über der Wasseroberfläche dar (Phillips [6]). Das bei Phillips durch ein Autokorrelationsintegral ausgedrückte Ergebnis läßt sich durch Einführen des von Wellenzahl und Frequenz abhängigen Spektrums $P(\mathfrak{K}, \omega)$ der Druckschwankungen p in anschaulichere Form bringen:

$$\alpha(\mathfrak{K}) = \frac{\pi}{4k} P(\mathfrak{K}, -\omega_k) \quad \text{mit} \quad \omega_k = \sqrt{gk}. \quad (5)$$

$P(\mathfrak{K}, \omega) dk_x dk_y d\omega$ ist der Anteil, den die Druck-Fourierkomponenten $A(\mathfrak{K}', \omega') e^{i[(\mathfrak{K}' \mathfrak{r}) + \omega' t]}$, deren Wellenzahlen $\mathfrak{K}'$ bzw. Kreisfrequenzen ω' in den Intervallen $k_x \leq k_x' \leq k_x + dk_x$, $k_y \leq k_y' \leq k_y + dk_y$, bzw. $\omega \leq \omega' \leq \omega + d\omega$ liegen, zum mittleren Quadrat des Drucks beitragen.

Der zweite Term auf der rechten Seite berücksichtigt die eventuell hinzukommende Anfachung infolge von Sheltering, Instabilität im Sinne von J. W. Miles [3] usw. Im Gegensatz zum ersten Term, der nur von der Turbulenz des Windes abhängt, sind diese Effekte (zumindest in erster Näherung) dem Wellenspektrum proportional.

Der dritte Term stellt die Dissipation infolge der turbulenten Viskosität ν_t dar (die zähe Reibung ist vernachlässigbar).

Der Term N_1 repräsentiert den nichtlinearen Einfluß der Brechung. Zur Zeit läßt sich über diesen Term wenig aussagen. Es ist wahrscheinlich, daß er ebenso wie die Reibungsterme vorwiegend im kurzwelligen Bereich des Spektrums von Bedeutung ist, jedoch wesentlich stärker als linear mit dem Spektrum zunimmt. Für unsere jetzige, mehr prinzipielle Untersuchung ist die exakte Form dieses Terms ohne Belang.

Der Term N_2 repräsentiert schließlich den Einfluß der nichtlinearen Wechselwirkungen zwischen verschiedenen Komponenten des Spektrums. Er lautet:

$$N_2 = \int\int\int\int_{-\infty}^{+\infty} F(\mathfrak{K}') F(\mathfrak{K}'') F(\mathfrak{K}' + \mathfrak{K}'' -$$
$$- \mathfrak{K}) T_1(\mathfrak{K}', \mathfrak{K}'', \mathfrak{K}' + \mathfrak{K}'' - \mathfrak{K}) dk_x' dk_y' dk_x'' dk_y''$$
$$- F(\mathfrak{K}) \int\int\int\int_{-\infty}^{+\infty} F(\mathfrak{K}') F(\mathfrak{K}'') T_2(\mathfrak{K}, \mathfrak{K}', \mathfrak{K}'') dk_x' dk_y' dk_x'' dk_y''$$
$$(6)$$

T_1 und T_2 sind komplizierte Austauschfunktionen, die hier nicht näher angegeben werden. Sie enthalten als Faktoren Diracsche δ-Funktionen der Form

$$\delta(\omega(\mathfrak{K}' + \mathfrak{K}'' - \mathfrak{K}) - \omega(\mathfrak{K}') - \omega(\mathfrak{K}'') + \omega(\mathfrak{K})).$$

Die der Einfachheit halber als vierfache Integrale dargestellten Ausdrücke auf der rechten Seite von (6) sind somit in Wirklichkeit nur dreifache Integrale über Hyperflächen der Form

$$\omega(\mathfrak{K}' + \mathfrak{K}'' - \mathfrak{K}) - \omega(\mathfrak{K}') - \omega(\mathfrak{K}'') + \omega(\mathfrak{K}) = 0$$

im $\mathfrak{K}' \cdot \mathfrak{K}''$-Raum ($\omega(\mathfrak{K}) = \omega_k = \sqrt{gk}$).

2. Formale Integration der Energiegleichung

Bezeichnen wir die Summe der lokalen Einwirkungen auf der rechten Seite von (3) kurz mit $L(\mathfrak{K}; \mathfrak{r}, t)$, so lautet die formale Lösung der Energiegleichung mit den zugehörigen Anfangs- und Randbedingungen:

$$F(\mathfrak{K}; \mathfrak{r}, t) = \int_{t_1}^{t} L(\mathfrak{K}; \mathfrak{r} - (t - t') \mathfrak{v}(\mathfrak{K}), t') dt' +$$
$$+ F(\mathfrak{K}; \mathfrak{r} - (t - t_1) \mathfrak{v}(\mathfrak{K}), t_1) \quad (7)$$

$$\text{mit } t_1 = \max \begin{cases} t_o \text{ (Begrenzung durch die Duration)} \\ t_r, \text{ wo } t_r = \max \mathfrak{r} \, (< t), \\ \text{für welches } \mathfrak{r} - (t - \mathfrak{r}) \mathfrak{v}(\mathfrak{K}) \text{ auf } \mathfrak{R} \text{ liegt.} \\ \text{(Begrenzung durch den Fetch)} \end{cases}$$

Gleichung (7) besagt, daß das Spektrum $F(\mathfrak{K}; \mathfrak{r}, t)$ bestimmt wird durch das Integral der lokalen Einwirkungen $L(\mathfrak{K})$ über den Weg, den eine Wellengruppe mit der Wellenzahl $\mathfrak{K}$, die zur Zeit t am Ort $\mathfrak{r}$ eintrifft, von einem bekannten Anfangspunkt $\mathfrak{r} - \mathfrak{v}(\mathfrak{K})(t - t_1)$ aus zurücklegt. Falls der zurückverlängerte Weg der Wellengruppe bis zur Anfangszeit t_o innerhalb des Gebietes $\mathfrak{G}$ bleibt, so ergibt sich der Anfangswert in (7) aus der Anfangsbedingung (1). Falls der zurückverlängerte Weg dagegen bereits zu einer Zeit $t_r > t_o$ auf den Rand $\mathfrak{R}$ stößt, so ergibt sich der Anfangswert in (7) aus der 1. Randbedingung (2) (Im Falle der 2. Randbedingung (3) müßte der Weg natürlich an $\mathfrak{R}$ reflektiert und weiter zurückverfolgt werden). Dieses Ergebnis ist kaum überraschend und hätte auch ohne explizite Aufstellung der Energiegleichung abgeleitet werden können. Dennoch ist (7) bisher nur für die Dünungsvoraussage angewandt worden [5]. Hier handelt es sich um das reine Dispersionsproblem mit $L(\mathfrak{K}) \equiv 0$, und (7) reduziert sich auf

$$F(\mathfrak{K}; \mathfrak{r}, t) = F(\mathfrak{K}; \mathfrak{r} - \mathfrak{v}(\mathfrak{K})(t - t_1), t_1). \quad (8)$$

Identifizieren wir im allgemeinen Fall $L(\mathfrak{K}) \neq 0$ das Gebiet $\mathfrak{G}$ mit einem scharf begrenzten Sturmgebiet, so erkennt man aus (7), daß zwischen dem Einfluß des Fetch und der Dauer des Sturms kein prinzipieller Unterschied besteht. Diese Größen machen sich in der gleichen Weise allein durch die Begrenzung der Einwirkungsdauer von $L(\mathfrak{K})$ bemerkbar. Wir werden daher auch später aus der Spektrumsvoraussage für ein unendlich ausgedehntes Sturmgebiet, das zur Zeit $t = t_o$ plötzlich entsteht, sofort die allgemeine Lösung für ein beliebiges (im Innern jedoch konstantes) Sturmgebiet mit zeitlich veränderlicher Umrandung ableiten können.

Die Gleichung (7) stellt zunächst nur eine formale Integration der Energiegleichung (4) dar; denn der Ausdruck $L(\mathfrak{K})$ hängt noch von dem Spektrum $F(\mathfrak{K})$ ab. Eine allgemeine Lösung dürfte wegen des sehr komplizierten nichtlinearen Terms N_2 nur auf numerischem Wege mit Hilfe elektronischer Rechenanlagen möglich sein. Im folgenden untersuchen wir daher lediglich zwei (allerdings wichtige) Grenzfälle: 1. die vollausgereifte Windsee, 2. die Entwicklungsphase, bei der die nichtlinearen Effekte noch vernachlässigbar sind.

3. Die ausgereifte Windsee

Unter der ausgereiften Windsee versteht man den Seegang, der sich für $t \to \infty$ bei einem unendlich ausgedehnten, konstanten Windfeld einstellt. In (4) verschwindet also die linke Seite. Für den relativ uninteressanten Fall eines schwachen Windfeldes, bei dem das Spektrum so klein bleibt, daß die nichtlinearen Terme vernachlässigt werden können, erhält man sofort

$$F(\mathfrak{K}) = \frac{\alpha(\mathfrak{K})}{4\,v_t\,k^2 - \beta(\mathfrak{K})} \quad (\beta(\mathfrak{K}) < 4\,v_t\,k^2 \text{ vorausgesetzt}). \quad (9)$$

Die allgemeine Lösung für den Fall, daß die nichtlinearen Terme nicht vernachlässigbar sind, können wir in diesem Rahmen nur qualitativ diskutieren. Die Austauschfunktionen T_1 und T_2 des nichtlinearen Terms N_2 sind beide positiv. Das erste Integral in (6) beschreibt die Energiezunahme des Spektrums an der Stelle $\mathfrak{K}$ durch die Wechselwirkungen zwischen den Wellenkomponenten an drei anderen Stellen des Spektrums. Die Zunahme ist unabhängig vom Wert des Spektrums an der Stelle $\mathfrak{K}$. Das zweite Integral stellt den Energieverlust des Spektrums $F(\mathfrak{K})$ durch unmittelbare Wechselwirkungen der Komponenten der Wellenzahl $\mathfrak{K}$ mit den Komponenten an jeweils zwei weiteren Stellen des Spektrums dar. Der Energieverlust ist dem Wert des Spektrums an der Stelle $\mathfrak{K}$ proportional. Die nichtlinearen Wechselwirkungen werden also die Tendenz haben, scharfe Spitzen des Spektrums zu glätten und die Energie gleichmäßiger über das Spektrum zu verteilen. Es ist zu erwarten, daß sich analog zum nichtlinearen Kaskadenprozeß der Turbulenz ein mittlerer Energiestrom ausbildet, der von den energiereichen langwelligen zu den energiearmen kurzwelligen Gebieten des Spektrums fließt. Im stationären Endzustand würde sich dann wahrscheinlich ein Gleichgewicht einstellen zwischen der Energiezufuhr $\alpha(\mathfrak{K}) + \beta(\mathfrak{K})\,F(\mathfrak{K})$ im langwelligen Gebiet, dem nichtlinearen Energietransport N_2 von längeren zu kürzeren Wellen, und den Energieverlusten $-4\,v_t\,k^2\,F(\mathfrak{K}) + N_1$ durch turbulente Reibung und Wellenbrechung, die vorwiegend im kurzwelligen Bereich erfolgen.

Der nichtlineare Energieaustausch ist der dritten Potenz des Spektrums und somit der sechsten Potenz der mittleren Wellensteilheit des Seegangs proportional. Wegen dieser sehr starken Abhängigkeit der nichtlinearen Wechselwirkungen von der mittleren Wellensteilheit wird der Übergang von der Entwicklungsphase, in der die nichtlinearen Effekte vernachlässigbar sind, zum stationären Endzustand des Seegangs für ein bestimmtes Spektralgebiet sehr rasch erfolgen. Durch Kombination der hier betrachteten Grenzfälle der linearen Entwicklungsphase und des nichtlinearen Endzustandes dürften sich somit auch im allgemeinen Fall brauchbare Seegangsvoraussagen erzielen lassen.

4. Die lineare Entwicklungsphase des Seeganges

Unter Vernachlässigung der nichtlinearen Terme läßt sich die Lösung der Energiegleichung auf Quadraturen zurückführen:

$$F(\mathfrak{K};\mathfrak{r},t) = \frac{1}{\gamma(\mathfrak{K};\mathfrak{r},t)} \int_{t_1}^{t} \gamma(\mathfrak{K};\mathfrak{r} - (t - t')\,\mathfrak{v}(\mathfrak{K}),t')$$

$$\alpha(\mathfrak{K};\mathfrak{r} - (t - t')\,\mathfrak{v}(\mathfrak{K}),t')\,dt' + F(\mathfrak{K};\mathfrak{r} - (t - t_1)\,\mathfrak{v}(\mathfrak{K}),t_1)$$

$$\text{mit } \gamma(\mathfrak{K};\mathfrak{r},t) = \exp \int_{t_1}^{t} [4\,v_t\,k^2 -$$

$$- \beta(\mathfrak{K};\mathfrak{r} - \mathfrak{v}(\mathfrak{K})\,(t - t')\,t')\,dt' \quad (10)$$

und t_1 wie in (7).

Mit geeigneten Annahmen über die Abhängigkeit der Koeffizienten $\alpha(\mathfrak{K})$ und $\beta(\mathfrak{K})$ von der Windstärke läßt sich aus (10) die Entwicklung des Spektrums für ein beliebiges Windfeld berechnen. Um einige prinzipielle Eigenschaften der Lösungen hervorzuheben, beschränken wir uns im folgenden jedoch auf den Fall $\alpha(\mathfrak{K})$, $\beta(\mathfrak{K})$ unabhängig von $\mathfrak{r}$, t. (10) wird dann:

$$F(\mathfrak{K};\mathfrak{r},t) = \frac{\alpha(\mathfrak{K})}{4v_t\,k^2 - \beta(\mathfrak{K})} (1 - e^{(\beta(\mathfrak{K}) - 4v_t\,k^2)(t - t_1)}) +$$

$$+ F(\mathfrak{K};\mathfrak{r} - (t - t_1)\,\mathfrak{v}(\mathfrak{K}),t_1) \quad (11)$$

Wir betrachten nun zwei spezielle Anfangs- und Randwertprobleme.

Fall 1: Das Gebiet $\mathfrak{G}$ ist unendlich; es ist $F_o(\mathfrak{K},\mathfrak{r}) \equiv 0$. Dies entspricht dem Fall eines plötzlich aufkommenden Windes über einem anfangs ruhigem Meer. Hier ist $t_1 = t_o$. Setzen wir $t_o = 0$, so wird (11):

$$F(\mathfrak{K};\mathfrak{r},t) = F_d(\mathfrak{K};t) = \frac{\alpha(\mathfrak{K})}{4v_t\,k^2 - \beta(\mathfrak{K})} (1 - e^{(\beta(\mathfrak{K}) - 4v_t\,k^2)\,t})$$

$$(12)$$

Für $\beta(\mathfrak{K}) < 4\,v_t\,k^2$ geht die Lösung für $t \to \infty$ in die stationäre Lösung (9) über. Gewöhnlich werden sich jedoch bereits vorher nichtlineare Effekte bemerkbar machen. Für $\beta(\mathfrak{K}) > 4v_t\,k^2$ nimmt das Spektrum exponentiell zu, und für $\beta(\mathfrak{K}) = 4v_t\,k^2$ erhält man die linear anwachsende Lösung

$$F_d(\mathfrak{K},t) = \alpha(\mathfrak{K})\,t \quad (13)$$

Fall 2: Das Gebiet $\mathfrak{G}$ wird durch zwei zur y-Achse parallele Geraden begrenzt, die sich mit der Geschwindigkeit U_r in x-Richtung fortbewegen (Bild 1). Innerhalb $\mathfrak{G}$ ist die Windgeschwindigkeit $U = \text{const}$; außerhalb $\mathfrak{G}$ ist $U = 0$, d. h. es ist $F_1(\mathfrak{K};\mathfrak{r},t) \equiv 0$.

Wird $t_o = -\infty$ gesetzt, so erhält man dann in einem mit $\mathfrak{G}$ mitbewegten Bezugssystem ein stationäres Problem. Führen wir in diesem System Koordinaten x_1 und x_2 gemäß Bild 1 ein, so wird (11):

$$F(\mathfrak{K};\mathfrak{r},t) = F_f(\mathfrak{K};x_i) = \begin{cases} F_d\left(\mathfrak{K};\dfrac{x_1}{v_x - U_r}\right) \text{für } v_x(\mathfrak{K}) > U_r \\[2ex] F_d\left(\mathfrak{K};\dfrac{x_2}{U_r - v_x}\right) \text{für } v_x(\mathfrak{K}) < U_r \end{cases}$$

$$(14)$$

wo F_d durch (12) gegeben ist.

Das Spektrum F_f für das beschränkte, sich mit der Geschwindigkeit U_r bewegende Windfeld läßt sich somit auf das Spektrum F_d für das unendliche, zur Zeit $t = 0$ plötzlich aufkommende Windfeld zurückführen. Dies gilt **allgemein** für ein konstantes Windfeld mit sich beliebig verändernder

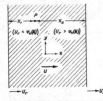

Bild 1 Begrenztes Windfeld mit Wanderungsgeschwindigkeit U_r und Windgeschwindigkeit U

Umrandung $\mathfrak{R}$. Die „Duration" t in F_d wird lediglich durch die Laufzeit τ (in unserem Spezialfall (14) also

$$\frac{x_1}{v_x(\mathfrak{R}) - U_r} \text{ bzw. } \frac{x_2}{U_r - v_x(\mathfrak{R})} \Big) \text{ ersetzt, die}$$

von der Wellengruppe der Wellenzahl $\mathfrak{R}$ benötigt wird, um vom Rand des Sturmgebiets ihren „Fetch" x_1 bzw x_2 bis zum betrachteten Punkt zu durchlaufen. Diese Zeit ist für jede Wellengruppe verschieden. Für $U_r > 0$ ist sie sogar für ein Spektralgebiet $(v_x(\mathfrak{R}) > U_r)$ von dem luvseitigen Fetch x_1 und für das andere Spektralgebiet $(v_x(\mathfrak{R}) < U_r)$ von dem leeseitigen Fetch x_2 abhängig. Für Wellenzahlen $\mathfrak{R}$ mit $v_x(\mathfrak{R}) = U_r$ wird die Laufzeit τ unendlich. Das Spektrum wird dort also (in der linearen Näherung) unendlich bzw., falls (9) zutrifft, ein hohes Maximum haben. Wir bezeichnen diese Erscheinung als G r u p p e n r e s o n a n z , im Gegensatz zur P h a s e n r e s o n a n z , die bei der Anfachung der Wellen durch die Druckschwankungen über der Wasseroberfläche eine entscheidende Rolle spielt.

Es ist aufschlußreich, das Zusammenwirken der Gruppen- und Phasenresonanz näher zu untersuchen. Wir beschränken uns dabei der Einfachheit halber auf den Fall eines linear zunehmenden Spektrums $F_d(\mathfrak{R}, t)$ nach (13). Zunächst fassen wir die Ergebnisse Phillips [6], die wir hier an Hand des Spektrums $P(\mathfrak{R}, \omega)$ etwas anschaulicher formulieren können, nochmals kurz zusammen. Die Zunahme des Spektrums $F(\mathfrak{R})$ ist der Spektraldichte der Druckschwankungen an der Resonanzstelle $(\mathfrak{R}, -\omega_k)$ proportional. An dieser Stelle haben die Fourierkomponenten $A(\mathfrak{R}, \omega) \exp[i(\mathfrak{R}\mathbf{r}) + i\omega t]$ des Drucks die gleiche Phasengeschwindigkeit wie die Oberflächenwellen gleicher Wellenzahl. Durch diese Resonanzerregung wird von den Druckwellen ständig Energie auf die Oberflächenwellen übertragen. Von den verstimmten Druckwellen mit Frequenzen $\omega \neq \pm \omega_k$ werden dagegen nur sehr kleine Oberflächenwellen konstanter Amplitude erzeugt. In einem natürlichen Windfeld entfällt nun der größte Teil der turbulenten Druckenergie auf Wellen, deren Phasengeschwindigkeit in Windrichtung nur wenig von der Windgeschwindigkeit U abweicht, d. h., die Wellen sind in einem mit der Windgeschwindigkeit bewegten Bezugssystem praktisch stehend. Im dreidimensionalen $\mathfrak{R}, \omega$-Raum wird die Spektraldichte $P(\mathfrak{R}, \omega)$ also auf die Umgebung der Ebene $\omega = -U k_x$ konzentriert sein. (U parallel zur x-Achse).

Beschränken wir uns zur Veranschaulichung der wesentlichen Zusammenhänge auf den Schnitt dieser Ebene mit der Ebene $k_y = 0$ (betrachten wir also nur Wellen, die sich in Windrichtung fortpflanzen), so liegt die maximale Dichte von $P(\mathfrak{R}, \omega)$ auf der Geraden $\omega = -U k_x, k_y = 0$. Das Maximum von $P(k, -\omega)$ auf der Geraden $\omega = U k_x, k_y = 0$, Bild 2)[3]. Zur Erregung der Oberflächenwellen trägt nun lediglich die Spektraldichte $P(k_x, -\omega_{k_x})$ längs der Eigenfrequenzkurve $\omega_{k_x} = \sqrt{g\,k_x}$ bei. Die Wellenerregung wird also dort am stärksten sein, wo die Eigenfrequenzen die Gerade der größten Spektraldichte schneidet, z. B. für $u = 10$ m/s bei $k = 0{,}1\ \text{m}^{-1}, \omega = 1\ \text{sec}^{-1}$ (Bild 2). Nach (5), (13) wird das Spektrum $F_d(k_x, t)$, also vermutlich in der Nähe dieser Wellenzahl $(k_x)_m$ ein Maximum aufweisen und in Richtung größerer und kleinerer k_x monoton abfallen. Der qualitative Verlauf von $F_d(k_x, t)$ (zu einer beliebig festgelegten Bezugszeit T) für $U = 10$ m/s ist in Bild 3 wiedergegeben. Die Kurve ließe sich für wirkliche Voraussagungszwecke an Beobachtungen anpassen, für die jetzige Betrachtung ist der Verlauf im einzelnen jedoch unwesentlich. Aus dieser angenommenen Kurve F_d sind nun nach (14) die Spektren $F_d(k_x, x_i)$ für den Fall begrenzten, mit konstanter Geschwindigkeit U_r wandernden Windfelds berechnet worden. Sämtliche Spektren sind auf den Fetch UT/2 bezogen. Die Laufzeit T für die der maximalen Spektraldichte $(F_d)_m$ zugeordneten Wellengruppe $(k_x)_m$. Für $U_r = 0$ haben die Spektren F_d und F_f ähnlichen Verlauf; es wird lediglich der kurzwellige Bereich von F_d wegen der relativ längeren Laufzeit der kurzen Wellen stärker betont. Für $U_r = -U/2$ ist F_f im Hauptbereich des Spektrums nur etwa halb so groß wie F_d, da die Laufzeit der Wellen bei $k_x = (k_x)_m$ um diesen Faktor verkürzt wird. In beiden Fällen wird die Laufzeit allein durch den luvseitigen Fetch x_1 begrenzt. Der Fall $U_r = U/2$ zeigt nun zum ersten Mal die Erscheinung der Gruppenresonanz. Für $k_x = (k_x)_m$ wird die Laufzeit und damit das Spektrum F_f unendlich. Die Anhebung des Spektrums macht sich auf beiden Seiten der singulären Stelle noch besonders bemerkbar, da die Stelle der Gruppenresonanz hier gerade mit der Phasenresonanz zusammenfällt. Im letzten Fall $U_r = U$ ist die Gruppenresonanzstelle ins langwellige Gebiet bei $k_x = (k_x)_m/4$ verschoben. Die Singularität beeinflußt dort nur ein relativ schmales Spektralgebiet, während sich das Spektrum im übrigen Gebiet nur unwesentlich von F_d unterscheidet. Dieses entspricht einer Singularität noch weiter nach links und F_f im Hauptspektralgebiet unterhalb der Kurve F_d liegen. In den Fällen $U_r > 0$ ist das Spektrum auf der linken Seite der singulären Stelle durch den luvseitigen Fetch x_1 und auf der rechten Seite durch den leeseitigen Fetch x_2 begrenzt. Bei der Anwendung der normierten Kurven in Bild 3 auf einen konkreten Fall werden die beiden Äste der Kurven daher mit — unter Umständen stark — verschiedenen Faktoren multipliziert. Im Fall $U_r > 0$ hat das Spektrum F_f somit ein im allgemeinen keine Ähnlichkeit mehr mit dem Spektrum F_d.

3) An Stelle der Geraden würde sich bei genauerer Berücksichtigung der Grenzschichteigenschaften der Luftströmung wahrscheinlich eine leicht nach unten gekrümmte Kurve ergeben.

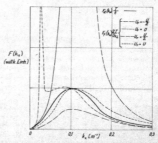

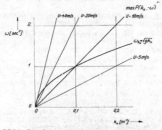

Bild 2 Geraden maximaler Spektraldichte $P(k_x, -\omega)$ der Druckschwankungen bei verschiedenen Windgeschwindigkeiten U

Bild 3 Seegangsspektren $F_f(\mathfrak{R}_x, x_i)$ in einem begrenzten Windfeld mit Windgeschwindigkeit $U = 10$ m/s bei verschiedenen Wanderungsgeschwindigkeiten U_r (qualitativ).

5. Wellenbildung bei endlicher Wassertiefe

Die Betrachtungen der letzten Abschnitte lassen sich unmittelbar auf den Fall endlicher Wassertiefe übertragen. Die Beziehungen für Phasengeschwindigkeit, Gruppengeschwindigkeit usw. müssen hierzu lediglich durch die entsprechenden Beziehungen bei endlicher Wassertiefe ersetzt werden. In den meisten der naheliegenden Anwendungsfälle der Wellenbildung in Küstennähe oder in Binnengewässern ergibt die endliche Tiefe jedoch nur geringe Korrekturen, da die von der Tiefe abhängigen längeren Wellen die betreffenden Gewässer im allgemeinen so schnell durchlaufen, daß sie der veränderten Windeinwirkung, Dämpfung usw. nur kurzzeitig ausgesetzt sind. (Die unmittelbaren Einflüsse der veränderlichen Wassertiefe auf die Welleneigenschaften werden hier natürlich nicht betrachtet.) Eine Ausnahme bildet jedoch der Extremfall des Neusiedler Sees, der bei einer Oberfläche von einigen Hundert Quadratkilometern eine durchschnittliche Wassertiefe von nur ca. 1 m aufweist. Trotz seiner geringen Tiefe ist dieser See für seine außerordentlich starke, bereits bei geringen Windstärken entstehende Wellenbildung bekannt. An Hand unserer vorhergehenden Betrachtungen können wir diese nun sehr einfach erklären. Im wesentlichen läßt sich die Wellenbildung auf zwei Ursachen zurückführen:

1. Bis auf relativ kurze Wellen ist die Phasengeschwindigkeit für sämtliche Wellen allein durch die Wassertiefe h bestimmt: $c = \sqrt{gh}$. Die Eigenfrequenzkurve $\omega_{k_x} = \sqrt{gk_x}$ in Bild 2 ist somit durch die Gerade $\omega_{k_x} = k_x \sqrt{gh}$ zu ersetzen. Bei einer Windgeschwindigkeit $U = c$ fällt nun diese Eigenfrequenzkurve gerade mit der Geraden der maximalen Spektraldichte der Druckschwankungen zusammen. Im Gegensatz zu den Meereswellen, bei denen stets nur ein schmaler Spektralbereich der Druckschwankungen zur Wellenanfachung beiträgt, werden in diesem Fall Oberflächenwellen von dem gesamten Wellenlängenbereich des Druckspektrums angeregt. Bei einer mittleren Seetiefe von 1 m tritt diese Resonanz bereits bei der sehr kleinen Windgeschwindigkeit von 3,1 m/s auf.

2. Wegen der kleinen Fortpflanzungsgeschwindigkeit ist die Laufzeit $\tau(\mathfrak{K})$ für sämtliche Wellen sehr groß. Während auf anderen Binnenseen vergleichbarer Größe die Ausbildung der energiereichen, langwelligen Komponenten wegen ihrer relativ großen Gruppengeschwindigkeit und des begrenzten Fetches stark unterdrückt wird, erreichen die Gruppengeschwindigkeiten sämtlicher Wellen in diesem Fall höchstens die kritische Geschwindigkeit $c = \sqrt{gh}$.

Zusammenfassung

Aus der Untersuchung der Energiegleichung für das Seegangsspektrum hat sich folgendes Bild der Seegangsentstehung ergeben:

1. Das Spektrum $F(\mathfrak{K})$ am Ort $\mathfrak{r}$ zur Zeit t wird durch ein Wegintegral über die Summe der lokalen Einwirkungen $L(\mathfrak{K}; \mathfrak{r}, t)$ bestimmt. Das Integral wird ausgeführt über den Weg $\mathfrak{r}' = \mathfrak{r} - (t - t') \mathfrak{v}(\mathfrak{K})$, den eine Wellengruppe der Wellenzahl $\mathfrak{K}$, die am Ort $\mathfrak{r}$ zur Zeit t eintrifft, von einem Anfangspunkt $\mathfrak{r}_1 = \mathfrak{r} - (t - t_1) \mathfrak{v}(\mathfrak{K})$ aus zurücklegt. Der Anfangswert des Spektrums am Ort $\mathfrak{r}_1$ zur Zeit t wird als bekannt vorausgesetzt. Als Anfangswert (gewöhnlich $F_1 = 0$) kann entweder ein Wert auf dem Rand $\mathfrak{R}$ oder ein zur Anfangszeit t_1, bekannter Spektralwert im Innern des betrachteten Seegebiets $\mathfrak{G}$ in Frage kommen. Das Spektrum ist im ersten Fall durch den Fetch, im zweiten Fall durch die Duration bestimmt. Diese Zusammenhänge verstehen sich fast von selbst auf Grund der Fortpflanzungseigenschaften von Wellen-

gruppen. Dennoch seien sie hier nochmals betont, da sämtliche bisher vorgeschlagenen empirischen Voraussagungsformeln, soweit sie den Einfluß des Fetches und der Duration berücksichtigen, hiermit in Widerspruch stehen.

2. Die Ausführung der Integration dürfte im allgemeinen Fall wegen des komplizierten Terms N_2, der den Einfluß der nichtlinearen Wechselwirkungen zwischen verschiedenen Spektralkomponenten wiedergibt, nur mit Hilfe elektronischer Rechenanlagen möglich sein. Hierzu fehlt ferner noch eine Abschätzung des den Einfluß der Wellenbrechung darstellenden nichtlinearen Terms N_1. Für die Entwicklungsphase des Seegangs, in der die nichtlinearen Terme des Spektrums vernachlässigbar sind, läßt sich die Energiegleichung jedoch durch Quadratur lösen. Der Fall eines konstanten Windfeldes mit variabler Berandung kann auf den Fall eines zur Zeit t = 0 plötzlich entstehenden, unendlich ausgedehnten, konstanten Windfeldes zurückgeführt werden. Die Spektren für den ersten Fall sind im allgemeinen stark von der Wanderungsgeschwindigkeit des Gebiets abhängig. Bewegt sich das Windfeld in Windrichtung, so werden die Wellen, deren Gruppengeschwindigkeit mit der Wanderungsgeschwindigkeit des Windfeldes zusammenfällt, und die sich somit sehr lange im Windfeld aufhalten, besonders stark geprägt (Gruppenresonanz). Die lokale Anfachung durch Windkräfte ist dagegen vorwiegend auf Wellen beschränkt, deren Phasengeschwindigkeit mit der Windgeschwindigkeit zusammenfällt und einer Phasenresonanzbedingung genügen. Fällt die Gruppenresonanzstelle gerade mit der Phasenresonanzstelle zusammen, d. h., ist die Wanderungsgeschwindigkeit eines Windgebiets gerade halb so groß wie die Windgeschwindigkeit selber, so wird sich selbst bei relativ kleiner Ausdehnung des Windgebietes eine sehr starke Windsee ausbilden.

3. Obwohl die Integration der Energiegleichung nur für einige charakteristische Beispiele durchgeführt wurde, läßt sich die Entwicklungsphase des Seegangs im Prinzip für jedes beliebige, als Funktion des Orts und der Zeit vorliegende Windfeld ermitteln. Hierzu müßte allerdings die Abhängigkeit der Koeffizienten $\alpha(\mathfrak{K})$, $\beta(\mathfrak{K})$ und v_t — die letzten beiden sind in der Entwicklungsphase jedoch voraussichtlich vernachlässigbar — von den Windverhältnissen durch Messung der turbulenten Druckschwankungen oder durch Vergleich von berechneten und gemessenen Seegangsspektren näher ermittelt werden.

(Eingegangen am 28. Oktober 1960)

Schrifttum

[1] Darbyshire, J., 1955: An investigation of storm waves in the North Atlantic Ocean. Proc. Roy. Soc. A, 230, 560.

[2] Darbyshire, J., 1956: An investigation into the generation of waves when the fetch of the wind is less than 100 miles. Quart J. Roy. Meteorol. Soc., Ldn. 82, no. 354, 461—468.

[3] Miles, J. W., 1957: On the generation of surface waves by shear flows, J. Fluid Mech. 3, 185—204.

[4] Neumann, G., 1953: On ocean wave spectra and a new method for forecasting wind generated sea. U.S. Beach Erosion Bd., Techn. Mem. No. 43, Dec. 42pp.

[5] Pierson, W. J. Neumann, G., und James, R. W., 1955: Practical methods for observing and forecasting ocean waves by means of wave spectra and statistics. U. S. Navy Dept., Hydrogr. Off. Pub. no. 603, Wash.

[6] Phillips, O. M., 1957: On the generation of waves by turbulent wind, J. Fluid Mech. 2, 417—445.

[7] Phillips, O. M., 1960: On the dynamics of unsteady gravity waves of finite amplitude, Part 1. The elementary interactions. J. Fluid Mech. 9, Part 2, 193—217.

[8] Taylor, G. I., 1953: An experimental study of standing waves, Proc. Roy. Soc. A, 218, 44.

3.2   Remote Sensing[4]

From the beginning of his career, Hasselmann has been working on ocean waves. His Ph.D. thesis from 1957 dealt with the "Propagation of the von Schmidt head waves". He later published 3 papers in the Journal of Fluid Mechanics "On the nonlinear energy transfer in gravity-wave spectrum, parts 1–3" in 1962–1963 [6, 8, 9]. When he was working with Walter Munk at IGPP, Scripps, he co-authored a paper with Munk and others about the "Propagation of ocean swell across the Pacific", which was published in 1966 [18]. By 1970, Hasselmann and M. Schieler had already published a "remote sensing paper" in which they discussed "Radar backscatter from the sea surface" [26]. Hasselmann and his colleagues [45–49] published several important papers in 1978, which concerned radar measurements of wind and waves, which were followed by several papers about the same topic over the following years. This culminated in the **international Marine Remote Sensing Experiment**, MARSEN, in the North Sea between the 16th of July to the 15th of October 1979 whose objectives were: "(1) to investigate the use of remote sensing techniques for oceanographic applications and (2) to utilise remote sensing techniques in combination with in-situ oceanographic measurements to investigate oceanic processes in finite-depth water in the near-shore zone". MARSEN was a well-integrated experiment in which six remote sensing aircraft took part including the NASA CV-990 with the JPL SAR with Omar Shemdin. 60 scientists from 6 countries took part in this very important experiment which was headed up by Hasselmann [67]. The results of the experiment were set out in several papers, 14 of which were published in a Special Issue of the JGR in 1983. One very important paper from the MARSEN Experiment entitled "Theory of SAR ocean wave imaging: A MARSEN view" appeared in the JGR in 1985 and was published by Hasselmann's international team. The paper included a proposal for a new SAR imaging model, which would be fundamental for SAR imaging of the ocean surface from satellites SARs in the future [75].

Hasselmann had already become a key member of the **ESA High Level Advisory Committee** (EOAS) by 1980 (see below). Working with ESA, Klaus Hasselmann and his wife Susanne published a fundamental paper "On the nonlinear mapping of an ocean wave spectrum onto a SAR image spectrum and its inversion" [102]. A facsimile of this paper is presented below as an example of one of the major contributions the Hasselmann family made to the future of the retrieval of the ocean wave spectrum from the ERS-1 C

---

[4] Prepared by Ola M. Johannessen, Guy Duchossois and Evert Attema.

band SAR on the global scale. Klaus and Susanne continued to contribute to the field of global SAR ocean wave spectrum research, but they drifted more and more into climate research. It appeared that their final contribution to ERS SAR Wave Mode was an extensive review entitled "The ERS SAR wave mission mode: A breakthrough in global ocean wave observations", which was published in "ERS missions –20 years of Observing Earth, ESA SP-1326.2013" by Klaus Hasselmann as lead author and 15 co-authors [176].

In the latter half of 1970 Hasselmann was invited to join ESA's thematic **Scatterometer Expert Group-SEG** and later, in 1981, he became a member of the **High-level Earth Observation Advisory Committee** (EOAC), which was founded by the ESA Director General. Of course, he was invited because he had seen the opportunity of using SAR from aeroplanes or satellites for ocean, wind and wave observations and had headed up the MARSEN Experiment in 1979. This expert group and committee provided outstanding scientific support for and made recommendations to ESA in various areas such as the definition of priority mission objectives, payload composition priorities, instrument performance specifications, in-orbit calibration requirements, the development of data processing algorithms, and geophysical product validation approaches. As a key contributor to this ESA Expert groups, Hasselmann played a major role in terms of the development of the mission objectives and the choice of the ERS-1 payload.

EOAC was given the following mandate by the ESA DG:

- To review and, if necessary, revise, the mission objectives of the European remote-sensing satellite programme as defined in the 1970s.
- To put forward an optimal configuration for the payload of the first ERS-1 mission.

The initial mission objectives, defined in the 1970s, had focused on the commercial and operational exploitation of remote-sensing applications. However, at the beginning of the 1980s, these objectives began to evolve within the Earth observation community, with the advent of worldwide programmes to study the oceans and the climate, such as the World Climate Research Programme (WCRP), the World Ocean Circulation Experiment (WOCE), and the Tropical Ocean Global Atmosphere (TOGA), which sought to answer the increasing concerns of the world scientific community, political decision-makers, and the general public over the issues of climate change and possible interactions with human activities. The new situation also required a deeper scientific understanding of the climate system, and hence also the main components of this system, namely the oceans, the polar regions, the continental land masses, and the atmosphere including the interactions between them—a field in which Hasselmann was also an expert. The EOAC recommended the following payload for **ERS-1**.

- The Active Microwave Instrumentation (AMI), combining a SAR mode and a wind scatterometer mode in the C band. The SAR wave mode was to determine the wave spectrum from 5×5 km mini-images collected globally every 200 km along the ground track of the orbit which was the result of Hasselmann's involvement. It would also collect high-resolution SAR images (25 m resolution) above continental and coastal regions, and polar ice caps.
- A radar altimeter operating in the Ku band.
- A laser retro-reflector system for precise restitution of the satellite orbit.

Hasselmann also contributed to the selection of an "unusual" orbit scenario combining several successive orbit cycle periods (3 days, 35 days and twice 168 days repeat cycles), which would satisfy the various research communities (ice, ocean circulation, SAR land imagery, geodesy).

Following the launch of ERS-1, the exploitation of the resulting data via complex processing algorithms, some from the Hasselmann team dealing with the global spectrum of waves, led to the organisation of many ERS-1 symposia by ESA with ever increasing participant numbers (400 participants in Cannes in 1992, 500 in Hamburg in 1993, 700 in Florence in 1997…) and with specialised workshops on downstream application demonstrations (200 participants in Toledo in 1994, London in 1995 and Zurich in 1996…). These ERS-1 symposia provided Hasselmann with opportunities to present the results of his team's work on wave mechanisms and global ocean wave spectra retrieved via the 5×5 km SAR images [e.g., 108, 115, 123, 124].

Hasselmann's expertise and ability to analyse and propose solutions to issues raised for ERS-1 was exceptional. He was the object of a general admiration by the entire ERS-1 team and was invited by ESA to attend the successful launch of ERS-1 in July 1991 in Kourou, Guyana. This was an opportunity for ESA to thank him for his dedication and valuable contributions to the success of this mission.

ERS-1 and its successor ERS-2, which was launched in 1995, paved the way to the successful Envisat mission, which was launched in 2002. Together, these three missions provided some 20 years of continuous data as recalled during ERS' 20th anniversary celebration at ESRIN Frascati in 2011 which was attended by some of the pioneers including Hasselmann [176]. These early missions were the precursors of the current joint ESA-EU Copernicus programme and the Sentinel mission series, making Europe a world leader in Earth observation and environment monitoring.

As previously mentioned, Hasselmann was an extremely important contributor during the early days of the ERS development. He was a fast talker with strong opinions. Not everybody could follow all of his complicated theories. Within his own scientific "bubble" he may not have been accustomed to much opposition in a debate, but he would always be open to accepting the opinions of his opponents if supported by correct theories and/or empirical evidence.

In addition to the political support for the ERS mission, Hasselmann was also very important within the scientific community. This was badly needed because, in the early days, reactions within the ERS scientific user community were very negative and even hostile. Today, following decades of successful application development, all opposition has clearly vanished.

Hasselmann's dedication is demonstrated by the following anecdote: in a SEG meeting he complained about the slow speed and high cost of industry studies, something "he could do in a couple of days with some of his students". ESA said: "great, let's do it. You have a week after which we'll come to your Institute on Friday to review the results". We found Hasselmann in his office submerged in paperwork, computer printouts, and graphics—but not quite with a conclusive answer despite his own efforts as well as those of his wife and some students who had reportedly spent several days and nights carrying out the research.

The SEG was a special group which included experts from ESA and scientific institutions as well as from industry. To avoid the complications involved in defining formal responsibilities, industry was no longer represented in the C/D phase. All members, especially of the SEG, were pioneers who had never been involved in a similar project before. The SEG host would normally present issues to the team asking for answers and/or recommendations. The SEG, including Hasselmann, actively participated in the discussions about such things as the required image size and the tracking distance between them needed to calculate the global ocean wave spectra.

Scatterometer Expert Group (SEG) meeting with industry representatives in 1980: from left to right David Offiler (UK Met Office), Tim Tucker (UK National Oceanography Centre), Werner Alpers (University of Hamburg), Evert Attema (Technical University Delft—later with ESA), Gert Dieterle (ESA), Alf Long (ESA), Gerbrand Komen (KNMI), Klaus Hasselmann (Max Planck Institute for Meteorology), Laurence Gray (CCRS), Juan Guijarro (ESA), Dave Lancashire (formerly Marconi Space Ltd., currently Airbus Defence and Space)

JOURNAL OF GEOPHYSICAL RESEARCH, VOL. 96, NO. C6, PAGES 10,713–10,729, JUNE 15, 1991

On the Nonlinear Mapping of an Ocean Wave Spectrum Into a Synthetic Aperture Radar Image Spectrum and Its Inversion

Klaus Hasselmann and Susanne Hasselmann

Max-Planck-Institut für Meteorologie, Hamburg, Germany

A new, closed nonlinear integral transformation relation is derived describing the mapping of a two-dimensional ocean wave spectrum into a synthetic aperture radar (SAR) image spectrum. The general integral relation is expanded in a power series with respect to orders of nonlinearity and velocity bunching. The individual terms of the series can be readily computed using fast Fourier transforms. The convergence of the series is rapid. The series expansion is also useful in identifying the different contributions to the net imaging process, consisting of the real aperture radar (RAR) cross-section modulation, the nonlinear motion (velocity bunching) effects, and their various interaction products. The lowest term of the expansion with respect to nonlinearity order yields a simple quasi-linear approximate mapping relation consisting of the standard linear SAR modulation expression multiplied by an additional nonlinear Gaussian azimuthal cutoff factor. The cutoff scale is given by the rms azimuthal (velocity bunching) displacement. The same cutoff factor applies to all terms of the power series expansion. The nonlinear mapping relation is inverted using a standard first-guess wave spectrum as regularization term. This is needed to overcome the basic 180° mapping ambiguity and the loss of information beyond the azimuthal cutoff. The inversion is solved numerically using an iteration technique based on the successive application of the explicit solution for the quasi-linear mapping approximation, with interposed corrections invoking the full nonlinear mapping expression. A straightforward application of this technique, however, generally yields unrealistic discontinuities of the best fit wave spectrum in the transition region separating the low azimuthal wave number domain, in which useful SAR information is available and the wave spectrum is modified, from the high azimuthal wave number region beyond the azimuthal cutoff, where the first-guess wave spectrum is retained. This difficulty is overcome by applying a two-step inversion procedure. In the first step the energy level of the wave spectrum is adjusted, and the wave number plane rotated and rescaled, without altering the shape of the spectrum. Using the resulting globally fitted spectrum as the new first-guess input spectrum, the original inversion method is then applied without further constraints in a second step to obtain a final fine-scale optimized spectrum. The forward mapping relation and inversion algorithms are illustrated for three Seasat cases representing different wave conditions corresponding to weakly, moderately, and strongly nonlinear imaging conditions.

1. Introduction

Sea state can be completely characterized statistically by the two-dimensional directional wave spectrum $F(\mathbf{k})$ describing the distribution of wave energy with respect to wave propagation wave number $\mathbf{k}$. All statistical properties of an ocean wave field at any given location and time can be derived from this function. Unfortunately, however, the two-dimensional wave spectrum has proved notoriously difficult to measure. Verifications of wave models, which routinely compute the space-time distributions of $F(\mathbf{k})$, have therefore been based largely on one-dimensional frequency spectra derived from wave buoys or wave staffs. More recently, directional wave buoys, which provide at least some integrated information on the directional distribution, have been more frequently deployed, but even these data have been limited to relatively sparsely distributed locations. Measurements of the full two-dimensional spectrum have been obtained only at selected sites and for restricted time periods using large wave-staff arrays [e.g., *Donelan et al.*, 1985], or special remote sensing systems onboard aircraft [*Plant*, 1987], such as Radar Ocean-Wave Spectrometer (ROWS) [*Jackson*, 1981; *Jackson et al.*, 1985a, b], surface contouring radar [*Kenney et al.*, 1979; *Walsh et al.*, 1985, 1987, 1991], or stereophotography [*Holthuijsen*, 1983].

Paper number 91JC00302.
0148-0227/91/91JC-00302$05.00

The sparsity of directional wave data (together with the limited availability even of one-dimensional frequency spectra or integrated wave height data) has not only handicapped the verification of wave models, but has also deterred wave modelers from seriously addressing the problem of assimilating wave data into their models.

This situation could change dramatically in the 1990s. We look forward in this decade to extensive, in some cases continuous, global measurements of the two-dimensional wave spectrum from synthetic aperture radars (SARs) flown on satellites such as ERS 1 and 2, Radarsat, shuttle missions, and polar platforms. These data will be augmented by global significant wave height measurements from radar altimeters on ERS 1, 2, Topex-Poseidon, Spinsat, and other satellites. Furthermore, global sea surface wind data from satellite scatterometers and altimeters will provide improved wind fields as input for wave models. The simultaneous assimilation of these wind and wave data into global wave models and atmospheric forecast models offers exciting new opportunities and perspectives for wave modelers, but also presents major challenges to the wave- and weather-forecasting community.

This is exemplified by the SAR wave data. The potential of space-borne SARs for imaging two-dimensional ocean wave fields from space has been convincingly demonstrated by Seasat (cf. *Alpers* [1983], *Beal et al.* [1983], and other papers in the Seasat issue, *Journal of Geophysical Research*, volume 88, 1983) and the shuttle SIR-B mission [cf. *Alpers et al.*, 1986; *Brüning et al.*, 1988; *Monaldo and Lyzenga*, 1988].

10,714   HASSELMANN AND HASSELMANN: MAPPING OF OCEAN WAVE SPECTRUM

The theory of the SAR imaging of a moving ocean wave surface is now also rather well understood (cf. MARSEN SAR review [K. *Hasselmann et al.*, 1985], referred to in the following as MSR). The theory has been verified in a number of field experiments with air-borne SARs as well as in Seasat and SIR-B hindcast studies. Nevertheless, the routine interpretation and application of SAR wave data is still generally regarded as a major unresolved problem because of the inherent complexities of the basically nonlinear imaging process.

A fundamental difficulty is that not all of the wave spectral information is mapped into the SAR image plane. Since SAR images provide only a snapshot of the instantaneous sea surface, they can determine the wave propagation direction only to within a sign. (The ambiguity can be removed, however, by correlating successive looks of the same scene [cf. *Rosenthal et al.*, 1989; *Vachon and Raney*, 1989]. A more serious loss of information is incurred by the nonlinear distortion induced by motion effects. These result in an azimuthal high wave number cutoff of the spectrum. The nonlinearities also produce significant shifts of the spectral peak and other distortions of the spectrum [*Alpers and Brüning*, 1986; *Brüning et al.*, 1988, 1990].

Most of these limitations, if properly understood, can be suitably dealt with if the SAR ocean wave image data are assimilated into a wave model. However, this is feasible only if there exists an accurate, operationally feasible method of computing the SAR image spectrum for a given wave spectrum, together with an associated technique for inverting the mapping relation. An essential first step toward the application of SAR wave data in wave models must therefore be the derivation of an efficient and accurate method for computing the mapping from a surface wave spectrum into a SAR surface image spectrum. Subsequently, a method must be devised for dealing with the loss of information incurred in the forward mapping relation and recovering the wave spectrum from the measured SAR image spectrum. Both of these questions are addressed in this paper.

Since the basic imaging mechanisms are known, a straightforward method of solving the forward mapping problem is to compute the SAR image for any given instantaneous realization of the sea surface pixel-by-pixel for each (moving) scattering element of the surface. Monte Carlo computations of the spectrum of the SAR image for a given two-dimensional wave spectrum using an ensemble of such deterministic mapping computations have been carried out by *Alpers* [1983], *Alpers et al.* [1986], *Alpers and Brüning* [1986], *Lyzenga* [1986], *Brüning et al.* [1988, 1990], and *Hasselmann et al.* [1988]. The transformation of the input wave spectrum into the output SAR image spectrum is determined in this method by generating an ensemble of individual surface wave field realizations for the prescribed surface wave spectrum, computing the image pixel-by-pixel for each realization, carrying out the Fourier transform for each SAR image, and finally forming the SAR image variance spectrum by averaging the modulus squared Fourier amplitudes over the ensemble of realizations.

In this paper we follow an alternative approach to develop a new, closed nonlinear integral mapping relation directly for the spectra. The closed relation circumvents the need for deterministic pixel-by-pixel transformation computations of individual images and is free of the inherent statistical sampling uncertainty of the "brute force" Monte Carlo approach.

Although the transformation is strongly nonlinear, the notorious closure problem of strongly nonlinear statistical systems does not arise, as the Gaussian property of the input wave field is not affected by the nonlinearities. This enables the expectation values of all higher-order nonlinear functions of the input wave variables occurring in the general relation for the output image spectrum to be reduced to closed expressions of the input wave spectrum.

The final integral transformation relation can be expanded in a Fourier transform series, which can be rapidly evaluated using fast Fourier transform (FFT) algorithms. The expansion also provides useful insights into the contributions to the net imaging process from the separate cross-section modulation and surface motion terms, together with their various interaction combinations.

Finally, with the availability of a closed, noise free, rapidly computable transformation expression, it is now possible to address the inverse problem of deriving the wave spectrum from the SAR spectrum. Because of the loss of information beyond the azimuthal cutoff and the 180° ambiguity, a rigorous inverse mapping solution does not exist. However, regularization can be achieved in the usual manner by minimizing a cost function which penalizes not only the deviation between the observed and predicted SAR spectrum, but also the deviation between the modified wave spectrum and a first-guess wave spectrum. The iterative inversion method developed in this paper generally converges within three or four iterations. The computations should be sufficiently rapid for application in an operational SAR data assimilation system.

The paper is structured as follows: section 2 reviews the different imaging mechanisms and defines notation. The basic nonlinear mapping relation is derived in section 3. Section 4 describes the inversion method. The results of sections 3 and 4 are illustrated in section 5 for three selected SAR images from Seasat, taking as first-guess input spectra the wave spectra derived from a global wave hindcast using the WAM third generation wave model [*WAMDIG*, 1988]. Section 6, finally, summarizes the principal results and conclusions of the study. An appendix describes the generalization of the pure velocity bunching theory presented in section 3 to higher order processes such as acceleration smearing.

2. SAR IMAGING OF OCEAN WAVES

After many years of debate, a rather wide consensus has emerged regarding the principal mechanisms governing the imaging of a moving ocean wave surface by a SAR (cf. MSR). The backscattered return may be represented generally as a superposition of the statistically phase uncorrelated returns from a continuous ensemble of small-scale backscattering surface elements. Although open questions still remain regarding finer details of the backscattering mechanism, it is generally agreed that in the range of incidence angles between 20° and 60° typical for satellite and most research aircraft SAR operations, the backscattered return from each surface element (facet) is dominated by Bragg scattering from short ripple waves. The ripple waves in turn are modulated in their orientation, energy, and motion by longer waves, thereby enabling the SAR to image normal wind-generated ocean waves.

The three basic modulation processes arise through (1) the change in the local incidence angle (tilt) of the facet through the long wave slope, (2) the hydrodynamic interaction between short and long waves, which modulates the energy and wave number of the short Bragg scattering waves, and (3) the advection of the backscattering facet by the long wave orbital velocity, which produces a Doppler shift in the return signal and induces an azimuthal displacement of the scattering element in the image plane.

For all processes, it can be assumed that to first order the backscattering ripple waves and the modulating ocean waves are widely separated in wavelength scale. On the basis of this two-scale description, a rather complete theory of SAR imaging of a random ocean wave field can be developed [Alpers and Rufenach, 1979; Swift and Wilson, 1979; Valenzuela, 1980; Alpers et al., 1981; Raney, 1981; Tucker, 1985; MSR].

An important feature of the theory is that SAR imaging is typically nonlinear. Although the hydrodynamic and tilt modulation can usually be approximated as linear processes, the so called velocity bunching mechanism associated with the orbital motion of the long waves is frequently strongly nonlinear, particularly for wind seas and short waves. Velocity bunching arises through the variations in the azimuthal displacements of the imaged backscattering elements induced by the variations of the orbital velocity within the long wave field. The alternate bunching and stretching of the apparent scatterer distribution in the image plane produces an image of the long waves, even in the hypothetical case that the backscattering cross section itself is not modulated. When the displacements are small compared with the characteristic wavelength of the long waves, the mechanism can be treated as a linear process, characterized by a velocity bunching modulation transfer function (MTF), in analogy with the hydrodynamic and tilt MTFs. For larger displacements, however, the process becomes nonlinear, and when the displacements significantly exceed a wavelength (for example, for short wind waves traveling in the azimuthal direction), the image can become completely smeared out.

Since the velocity bunching nonlinearity normally strongly dominates over any nonlinearities of the tilt or hydrodynamic modulation, we shall simply ignore the latter to avoid unnecessarily complicating the analysis, although higher order tilt and hydrodynamic modulation terms (to the extent that they are known) can in principle be readily included [cf. Hasselmann et al., 1990].

In the following it will be convenient to regard the SAR wave image as produced by two consecutive imaging processes: the frozen surface (or real aperture radar, RAR) imaging mechanism, governed by the hydrodynamic and tilt modulation, and the additional motion effects, which are specific to a SAR and do not affect the RAR image.

The Frozen Surface Contribution

In the framework of linear modulation theory, the surface elevation $\zeta(\mathbf{r}, t)$ and the variations of the local (specific) backscattering cross section $\sigma(\mathbf{r}, t)$ sensed by a RAR may both be represented as a superposition of propagating wave components,

$$\zeta(\mathbf{r}, t) = \sum_{\mathbf{k}} \zeta_{\mathbf{k}} \exp{(i[\mathbf{k} \cdot \mathbf{r} - \omega t])} + \text{complex conjugate} \quad (1)$$

$$\sigma(\mathbf{r}, t) = \bar{\sigma}\left\{ 1 + \left[\sum_{\mathbf{k}} m_{\mathbf{k}} \exp{i(\mathbf{k} \cdot \mathbf{r} - \omega t)} + \text{c.c.} \right] \right\} \quad (2)$$

where $\bar{\sigma}$ denotes the spatially averaged specific cross section, $\omega = (gk)^{1/2}$ is the gravity wave frequency, and the cross-section modulation factor $m_{\mathbf{k}}$ and wave amplitude $\zeta_{\mathbf{k}}$ are linearly related through the RAR modulation transfer function $T_{\mathbf{k}}^R$,

$$m_{\mathbf{k}} = T_{\mathbf{k}}^R \zeta_{\mathbf{k}} \quad (3)$$

(Note that the MTFs refer here to the wave height components, $\zeta_{\mathbf{k}}$ and not, as often defined, to the wave slope $|\mathbf{k}|\zeta_{\mathbf{k}}$.)

A discrete Fourier sum notation has been chosen rather than continuous Fourier, or more rigorously, Fourier-Stieltjes, integrals, as we shall be considering later derivatives with respect to individual Fourier components. The discrete representation avoids the rather cumbersome functional derivative notation required for continuous integrals. The transition to continuous integrals is carried out at the end of the analysis.

For the general theory presented in the next section, $T_{\mathbf{k}}^R$ need not be further specified. However, for later numerical applications, $T_{\mathbf{k}}^R$ needs to be evaluated in more detail by decomposition into its tilt and hydrodynamic modulation components,

$$T_{\mathbf{k}}^R = T_{\mathbf{k}}^t + T_{\mathbf{k}}^h \quad (4)$$

For a Phillips k^{-4} high wave number spectrum, the tilt MTF can be approximated for large dialectic constants (which for seawater are of the order of 80), by the expressions [cf. Wright, 1968; Lyzenga, 1986]

$$T^t(\mathbf{k}) = 4ik_l \cot \theta (1 + \sin^2 \theta)^{-1} \quad \theta \le 60° \quad (5)$$

$$T^t(\mathbf{k}) = 8ik_l(\sin 2\theta)^{-1}$$

for vertical polarization (VV) and for horizontal polarization (HH), respectively, where θ is the radar incidence angle and k_l the component of the incident wave number vector in the radar look direction.

The hydrodynamic MTF can be derived from a two-scale model of hydrodynamic short wave–long wave interactions. A simple relaxation type source term, characterized by a damping factor μ, is normally introduced to describe the response of the short waves to the long wave modulation [cf. Keller and Wright, 1975]. Feindt [1985] found that a better agreement with laboratory measurements could be obtained by including an additional feedback term, characterized by a complex feedback factor $Y_r + iY_i$, representing the long wave modulation of the wind input to the short waves. This yields a hydrodynamic MTF

$$T_{\mathbf{k}}^h = \frac{\omega - i\mu}{\omega^2 + \mu^2} (4.5)k\omega\left(\frac{k_y^2}{k^2} + Y_r + iY_i\right) \quad (6)$$

Coordinates are chosen such that the x axis points in the SAR flight (azimuthal) direction and the y axis forms a right-handed coordinate system with x (thus y points in the positive or negative look direction l for a left or right looking SAR, respectively).

To first order, both RARs and SARs produce quasi-instantaneous images of the surface at a fixed time, $t = 0$, say. We shall ignore for simplicity the distortion effects due to the fact that a side looking radar does not, in fact, take an instantaneous snapshot but builds up an image from a sequence of consecutively imaged range strips. Thus, moving waves are imaged with slightly Doppler displaced "wave numbers of encounter." This straightforward geometric effect applies equally for a RAR and a SAR and should be distinguished from the SAR motion effects summarized in the following subsection. It can be important for more slowly moving aircraft SARs but is generally negligible for spaceborne SAR imagery. In the same spirit, we shall ignore effects arising from the time delay between individual images in multilook images, which can be used, for example, to gain information on the wave propagation direction, as mentioned above.

Since both RARs and SARs produce quasi-instantaneous images of the surface at time $t = 0$, say, the Fourier decomposition of the image modulation intensity $I(r)$ (after subtraction of the mean) has the form

$$I(\mathbf{r}) = \sum_{\mathbf{k}} I_{\mathbf{k}} \exp (i\mathbf{k} \cdot \mathbf{r}) \tag{7}$$

where

$$I_{\mathbf{k}} = (I_{-\mathbf{k}})^* \tag{8}$$

The structural difference between the Fourier representations (1), (2), and (7) is sometimes overlooked. In contrast to the standard two-dimensional Fourier form (7), equations (1) and (2) really represent three-dimensional wave number–frequency spectra. They can be represented as two-dimensional distributions, however, because the frequencies are constrained to lie on the two free gravity wave dispersion surfaces $\omega = \pm \sqrt{gk}$. Thus in (1) and (2) Fourier components of opposite sign in $\mathbf{k}$ represent different waves traveling in opposite directions, and are not related, in contrast to (7), where the components are related through (8). (For the same reason, (1) and (2) include a second explicit complex conjugate sum, whereas in (7) the complex conjugate wave number pairs are already included implicitly in the single sum over positive and negative $\mathbf{k}$.)

For a RAR, the image intensity is directly proportional to the specific cross section. Thus if the image modulation is normalized by the mean image intensity, we have

$$I^R(\mathbf{r}) = \sigma(\mathbf{r}, 0)/\bar{\sigma} - 1 \tag{9}$$

and (1)–(3), (7), and (8) yield

$$I_{\mathbf{k}}^R = T_{\mathbf{k}}^R \zeta_{\mathbf{k}} + (T_{-\mathbf{k}}^R \zeta_{-\mathbf{k}})^* \tag{10}$$

We have not explicitly introduced the RAR (SAR) system transfer function into (9) and (10). This may be represented simply as an additional multiplicative factor in the right-hand side of (10). We shall assume in the following that the system transfer function has already been incorporated in the definition of $T_{\mathbf{k}}^R$.

We have also not considered clutter effects. To first order, these may be represented simply as an additional clutter noise spectrum superimposed on the image spectrum considered here [cf. *Alpers and Hasselmann*, 1982].

In terms of the ocean wave and RAR image variance spectra $F_{\mathbf{k}}$, $P_{\mathbf{k}}^R$, defined by

$$\langle \zeta^2 \rangle = \sum_{\mathbf{k}} F_{\mathbf{k}} = 2 \sum_{\mathbf{k}} \langle \zeta_{\mathbf{k}}^* \zeta_{\mathbf{k}} \rangle \tag{11}$$

$$\langle I^{R^2} \rangle = \sum_{\mathbf{k}} P_{\mathbf{k}}^R = \sum_{\mathbf{k}} \langle I_{\mathbf{k}}^{R*} I_{\mathbf{k}}^R \rangle \tag{12}$$

where the angle brackets denote ensemble means, the linear relation (10) yields

$$P_{\mathbf{k}}^R = \frac{1}{2} \{ |T_{\mathbf{k}}^R|^2 F_{\mathbf{k}} + |T_{-\mathbf{k}}^R|^2 F_{-\mathbf{k}} \} \tag{13}$$

Motion Effects

We consider now the modification of the frozen image induced by the surface motion. This is normally described by two effects: the azimuthal displacement ξ of the apparent position of a backscattering element in the image plane, and an azimuthal smearing or broadening Δx of the image of the (theoretically infinitesimal) backscattering element.

According to standard SAR imaging theory, the azimuthal displacement ξ of the backscattering element is proportional to the range component v of the long wave orbital velocity with which the backscattering element is advected,

$$\xi = \beta v \tag{14}$$

where

$$\beta = (\text{slant range } \rho)/(\text{SAR platform velocity } U) \tag{15}$$

The orbital velocity v is defined here as the time average over the period during which the scattering element is viewed by the SAR. Normally, the SAR illumination time is small compared with the wave period, so that to first order, v may be set equal to the instantaneous orbital velocity in the center of the viewing window.

From classical surface wave theory [*Lamb*, 1932],

$$v = \sum_{\mathbf{k}} T_{\mathbf{k}}^v \zeta_{\mathbf{k}} \exp (i\mathbf{k}r) + \text{c.c.} \tag{16}$$

where the range velocity transfer function is given by

$$T_{\mathbf{k}}^v = -\omega \left(\sin \theta \, \frac{k_l}{|k|} + i \cos \theta \right) \tag{17}$$

We have neglected for simplicity the small additional Doppler shift due to the finite phase velocity of the Bragg scattering ripples. This can be readily included in the theory but encumbers the notation. We shall also neglect the Doppler shifts due to the dynamics of the ripple waves, which are smaller than the phase velocity terms by another order of magnitude, and would appear formally in a smearing term (cf. MSR).

The smearing term Δx is normally represented as the sum of a second-order acceleration term and a velocity spread term (cf. MSR).

The acceleration smearing arises through the variation of the instantaneous orbital velocity during the SAR viewing interval. This yields slightly different effective displacements ξ for the beginning and end of the SAR illumination period. The term is generally an order of magnitude smaller than the velocity spread term [cf. *Alpers and Rufenach*, 1979; *Alpers et al.*, 1981; MSR]. To simplify the presentation, it will not

124 H. von Storch

be considered in this section. However, the extension of the theory to include this effect is basically straightforward and is presented in the appendix.

The velocity spread term is identical in physical origin to the azimuthal displacement term. It is generally introduced as a separate term primarily for conceptual purposes [cf. *Tucker*, 1985; MSR], although the distinction also has important practical implications for Monte Carlo simulations. In the present theory, however, there is no need to treat the velocity spread term separately from the general velocity bunching formalism. The term has nevertheless played some role in the discussion of the azimuthal cutoff of SAR images, which will be considered in the following section, and is therefore briefly described here.

The velocity spread term arises through the introduction of a second separation scale, the SAR resolution scale L_{SAR}, into the SAR imaging model. The scale L_{SAR} is typically of order 20 m and is therefore an order of magnitude larger than the basic separation scale L_{hyd} of the standard hydrodynamic interaction and Bragg backscattering two-scale model. The scale L_{hyd} lies between the wavelength of the Bragg scattering ripples and the long waves and is thus generally of order 1 m. Since the SAR is unable to distinguish between individual backscattering facets within a SAR resolution cell, the entire ensemble of backscattering facets within a resolution cell is mapped into a single image pixel. The mean azimuthal displacement of the pixel is given by $\bar{\xi} = \beta \bar{v}$, where $\bar{v}$ is the mean orbital velocity of the cell facet ensemble. The deviations $\xi - \bar{\xi}$ of the individual facet displacements relative to this mean value then produce the "velocity spread" smearing of the image of the resolution cell.

In Monte Carlo simulations, this subresolution smearing can be treated as a statistical process which can be represented simply as an effective degradation of the SAR system MTF. The long wave spectrum can then be subdivided at the scale L_{SAR}, and only waves with scales greater than L_{SAR} need be included explicitly in the simulation. The mapping computations can therefore be carried out at the relatively coarse resolution of the SAR rather than at the much finer resolution scale L_{hyd} of the backscattering facets.

In the present analysis, however, the subdivision of the wave spectrum at the SAR separation scale L_{SAR} is unnecessary, since the theory can be carried through uniformly up to the high wave number cutoff $(L_{hyd})^{-1}$ of the backscattering-hydrodynamic two-scale model. We may therefore regard the SAR image directly as the superposition of the (statistically independent) images of all subresolution scale backscattering facets, without clustering these elements together to larger entities of the dimension of the SAR resolution cell.

The relation between the SAR and RAR images in the present "pure velocity bunching" model is obtained by simply mapping each facet at position $\mathbf{r}'$ into its corresponding position $\mathbf{r} = \mathbf{r}' + \xi(\mathbf{r}')$ in the image plane,

$$\hat{I}^S(\mathbf{r}) = \int \hat{I}^R(\mathbf{r}')\delta[\mathbf{r} - \mathbf{r}' - \xi(\mathbf{r}')]\, d\mathbf{r}' \quad (18)$$

where $\xi = \mathbf{a}\xi$, $\mathbf{a}$ denotes the unit vector in the azimuthal (x) direction, and

$$\hat{I}^{S,R}(\mathbf{r}) = 1 + I^{S,R}(\mathbf{r}) \quad (19)$$

is the total normalized image intensity.

Integrating out the δ function, (18) yields

$$\hat{I}^S(\mathbf{r}) = \left\{ \hat{I}^R(\mathbf{r}')\left|\frac{d\mathbf{r}'}{d\mathbf{r}}\right| \right\}_{\mathbf{r}' = \mathbf{r} - \xi(\mathbf{r}')} \quad (20)$$

The Jakobian velocity bunching factor

$$\left|\frac{d\mathbf{r}'}{d\mathbf{r}}\right| = \left|1 + \frac{d\xi'(\mathbf{r}')}{d\mathbf{r}'}\right|^{-1} \quad (21)$$

represents the variation in the effective density of backscattering elements in the image plane due to the compression or stretching of the originally homogeneous distribution of facets. As pointed out, this enables the SAR to image ocean waves even in the hypothetical situation in which the RAR transfer function vanishes, i.e., $I^R(\mathbf{r}) = 0$.

For the case

$$\left|\frac{\partial \xi'}{\partial \mathbf{r}'}\right| \ll 1 \quad (22)$$

(21) can be expanded in a power series and truncated after the linear term. Equations (20) and (21) yield then for the SAR image amplitude spectrum, applying (14) and (16),

$$I_{\mathbf{k}}^S = I_{\mathbf{k}}^R + [T_{\mathbf{k}}^{vb}\zeta_{\mathbf{k}} + (T_{-\mathbf{k}}^{vb}\zeta_{-\mathbf{k}})^*] \quad (23)$$

where the velocity bunching modulation transfer function

$$T_{\mathbf{k}}^{vb} = -i\beta k_x T_{\mathbf{k}}^v$$

$$= -\beta k_x \omega (\cos\theta - i\sin\theta\, k_l/k) \quad (24)$$

Thus in the linear approximation

$$I_{\mathbf{k}}^S = T_{\mathbf{k}}^S \zeta_{\mathbf{k}} + (T_{-\mathbf{k}}^S \zeta_{-\mathbf{k}})^* \quad (25)$$

and the image variance spectrum is given by

$$P_{\mathbf{k}}^S = |T_{\mathbf{k}}^S|^2 \frac{F_{\mathbf{k}}}{2} + |T_{-\mathbf{k}}^S|^2 \frac{F_{-\mathbf{k}}}{2} \quad (26)$$

with the net SAR imaging modulation transfer function

$$T_{\mathbf{k}}^S = T_{\mathbf{k}}^R + T_{\mathbf{k}}^{vb} \quad (27)$$

The condition (22) is generally satisfied for swell. However, in many situations, in particular for short wind seas, the inequality does not hold or is even reversed [cf. MSR; *Brüning et al.*, 1990]. Moreover, even for sea states for which (22) applies in a spectrally averaged sense, the inequality breaks down for short azimuthally propagating waves. Thus in all cases, (18) and (20) represent a strongly nonlinear transformation either for all or at least part of the spectrum, and we must address the general problem of deriving the fully nonlinear transformation relation between the surface wave spectrum and the SAR image spectrum.

3. GENERAL NONLINEAR MAPPING RELATION

To determine the dependence of the SAR image Fourier components $I_{\bar{k}}^S$ on the wave Fourier components in the general nonlinear case, we first apply a Fourier transform to the basic mapping relation (20):

10,718   HASSELMANN AND HASSELMANN: MAPPING OF OCEAN WAVE SPECTRUM

$$I_k^S = \frac{1}{A} \int d\mathbf{r} \, \exp(-i\mathbf{k} \cdot \mathbf{r}) \left\{ I^R(\mathbf{r}') \left| \frac{d\mathbf{r}'}{d\mathbf{r}} \right| \right\}_{\mathbf{r}' = \mathbf{r} - \xi(\mathbf{r}')}$$

$$= \frac{1}{A} \int d\mathbf{r}' I^R(\mathbf{r}') \exp\{-i\mathbf{k} \cdot [\mathbf{r}' + \xi(\mathbf{r}')]\} \qquad (28)$$

Here A denotes the finite rectangular area of the sea surface corresponding to the discrete Fourier representation (in the final expression, $A \to \infty$).

Substituting the Fourier representation (7), (10) for I^R into (28), we obtain

$$I_k^S = \frac{1}{A} \int d\mathbf{r}' \left\{ 1 + \sum_{\mathbf{k}'} (T_{\mathbf{k}'}^R \zeta_{\mathbf{k}'} + T_{-\mathbf{k}'}^{R*} \zeta_{-\mathbf{k}'}^*) \right.$$

$$\left. \cdot \exp i\mathbf{k}' \cdot \mathbf{r} \right\} \exp[-i\mathbf{k} \cdot \mathbf{r}' - i\mathbf{k} \cdot \xi(\mathbf{r}')] \qquad (29)$$

This yields for the SAR image variance spectrum

$$P_k^S = \langle I_k^S \cdot (I_k^S)^* \rangle$$

$$= \left\langle A^{-2} \iint d\mathbf{r}' \, d\mathbf{r}'' \right.$$

$$\cdot \exp\{-i\mathbf{k} \cdot (\mathbf{r}' - \mathbf{r}'') - i\mathbf{k} \cdot [\xi(\mathbf{r}') - \xi(\mathbf{r}'')]\}$$

$$\cdot \left\{ 1 + \sum_{\mathbf{k}'} (T_{\mathbf{k}'}^R \zeta_{\mathbf{k}'} + T_{-\mathbf{k}'}^{R*} \zeta_{-\mathbf{k}'}^*) e^{i\mathbf{k}' \cdot \mathbf{r}'} \right\}$$

$$\cdot \left. \left\{ 1 + \sum_{\mathbf{k}''} (T_{\mathbf{k}''}^{R*} \zeta_{\mathbf{k}''}^* + T_{-\mathbf{k}''}^R \zeta_{-\mathbf{k}''}) e^{-i\mathbf{k}'' \cdot \mathbf{r}''} \right\} \right\rangle \qquad (30)$$

The nonlinearity of this integral expression appears solely in the factor

$$N_k = \exp\{-i\mathbf{k} \cdot [\xi(\mathbf{r}') - \xi(\mathbf{r}'')]\} \qquad (31)$$

The term occurs in the following mean product combinations:

$$E_{\mathbf{k}'}^a = \langle N_k \cdot \zeta_{\mathbf{k}'} \rangle \qquad E_{\mathbf{k}'}^b = \langle N_k \cdot \zeta_{-\mathbf{k}'}^* \rangle \qquad (32)$$

$$E_{\mathbf{k}'\mathbf{k}''}^{aa} = \langle N_k \cdot \zeta_{\mathbf{k}'} \zeta_{\mathbf{k}''} \rangle \qquad E_{\mathbf{k}'\mathbf{k}''}^{ab} = \langle N_k \cdot \zeta_{\mathbf{k}'} \zeta_{-\mathbf{k}''}^* \rangle \qquad (33)$$

$$E_{\mathbf{k}'\mathbf{k}''}^{ba} = \langle N_k \cdot \zeta_{-\mathbf{k}'}^* \zeta_{\mathbf{k}''} \rangle \qquad E_{\mathbf{k}'\mathbf{k}''}^{bb} = \langle N_k \cdot \zeta_{-\mathbf{k}'}^* \zeta_{-\mathbf{k}''}^* \rangle \qquad (34)$$

To evaluate these expressions, we decompose N_k into a sum over a set of terms each of which is composed of two statistically independent factors. The first factor consists of an infinitesimal expression containing the specific wave components which appear in the products (32)–(34). The second factor contains the remaining components of the wave field. Since for a Gaussian wave field all wave components are statistically independent, the second factor in each term is statistically independent of the first factor, and the mean products (32)–(34) can therefore be immediately evaluated.

The Fourier representation of the azimuthal displacement difference

$$\Delta\xi = \xi(\mathbf{r}') - \xi(\mathbf{r}'') \qquad (35)$$

appearing in the exponent of N_k may be written in the form

$$\Delta\xi = \sum_{\mathbf{k}''} (K_{\mathbf{k}''} \zeta_{\mathbf{k}''} + \text{c.c.}) \qquad (36)$$

where

$$K_{\mathbf{k}''} = \beta T_{\mathbf{k}''}^v (e^{i\mathbf{k}'' \cdot \mathbf{r}'} - e^{i\mathbf{k}'' \cdot \mathbf{r}''}) \qquad (37)$$

(cf. (14), (16)). Splitting off from the sum (36) the subset S of infinitesimal wave components which appear in the products (32)–(34), and denoting the residual sum Σ' by R, we have

$$\Delta\xi = (K_{\mathbf{k}'}^* \zeta_{\mathbf{k}'}^* + K_{\mathbf{k}'}^* \zeta_{\mathbf{k}'}^* + K_{-\mathbf{k}'} \xi_{-\mathbf{k}'} + K_{-\mathbf{k}'} \zeta_{\mathbf{k}''} + \text{c.c})$$

$$+ \sum_{\mathbf{k}''}' (K_{\mathbf{k}''} \zeta_{\mathbf{k}''} + \text{c.c})$$

$$= S + R \qquad (38)$$

Since S is infinitesimal, we may expand $N_k = \exp(-ik_x\Delta\xi)$ in a Taylor series

$$N_k = e^{-ik_x R} \left(1 - ik_x S - k_x^2 \frac{S^2}{2} + \cdots \right) \qquad (39)$$

The rest sum R contains only wave components which are statistically independent of the wave component factors X appearing in (32)–(34). Thus the expectation values in these expressions may be factorized in the form

$$\langle N_k \cdot X \rangle = \langle e^{-ik_x R} \rangle \left(\langle X \rangle - ik_x \langle SX \rangle - k_x^2 \frac{\langle S^2 X \rangle}{2} \right) \qquad (40)$$

The first term $\langle X \rangle$ in (40) vanishes, since $\langle \zeta_{\mathbf{k}'} \rangle = 0$, $\langle \zeta_{\mathbf{k}'} \zeta_{\mathbf{k}''} \rangle = 0$, $\langle \zeta_{\mathbf{k}'} \zeta_{-\mathbf{k}''}^* \rangle = 0$ (except for the subset $\mathbf{k}' + \mathbf{k}'' = 0$, which has zero integral measure in the limit of a continuous spectrum). The second term $\langle SX \rangle$ is proportional to the wave spectrum, while the third term $\langle S^2 X \rangle$ represents a quadratic wave spectral product. Since X is either linear (32) or quadratic ((33) and (34)), only the first two terms in the expansion (39) contribute to (40). (In the general theory including acceleration smearing presented in the appendix, however, the full expansion (39) is needed.)

The first factor $\langle e^{-ik_x R} \rangle$ in (40) may now be replaced again by $\langle N_k \rangle$, since the two expressions differ only by the negligible infinitesimal components S. The term can be evaluated by again making use of the Gaussian property of the wave field. Since $\Delta\xi$ is a linear function of the wave field, the variable is normally distributed, and one obtains by direct integration over the probability distribution

$$\langle N_k \rangle = \langle \exp(-ik_x \Delta\xi) \rangle = \exp(-k_x^2 \langle \Delta\xi^2 \rangle / 2) \qquad (41)$$

From (36), we have further

$$\langle \Delta\xi^2 \rangle = 2\beta^2 \int |T_k^v|^2 F(\mathbf{k})[1 - \cos \mathbf{k}(\mathbf{r}' - \mathbf{r}'')] \, d\mathbf{k}$$

$$= 2\xi'^2[1 - \langle v^2 \rangle^{-1} f^v(\mathbf{r}' - \mathbf{r}'')] \qquad (42)$$

where

$$f^v(\mathbf{r}) = \langle v(\mathbf{x} + \mathbf{r})v(\mathbf{x}) \rangle = \int F(\mathbf{k}) |T_k^v|^2 e^{i\mathbf{k} \cdot \mathbf{r}} \, d\mathbf{k} \qquad (43)$$

is the orbital velocity covariance function and

$$\xi'^2 = \langle \xi^2 \rangle = \beta^2 \langle v^2 \rangle = \beta^2 \int |T_k^v|^2 \, F(k) \, dk \quad (44)$$

is the mean square azimuthal displacement of a scattering element. We have introduced at this point the continuous spectral notation

$$F(k) = \frac{F_k}{\Delta k} = (2\pi)^{-2} A F_k \quad (45)$$

After some straightforward algebra to evaluate the mean products within the parentheses in (40), equation (30) yields, together with (42)–(44), the closed nonlinear spectral transform expression

$$P^S(k) = (2\pi)^{-2} \exp\left[-k_x^2 \xi'^2\right]$$

$$\cdot \int d\mathbf{r} e^{-i\mathbf{k}\cdot\mathbf{r}} \, \exp\left[k_x^2 \xi'^2 \langle v^2 \rangle^{-1} f^v(\mathbf{r})\right]$$

$$\times \{1 + f^R(\mathbf{r}) + i k_x \beta [f^{Rv}(\mathbf{r}) - f^{Rv}(-\mathbf{r})]$$

$$+ (k_x \beta)^2 [f^{Rv}(\mathbf{r}) - f^{Rv}(0)][f^{Rv}(-\mathbf{r}) - f^{Rv}(0)]\}$$

$$(46)$$

where $P^S(k) = P_k^S / \Delta k$ and

$$f^R(\mathbf{r}) = \langle I^R(\mathbf{x} + \mathbf{r}) I^R(\mathbf{x}) \rangle$$

$$= \frac{1}{2} \int \{F(k)|T_k^R|^2 + F(-k)|T_{-k}^R|^2\} e^{i\mathbf{k}\cdot\mathbf{r}} \, dk \quad (47)$$

$$f^{Rv}(\mathbf{r}) = \langle I^R(\mathbf{x} + \mathbf{r}) v(\mathbf{x}) \rangle$$

$$= \frac{1}{2} \int \{F(k) T_k^R (T_k^v)^* + F(-k)(T_{-k}^R)^* T_{-k}^v\} e^{i\mathbf{k}\cdot\mathbf{r}} \, dk \quad (48)$$

represent the autovariance function of the RAR image intensity $I^R(\mathbf{x})$ and the covariance function of $I^R(\mathbf{x})$ and $v(\mathbf{x})$, respectively.

Apart from the second exponential factor, the integral (46) has the form of a Fourier transform. It can be reduced to a series of Fourier transforms by expanding the exponential in a Taylor series,

$$\exp\left[k_x^2 \xi'^2 \langle v^2 \rangle^{-1} f^v(r)\right]$$

$$= [1 + k_x^2 \xi'^2 \langle v^2 \rangle^{-1} f^v(r) + \cdots] \quad (49)$$

This yields a spectral series expansion of the form

$$P^S(k) = \exp\left(-k_x^2 \xi'^2\right) \sum_{n=1}^{\infty} \sum_{m=2n-2}^{2n} (k_x \beta)^m P_{nm}^S(k) \quad (50)$$

where the index n indicates the nonlinearity order with respect to the input wave spectrum and the index m the order with respect to the velocity bunching parameter β (which is seen to occur always in combination with the azimuthal wave number k_x). Explicitly, the spectral expansion terms are given by

$$P_{n,2n}^S = \Omega_n \left\{ \frac{f^v(r)^n}{n!} \right\} \quad (51)$$

$$P_{n,2n-1}^S = \Omega_n \left\{ \frac{i[f^{Rv}(\mathbf{r}) - f^{Rv}(-\mathbf{r})] f^v(\mathbf{r})^{n-1}}{(n-1)!} \right\} \quad (52)$$

$$P_{n,2n-2}^S = \Omega_n \left\{ \frac{1}{(n-1)!} f^R(\mathbf{r}) f^v(\mathbf{r})^{n-1} \right.$$

$$+ \frac{1}{(n-2)!} [f^{Rv}(\mathbf{r}) - f^{Rv}(0)]$$

$$\cdot [f^{Rv}(-\mathbf{r}) - f^{Rv}(0)] f^v(\mathbf{r})^{n-2} \quad (53)$$

where Ω_n is the Fourier transform operator

$$\Omega_n = (2\pi)^{-2} \int d\mathbf{r} \, \exp\left(-i\mathbf{k}\cdot\mathbf{r}\right) \quad (54)$$

(for the integers 0, −1 the factorial function is defined as $0! = 1$ and $[(-1)!]^{-1} = 0$).

We have left out a term $P_{0,0}$ in the sum representing an irrelevant δ function contribution at $k = 0$ associated with the mean image intensity.

An expansion with respect to nonlinearity only can be obtained by summing over the velocity bunching index m for fixed nonlinearity order n,

$$P^S(k) = \exp\left(-k_x^2 \xi'^2\right)(P_1^S(k) + P_2^S(k)$$

$$+ \cdots + P_n^S(k) + \cdots) \quad (55)$$

The linear term P_1^S is found to be identical (as it must be) with the linear SAR spectrum of (26).

It should be noted that the terms $P_n^S(k)$ in (55) do not represent the direct expansion of $P^S(k)$ in powers of the input wave spectrum, as the common (nonlinear) azimuthal cutoff factor $\exp\left(-k_x^2 \xi'^2\right)$ has been taken out of the sum. This is an important feature of the theory.

The first term in the expansion (55) yields the quasi-linear approximation

$$P_{ql}^S(k) = \exp\left(-k_x^2 \xi'^2\right) P_1^S(k) \quad (56)$$

The significance of the azimuthal cutoff factor is well illustrated by this term. The approximation $P_1^S(k)$ of purely linear SAR imaging theory, without the cutoff factor, always breaks down for high azimuthal wave numbers, even for very low waves, since real wave spectra (and therefore also the computed linear SAR image spectra) fall off as a power law at high wave numbers, rather than exponentially, as required by (55). In contrast, the lowest order quasi-linear approximation, including the azimuthal cutoff factor, represents a uniformly valid approximation for the entire spectrum.

In a hindcast study of 34 Seasat SAR spectra covering a wide variety of sea states (C. Brüning et al., manuscript in preparation, 1991), it was found that the quasi-linear form (56) yielded an acceptable first-order description of the SAR spectrum for about half of the cases analyzed and captured the qualitative features of the spectrum (although with displaced peaks, etc.) in all cases. The robustness of the quasi-linear approximation will be used in the next section to develop an iterative scheme for inverting the fully nonlinear transformation (55).

The decomposition of the quasi-linear spectrum P_{ql}^S into its contributions of different velocity bunching order yields

10,720   HASSELMANN AND HASSELMANN: MAPPING OF OCEAN WAVE SPECTRUM

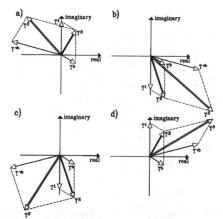

Fig. 1.   Orientations of the tilt, hydrodynamic, velocity bunching, and net RAR and SAR MTFs in the complex plane for a given wave component and different sign combinations of the SAR flight and look directions: (*a*) $k_x > 0$, $k_y > 0$, $k_l > 0$ (left looking SAR); (*b*) change in sign of flight direction ($k_x < 0$, $k_y < 0$, $k_l < 0$); (*c*, *d*) same as Figures 1*a* and 1*b* for a right looking SAR ($k_l = -k_y$).

$$P_{ql}^S = P_{ql}^R + P_{ql}^{int} + P_{ql}^{vb} \qquad (57)$$

where the pure RAR spectrum P_{ql}^R, the pure quasi-linear velocity bunching spectrum P_{ql}^{vb}, and the quasi-linear interference spectral term P_{ql}^{int} are given by

$$\begin{Bmatrix} P_{ql}^R \\ P_{ql}^{int} \\ P_{ql}^{vb} \end{Bmatrix} = \exp\left(-k_x^2 \xi'^2\right) \begin{Bmatrix} P_{10}^S \\ (k_x \beta) P_{11}^S \\ (k_x \beta)^2 P_{12}^S \end{Bmatrix} \qquad (58)$$

Applying (26) and (27), this may be written

$$P_{ql}^{\cdot\cdot} = H_k^{\cdot\cdot} \cdot \frac{F_k}{2} + H_{-k}^{\cdot\cdot} \cdot \frac{F_{-k}}{2} \qquad (59)$$

(for any index R, int, vb, or S), where

$$\begin{Bmatrix} H_k^R \\ H_k^{int} \\ H_k^{vb} \\ H_k^S \end{Bmatrix} = \exp\left(-k_x \xi'^2\right) \begin{Bmatrix} |T_k^R|^2 \\ (T_k^R T_k^{vb*} + T_k^{R*} T_k^{vb}) \\ |T_k^{vb}|^2 \\ |T_k^S|^2 \end{Bmatrix} \qquad (60)$$

The orientations in the complex plane of the various MTFs which determine the filter factors $H_k^{\cdot}$ are indicated in Figure 1. The MTFs apply for a given surface wave component and various sign combinations of the SAR look and flight directions. Figure 1*a* applies to a wave component traveling in the positive quadrant of the xy plane ($k_x > 0$, $k_y > 0$) and a left looking SAR ($k_l = k_y$). Figure 1*b* applies for the same left looking SAR viewing the same wave component but for a SAR platform flying in the opposite direction ($k_x < 0$, $k_y = k_l < 0$). Figures 1*c* and 1*d* correspond to Figures 1*a* and 1*b*, respectively, for a right looking SAR ($k_l = -k_y$). The resultant SAR MTF is seen to

be strongly dependent on the orientation of the SAR look and flight directions. This has been confirmed by aircraft SAR measurements [e.g., *Hasselmann et al.*, 1989].

The strong dependence on the viewing geometry is at first sight perhaps surprising, since the moduli of the individual velocity bunching, tilt, and hydrodynamic transfer functions are invariant with respect to the four viewing combinations shown in Figure 1. The modulus of the net RAR transfer function is also only weakly dependent on the look and flight direction (through the imaginary component of the hydrodynamic MTF, which, in contrast to the imaginary tilt MTF, is invariant with respect to the look and flight direction).

Since the moduli of the separate filter functions H_k^R and H_k^{vb} for pure RAR and pure velocity bunching imaging, respectively, are approximately or exactly independent of the sign combinations of the viewing geometry, the strong viewing geometry dependence of the net SAR filter function,

$$H_k^S = H_k^R + H_k^{int} + H_k^{vb} \qquad (61)$$

must come about through the interference filter function H_k^{int}.

This is illustrated by the plots of the four filter functions H_k^R, H_k^{int}, H_k^{vb} and H_k^S shown in Figure 2. The cutoff scale was chosen as $\xi' = 70$ m, or $k_x^{cutoff} = (\xi')^{-1} = 0.014$ m^{-1}, corresponding to the Seasat value $\beta = 113.5$ and a Pierson-Moskowitz [*Pierson and Moskowitz*, 1964] fully developed wind sea spectrum for a wind speed at 10 m height of 10 m/s ($\langle v^2 \rangle^{1/2} = 0.62$ m/s). The damping factor in T_k^{hyd} (equation (6)) was set at $\mu = 0.5$ s^{-1}, and no wind input modulation terms were included (as in the Seasat computations in section 5).

The filter function H_k^{vb} is seen to be exactly symmetrical with respect to a change in sign of k_x or k_y, the filter function H_k^R is exactly symmetrical with respect to a change in sign of k_x, and approximately symmetrical with respect to a change in sign of k_y, while the filter function H_k^{int} is exactly antisymmetrical with respect to the transformation $k_y \to -k_y$. The net filter function H_k^S is therefore approximately symmetrical with respect to a change in sign of k_y, but has pronounced asymmetries with respect to the transformation $k_x \to -k_x$.

The general structures of the filter functions shown in Figure 2 are independent of the parameters chosen. It will be useful to keep Figure 2 in mind later in discussing the origin of the various distortions and asymmetries found in computed and observed SAR image spectra.

As pointed out, the common azimuthal cutoff factor applies not only to the quasi-linear spectral terms but also to the entire series expansion (50) or (55). This has a useful practical implication. The azimuthal cutoff of an observed SAR spectrum is usually a relatively well-defined feature. Its experimental determination, independent of the details of the mapping process, yields an important integral property of the wave spectrum, the mean square orbital velocity (cf. (44)).

Beal et al. [1983], *Lyzenga* [1986], and *Monaldo and Lyzenga* [1986, 1988] have verified experimentally the proportionality of the azimuthal cutoff scale to the rms orbital range velocity component, or some related integral property of the wave field. Previously, this finding has been difficult to interpret theoretically. The SAR two-scale model using the SAR resolution scale as separation scale yields a cutoff

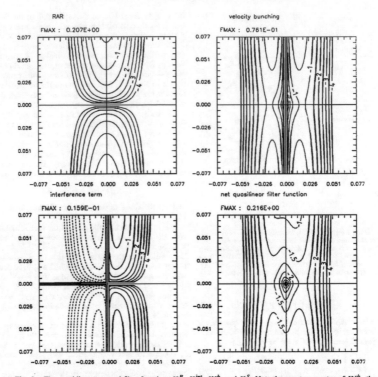

Fig. 2. The quasi-linear spectral filter functions $H_\mathbf{k}^R$, $H_\mathbf{k}^{int}$, $H_\mathbf{k}^{vb}$, and $H_\mathbf{k}^S$. Note the exact symmetry of $H_\mathbf{k}^{vb}$, the approximate symmetry of $H_\mathbf{k}^R$, and the antisymmetry of $H_\mathbf{k}^{int}$, yielding a nonsymmetrical net filter function $H_\mathbf{k}^S$. Isolines are logarithmically spaced relative to the maximum value, with two isolines per decade.

factor due to velocity spreading which is determined by the rms orbital velocity in the subresolution short wave spectral band between the SAR resolution scale and the facet scale. *Tucker* [1985] computed the contribution of this velocity spread smearing and obtained an azimuthal cutoff factor which was indeed identical to our form, but with ζ' replaced by the rms azimuthal displacement (the "velocity spread") of only the short subresolution scale waves. The present closed theory implies that the nonlinear velocity bunching mechanism in the "deterministic" wave number region, below the SAR resolution wave number, not only distorts the spectrum in this region, but must also contribute to the azimuthal cutoff. When this effect is combined with the velocity spread term, one obtains the simple result that the net azimuthal filtering can be represented by a common Gaussian cutoff factor which acts on all terms in the nonlinear spectral expansion.

We conclude this section by summarizing again the basic three computational steps needed to determine the SAR spectrum according to (50):

1. Computation of the three autovariance and covariance functions $f^v(\mathbf{r})$, $f^R(\mathbf{r})$, and $f^{Rv}(\mathbf{r})$ using the Fourier transform relations (43), (47), and (48).

2. Computation of the covariance product expressions appearing in (51)–(53).

3. Computation of the Fourier transforms (51)–(53). If only the final SAR spectrum is of interest, without regard for the separate contributions from different nonlinearity order, the covariance products of different nonlinearity order n for given velocity bunching order m (power of βk_x) can be collected together and Fourier transformed in a single operation.

Since only Fourier transforms are involved, the computations are rather fast (less than 1 s on a CRAY-2, for 128 × 128 pixel scene using full FFT representations). Good convergence was normally attained even for strongly nonlinear spectra with a truncation at nonlinearity order $n = 6$. The higher order terms contribute mainly to the resolution of the (normally not very important) details near the azimuthal cutoff (cf. section 5).

4. INVERSION

A unique formal inversion of the forward mapping relation cannot exist in view of the 180° ambiguity of the SAR image and the loss of information beyond the azimuthal cutoff. The standard procedure for resolving such underdetermined inversion problems is to introduce a regularization term which makes use of additional information from a first-guess wave spectrum $\hat{F}(\mathbf{k})$.

Following this approach, we define the optimal fit wave spectrum $F(\mathbf{k})$ as the spectrum which minimizes the cost function

$$J = \int [P(\mathbf{k}) - \hat{P}(\mathbf{k})]^2 \, d\mathbf{k} + \mu \int \left[\frac{[F(\mathbf{k}) - \hat{F}(\mathbf{k})]}{[B + \hat{F}(\mathbf{k})]} \right]^2 \, d\mathbf{k} \tag{62}$$

where $\hat{P}(\mathbf{k})$, $P(\mathbf{k})$ are the observed and fitted SAR spectra, respectively (the index S has been dropped, as there is now no need to distinguish between the SAR and RAR spectrum), μ is some suitably chosen weight reflecting the relative confidence ascribed to the observed SAR spectrum and the first-guess wave spectrum (which will normally be provided by a wave model), and a small positive constant B has been introduced into the normalizing denominator in the second integral to avoid numerical infinities for $\hat{F}(\mathbf{k}) = 0$ (formally, infinities in normalizing factors are acceptable, as they merely express an infinitely hard side condition).

Equation (62) can be generalized by introducing wave number dependent weights within the integrals or a nondiagonal metric. The rationale for such generalizations is normally provided by maximum likelihood considerations. Since the required input for these generalizations (the error covariance matrix for the combined data set $\hat{P}(\mathbf{k})$, $\hat{F}(\mathbf{k})$) cannot be readily estimated in our case, these options were not further pursued.

However, to enhance the agreement of the computed and observed SAR spectra in the neighborhood of the SAR spectral peaks, we considered also the alternative cost function

$$J' = \int [P(\mathbf{k}) - \hat{P}(\mathbf{k})]^2 \hat{P}(\mathbf{k}) \, d\mathbf{k}$$

$$+ \mu \int \left\{ \frac{[F(\mathbf{k}) - \hat{F}(\mathbf{k})]}{[B + \hat{F}(\mathbf{k})]} \right\}^2 \, d\mathbf{k} \tag{63}$$

with an additional factor $\hat{P}(\mathbf{k})$ in the first integral. It should be stressed that at this time both cost functions (62) and (63) should be regarded only as rather arbitrarily selected candidates which will need to be further tested and possibly modified in more extensive studies.

The solution of the general nonlinear variational problem

$$\frac{\delta J}{\delta F(\mathbf{k})} = 0 \tag{64}$$

was obtained by an iterative technique which made use of the approximate quasi-linear mapping relation (56), as follows.

Starting from a first estimate $F^1(\mathbf{k}) = \hat{F}(\mathbf{k})$, let $F^n(\mathbf{k})$, $P^n(\mathbf{k})$ represent the approximate solution after n iteration steps, where $P^n(\mathbf{k})$ is the associated SAR spectrum for the wave spectrum $F^n(\mathbf{k})$ in accordance with the fully nonlinear mapping relation (55),

$$P^n = M_{nl}(F^n) \tag{65}$$

Construct now an improved solution

$$F^{n+1} = F^n + \Delta F^n \tag{66}$$

by assuming, in a first step, that the increment ΔF^n leads to a modified SAR spectrum

$$P^{n+1} = P^n + \Delta P^n \tag{67}$$

for which the change ΔP^n is related to ΔF^n through the approximate quasi-linear transformation (equations (56) and (26)),

$$\Delta P^n(\mathbf{k}) = \frac{1}{2} \exp (-k_x^2 \xi_n'^2)[|T^S(\mathbf{k})|^2 \Delta F^n(\mathbf{k})$$

$$+ |T^S(-\mathbf{k})|^2 \Delta F^n(-\mathbf{k})] \tag{68}$$

Substituting these new estimates for F, P into (62), one obtains

$$J = \int [\Delta P^n - (\hat{P} - P^n)]^2 \, d\mathbf{k}$$

$$+ \mu \int [\Delta F^n - (\hat{F} - F^n)]^2 \, d\mathbf{k} \tag{69}$$

The solution of the variational equations for J with respect to ΔF^n, with ΔP^n given by (68), can be immediately found:

$$\Delta F^n = \frac{[A_{-\mathbf{k}}(W_{\mathbf{k}} \delta P + \mu \, \delta F_{\mathbf{k}}) - B_{\mathbf{k}}(W_{-\mathbf{k}} \delta P + \mu \, \delta F_{-\mathbf{k}})]}{[A_{\mathbf{k}} A_{-\mathbf{k}} - B_{\mathbf{k}}^2]} \tag{70}$$

where

$$\delta P = \hat{P}(\mathbf{k}) - P^n(\mathbf{k}) = \hat{P}(-\mathbf{k}) - P^n(-\mathbf{k}) \tag{71}$$

$$\delta F_{\mathbf{k}} = \hat{F}(\mathbf{k}) - F^n(\mathbf{k}) \tag{72}$$

$$A_{\mathbf{k}} = W_{\mathbf{k}}^2 + 2\mu \tag{73}$$

$$B_{\mathbf{k}} = W_{\mathbf{k}} W_{-\mathbf{k}} \tag{74}$$

and

$$W_{\mathbf{k}} = |T^S(\mathbf{k})|^2 \exp (-k_x^2 \xi'^2) \tag{75}$$

Having determined ΔF^n and the new wave spectrum F^{n+1}, the iteration step is then completed by computing the associated SAR spectrum, using the fully nonlinear transformation relation $P^{n+1} = M_{nl}(F^{n+1})$.

The technique can be applied equally well to the form (63). In computing the perturbations ΔF^n, ΔP^n, the additional factor P in the first integral in J is set equal in this case to P^n.

The iteration scheme was found to converge in all cases studied, including cases with strongly nonlinear and poor first guesses, provided μ was not chosen too small, namely, $\mu \gtrsim \mu_0$, where

$$\mu_0 = 0.1 \hat{P}_{max}^2 \tag{76}$$

In most applications, we chose $\mu = \mu_0$. The constant B was set at $0.01\hat{F}_{max}$.

In the Seasat cases discussed in the following section, the SAR calibration was not known. This can be readily accommodated in the inversion formalism by including the dependence on the unknown calibration factor explicitly in the expression for J and minimizing the cost function simultaneously with respect to both F and the unknown calibration factor. The minimization with respect to the calibration factor can be given analytically and was carried out after each iteration step.

Other unknown or poorly known parameters a_j (for example, in the hydrodynamic MTF) can be treated in the same way. If first-guess estimates $\hat{a}_j$ of these parameters exist, deviations from these values can be penalized by adding standard penalty terms to J of the form $\Sigma_j \mu_j(a_j - \hat{a}_j)^2$.

The basic inversion formalism can be generalized also in other ways. In practice, it was found that a straightforward application of the inversion method just described yielded wave spectra which successfully reproduced the observed SAR spectra, but were nonetheless clearly unrealistic. The inversion modified the wave spectrum within the wave number region $|k_x| < k_{x_x}^{cutoff}$ containing valid SAR information, but left the first-guess wave spectrum unchanged in the high azimuthal wave number region beyond the cutoff. Although entirely consistent with the intended operation of the cost function, the resultant spectra exhibited dislocations along the transitional azimuthal cutoff bands which were obviously spurious.

The origin of these problems is clearly the lack of dynamical constraints in the inversion formalism. In reality, the development of dislocated spectra is prevented by nonlinear wave-wave interactions, which maintain an approximately universal spectral shape in the wind sea region of the spectrum [cf. Hasselmann et al., 1973, 1976; Komen et al., 1984]. The difficulty would not arise if the inversion technique were imbedded in a general wind and wave data assimilation scheme in which all modifications of the wave spectrum were attributed to modifications in the wind field. These would automatically ensure dynamically consistent changes in the wave spectrum through the application of a wave model. This is the ultimate goal of the assimilation program of the Wave Modeling (WAM) Group. However, for the present intermediate level of inversion, without direct coupling to the wind field, some form of dynamical constraints are needed. These were introduced using the following simple two-stage procedure.

As the high wave number region of the spectra is strongly coupled to lower wave numbers through the nonlinear transfer, we restricted the modification of the spectrum in the first inversion stage to transformations which apply uniformly to the entire wave spectrum. The simplest such transformation is a rotation ϕ_0 in the wave number plane combined with scale changes A, B in the energy and wave number, respectively.

$$F'(\mathbf{k}) = AF(\mathbf{k}') \tag{77}$$

where

$$k_x' = B(k_x \cos \phi_0 - k_y \sin \phi_0)$$
$$k_y' = B(k_x \sin \phi_0 + k_y \cos \phi_0) \tag{78}$$

After minimizing the cost function with respect to the parameters ϕ_0, A, B, the original minimization procedure without constraints was then applied in a second stage.

The first stage normally yielded a close fit to the SAR spectral peak, while ensuring continuity of the overall spectral distribution. The second stage then provided further fine-scale adjustments within the azimuthal wave number band for which detailed SAR information was available. Since a reasonable first-order fit was achieved already in the first stage, the second stage no longer produced significant dislocations in the azimuthal cutoff region.

As pointed out earlier, the energy scale parameter A can be determined rather reliably (for given B, ϕ_0) from the observed azimuthal cutoff scale ζ', which is independent of the details of the SAR spectrum. In practice, the least squares minimization routine was therefore applied in the first stage only to the parameters ϕ_0 and B, while A was determined explicitly from ϕ_0, B, and ζ' using the relation (44) for the azimuthal cutoff scale.

5. SOME EXAMPLES FROM SEASAT

The computation of the forward transformation relation is illustrated in Figure 3 for a typical Seasat case. The case was selected together with the two other cases discussed in this section from a larger set of SAR image spectra analyzed in the course of a wave hindcast study using the WAM third-generation wave model [WAMDIG, 1988; Hasselmann et al., 1988].

The individual panels show the hindcast WAM wave spectrum, the SAR spectrum computed from the WAM spectrum, and some typical spectral terms of the nonlinear spectral expansion. The case is only weakly nonlinear, so that only little of the azimuthally traveling short wave energy is lost in the SAR image, while most of the wave energy propagating in the range direction is retained. The quasilinear approximation, consisting of the sum of the first three quasi-linear contributions is seen to yield a fairly good approximation of the fully nonlinear image spectrum.

The splitting of the single-wave spectral peak into two peaks in the SAR image spectrum is a common feature in SAR images of predominantly range traveling waves. It arises because the velocity bunching MTF, which normally dominates over the RAR MTF, vanishes in the range direction (cf. section 3 and Figure 2).

The asymmetry of the SAR response about the look direction due to the interference term (cf. Figure 2) is evident in the quasi-linear and fully nonlinear SAR spectra and in the interference term itself. In general, all terms with odd powers of $(k_x\beta)$ contribute to the asymmetry. As pointed out in section 3, the asymmetry is dependent on the look and flight directions, so that different SAR image spectra are obtained, for example, if the same wave field is viewed from the upwind or downwind direction [Hasselmann et al., 1989]. In comparing Figures 2 and 3, it should be noted that the SAR spectrum is formed from both positive and negative $\mathbf{k}$ contributions (cf. (26)). Thus in contrast to the filter functions of Figure 2, which apply for only one wave component, the spectra of Figure 3 are symmetrical with respect to the transformation $\mathbf{k} \to -\mathbf{k}$.

The higher order terms in the expansion are proportional to the product of a high power of $(k_x\beta)$ with the exponential azimuthal cutoff factor and are therefore normally strongly

10,724   HASSELMANN AND HASSELMANN: MAPPING OF OCEAN WAVE SPECTRUM

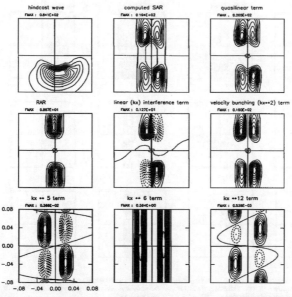

Fig. 3. Hindcast wave spectrum and computed (hindcast) SAR spectra for August 21, 0740, at 57°N, 9°W (top row). Also shown are various spectral expansion terms (see text). The x axis points in the satellite flight direction, and the y axis in the negative look direction (right looking SAR).

peaked along the azimuthal cutoff band. These terms do not contribute significantly to the structure of the SAR spectrum in the neighborhood of the spectral peak, so that the effective convergence of the series in the main part of the spectrum is rather rapid (see also Figure 5, discussed below).

Figure 4 shows a comparison of the observed SAR spectra and the computed wave and SAR spectra before and after inversion for the three Seasat cases. Through the combined effects of the azimuthal cutoff, the strongly varying modulation transfer functions (cf. Figure 2) and the nonlinear distortions, the observed and computed SAR spectra (first and third columns, respectively) show relatively little similarity with the hindcast first-guess wave spectra (second column). The first-guess SAR spectra computed from the hindcast wave spectra reproduce the azimuthally banded structure of the observed SAR spectra but nonetheless still show significant deviations.

The fourth and fifth columns show the best fit wave spectra and the associated computed SAR spectra derived by the inversion method. The agreement between the best fit and observed SAR spectra is now seen to be markedly improved. The parameters of the hydrodynamic MTF (6) were chosen in all cases as $\mu = 0.5 \, s^{-1}$ and $\gamma = 0$. This is consistent with field and laboratory measurements [cf. *Keller and Wright*, 1975; *Plant et al.*, 1983; *Feindt et al.*, 1986; *Schröter et al.*, 1986], but no attempt was made to optimize these parameters. The inversions were based on the peak-enhanced cost function (63).

The three cases were selected to illustrate different degrees of nonlinearity and different directions of wave propagation relative to the SAR look direction. The first case is weakly nonlinear, with predominantly range propagating waves. The second case is moderately nonlinear and represents a wave field propagating at an angle between the range and azimuthal directions. The third case, finally, is strongly nonlinear and was chosen also as an example of a more complex sea state, consisting of a superposition of swell and wind sea components propagating at nearly 90° relative to one another. The azimuthally propagating major swell component is seen to be almost entirely lost due to the azimuthal cutoff.

The individual modifications introduced into the best fit wave spectra through the two-step inversion procedure (summarized in Table 1) can be clearly recognized.

1. The spectra have been rotated and the wave number scales adjusted to reproduce the positions of the SAR spectral peaks.

2. The energy scales have been adjusted (together with the wave number scales) to reproduce the observed azimuthal cutoffs. This effect is evident in the changed azimuthal limits between the first-guess and best fit SAR spectra (columns 3 and 5 of Figure 4).

3. The subsequent modifications of the detailed structures of the spectra, acting separately on all components of the wave spectra, have resulted mainly in some sharpening of the spectral peaks, which were generally too broad in the

132   H. von Storch

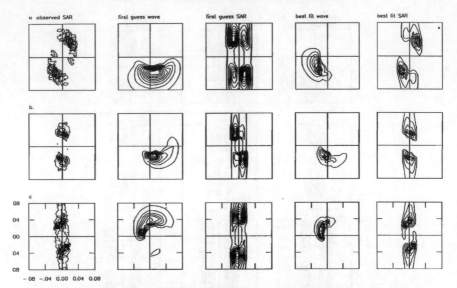

Fig. 4.   Observed Seasat SAR spectrum, first-guess (hindcast) wave spectrum, computed first-guess (hindcast) SAR spectrum, and best fit wave and SAR spectra for the case shown in Figure 3 (top row), for August 19, 0620, 60°N, 6°W (second row) and for September 30, 0000, 60°N, 9°W (bottom row).

original wave spectra. (The finite 30° resolution of the WAM model, together with the discrete interaction approximation of the nonlinear transfer source function [cf. S. *Hasselmann et al.*, 1985], are known to result in somewhat too broad peaks in the model spectra.)

The third example illustrates a basic limitation of the present simplified approach, in which a single linear coordinate transformation is applied to the entire spectrum in the first stage of the inversion. The coordinate transformation was governed in this case by the wind sea region of the spectrum, represented by the broad higher frequency peak in the range direction. The peak needed to be rotated about 30° to the left to achieve coincidence between the computed and observed SAR peaks. However, the rotation modified also the azimuthally propagating swell components of the original hindcast wave spectrum, although the swell hindcast was presumably not directly affected by errors in the wind sea hindcast. This deficiency could in principle have been overcome by considering a more sophisticated class of transformations. However, this was not pursued further, since, as has been pointed out, it is anticipated that the present two-stage inversion procedure will be ultimately superseded by a comprehensive data assimilation scheme in which the measured SAR spectrum is used to modify the wind field rather than the wave spectrum directly.

The convergence properties of the spectral expansion (50) for the three cases considered are indicated in Figure 5. Terms of different nonlinearity order n for a given velocity bunching order m (which appear in the same Fourier transform contribution) have been collected into a single term. The curves show the maximal spectral values for each mth-order spectrum of the expansion. Since these values, as already mentioned, tend to lie near the relatively unimportant azimuthal cutoff limits for the higher order expansion terms, the effective convergence is in fact better than implied by the figure. In practice, good convergence was achieved in all cases studied with a truncation of the series at $m = 12(n = 6, 7)$.

6.   CONCLUSIONS

The new closed, nonlinear integral transformation relation derived in this paper, together with its expansion in a

TABLE 1.   Inversion Parameters of Seasat SAR Spectra

Case	Seasat Swath	Latitude, deg N	Longitude, deg W	Date	Time, UT	ϕ_0, deg	B	A
a	794	57	9	Aug. 21	0740	64	1.2	0.75
b	762	60	6	Aug. 19	0620	33	1.1	1.6
c	1359	60	9	Sept. 30	0000	−28	1.5	1.3

HASSELMANN AND HASSELMANN: MAPPING OF OCEAN WAVE SPECTRUM

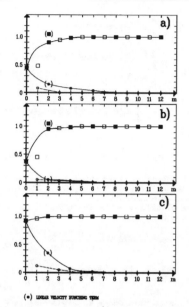

(●) LINEAR VELOCITY BUNCHING TERM

Fig. 5. Convergence of the spectral series expansions with respect to velocity bunching order m. Plotted are the maximal values of each spectrum. Squares denote the partial sum, dots the contribution of the last term in the sum. Solid and open points refer to even and odd m, respectively. The convergence in the neighborhood of the spectral peak of the net SAR spectrum is faster than indicated by the individual maximal values.

spectral series with respect to nonlinearity and velocity bunching order, presents a number of advantages.

1. It can be computed rather rapidly using fast Fourier transforms and is free of the statistical sampling errors of Monte Carlo methods. It should thus make feasible the operational processing of SAR wave images which will be obtained from future satellites such as ERS-1, 2, and Radarsat.

2. It provides a clearer insight into the imaging mechanism by identifying the different contributions from the RAR modulation and nonlinear velocity bunching processes and their various interference terms.

3. It yields a simple expression for the azimuthal cutoff in the form of a Gaussian filter factor which acts on all terms in the series expansion. The azimuthal cutoff scale is given by the rms azimuthal velocity bunching displacement. The observed azimuthal cutoff therefore yields a useful integral constraint on the wave spectrum.

4. The lowest order quasi-linear term of the spectral series expansion, consisting of the product of the standard linear SAR spectrum and the azimuthal cutoff factor, yields a useful first-order approximation of the fully nonlinear mapping relation.

5. It provides the necessary basis for inverting the transformation using standard inverse modeling techniques.

Because of the 180° ambiguity of the spectral mapping relation and the loss of information beyond the azimuthal cutoff, an inversion of the wave-to-image transformation will in general be feasible only if a first-guess wave spectrum is available. This implies that the interpretation and application of SAR wave image data will normally require the application of a wave model.

The inversion technique developed in this study provides an optimal wave spectral estimate for a given first-guess wave spectrum and observed SAR spectrum. The examples shown demonstrate that SAR ocean wave images can indeed provide valuable information to correct modeled wave spectra.

No attempt was made at this stage to correct for possible deficiencies in the wind field driving the wave model, which will normally be the principal cause of discrepancies between observed and predicted SAR spectra. The present inversion technique requires additional ad hoc assumptions to replace the dynamical constraints which would be introduced automatically if the inversion method were integrated in a general wind and wave data assimilation scheme. The inversion technique presented here should therefore be regarded only as an intermediate step toward the development of such a comprehensive data assimilation system.

APPENDIX: EXTENSION TO ACCELERATION SMEARING AND GENERAL DISPERSIVE MAPPING

The pure velocity bunching theory presented in section 3 is nondispersive: an infinitesimal cross-section element $\sigma(\mathbf{r}')$ $d\mathbf{r}'$, which in the absence of motion effects would correspond to an image element $\hat{I}^R(\mathbf{r}')\,d\mathbf{r}' = \sigma(\mathbf{r}')\,d\mathbf{r}'/\bar{\sigma}$, is mapped one-to-one into a displaced infinitesimal element

$$\hat{I}^S(\mathbf{r})\,d\mathbf{r} = \hat{I}^R(\mathbf{r}')\,d\mathbf{r}' \qquad (A1)$$

in the SAR image plane, where

$$r = \mathbf{r}' + \boldsymbol{\xi}(\mathbf{r}') \qquad (A2)$$

and $\boldsymbol{\xi} = \mathbf{a}\beta v$ (equation (14)).

In the general SAR theory of ocean wave imaging for arbitrarily moving scattering elements, the mapping is dispersive (cf. MSR): an infinitesimal element on the sea surface is mapped into a finite patch in the SAR image plane. The form (A1) must accordingly be generalized to the integral relation

$$\hat{I}^S(\mathbf{r}) = \int \hat{I}^R(\mathbf{r}')M(\mathbf{r} - \mathbf{r}';\,\mathbf{r}')\,d\mathbf{r}' \qquad (A3)$$

The mapping function $M(\mathbf{r} - \mathbf{r}';\,\mathbf{r}')$ represents a finite width distribution with respect to the primary spatial separation variable $\mathbf{r} - \mathbf{r}'$ and depends in general also on the details of the motion history of the scattering element at $\mathbf{r}'$. To a good approximation, the shape of M is given by the shape of the Doppler spectrum of the backscattered return signal (MSR).

In the nondispersive velocity bunching model, the Doppler spectrum is a line spectrum, and M therefore reduces to a δ function,

$$M(\mathbf{r} - \mathbf{r}';\,\mathbf{r}') \doteq \delta[\mathbf{r} - \mathbf{r}' - \boldsymbol{\xi}(\mathbf{r}')] \qquad (A4)$$

In the general case, the time dependence of the backscattering elements cannot be represented simply as the constant advection of a frozen scatterer, and the Doppler spectrum and mapping function M have finite width.

For Bragg scattering, the assumption of a constant advection velocity given by the instantaneous orbital velocity in the center of the viewing window nevertheless remains a good first approximation. The pure velocity bunching theory is therefore normally generalized by expanding the change in orbital velocity during the SAR viewing interval $-\tau/2 < t < \tau/2$ in a Taylor series,

$$v(\mathbf{r}, t) = v(\mathbf{r}, 0) + ta(\mathbf{r}, 0) + \cdots \quad (A5)$$

where $a(\mathbf{r}, 0) = [dv(\mathbf{r}, t)/dt]_{t=0}$ represents the orbital acceleration (in the range direction) in the center of the viewing window. It is assumed that τ is small compared with the wave period.

The linear drift of the orbital velocity during the SAR illumination time leads to an approximately uniform azimuthal smearing of the scattering element in the image plane over the interval $\xi - \beta|a|\tau/2 < x < \xi + \beta|a|\tau/2$. The mapping function in this case becomes

$$M^h(\mathbf{r} - \mathbf{r}'; \ \mathbf{r}') = \delta(y - y')H\left(\frac{x - x' - \xi(\mathbf{r}')}{\beta|a|\tau}\right)(\beta|a|\tau)^{-1} \quad (A6)$$

where $H(\eta)$ denotes the top hat function

$$H(\eta) = 1 \qquad |\eta| \leq \frac{1}{2} \quad (A7)$$

$$H(\eta) = 0 \qquad |\eta| > \frac{1}{2}$$

In place of (A6), a Gaussian distribution

$$M^g(\mathbf{r} - \mathbf{r}'; \ \mathbf{r}')$$

$$= (2\pi)^{-1/2}(\Delta x)^{-1} \exp\left\{\frac{[x - x' - \xi(\mathbf{r}')]^2}{2(\Delta x)^2}\right\}\delta(y - y') \quad (A8)$$

with the same rms width

$$\Delta x(\mathbf{r}') = \frac{\beta\tau|a(\mathbf{r}')|}{2(3)^{1/2}} \quad (A9)$$

as the top hat form (A6) is sometimes used [cf. *Alpers and Brüning*, 1986]. This simplifies the treatment of acceleration smearing within the framework of a more general analysis, including the effects of the antenna pattern and the matched filter and other SAR system characteristics.

The distinction between (A6) and (A8) is immaterial in the present context. We shall show that if the function M is known, regardless of the model used, the surface wave–SAR image spectral mapping relations can be derived as before in closed form.

Starting from the general form (A3), the expression (28) for the Fourier components of the SAR image now becomes (for $\mathbf{k} \neq 0$, so that $I(\mathbf{r})$ may be replaced by $\hat{I}(\mathbf{r})$)

$$I_{\mathbf{k}}^S = \frac{1}{A} \int d\mathbf{r}\, \hat{I}^S(\mathbf{r}) \exp(-i\mathbf{k} \cdot \mathbf{r})$$

$$= \frac{1}{A} \iint d\mathbf{r}\, d\mathbf{r}'\, \hat{I}^R(\mathbf{r}')M(\mathbf{r} - \mathbf{r}'; \ \mathbf{r}') \exp(-i\mathbf{k} \cdot \mathbf{r}) \quad (A10)$$

or

$$I_{\mathbf{k}}^S = \frac{1}{A} \int d\mathbf{r}'\, \hat{I}^R(\mathbf{r}')G_{\mathbf{k}}(\mathbf{r}') \exp(-i\mathbf{k} \cdot \mathbf{r}') \quad (A11)$$

where

$$G_{\mathbf{k}}(\mathbf{r}') = \int d\boldsymbol{\rho}\, M(\boldsymbol{\rho}; \ \mathbf{r}') \exp(-i\mathbf{k} \cdot \boldsymbol{\rho}) \quad (A12)$$

The nonlinearity of the imaging relation (A11) arises through the dependence of the weighting function $G_{\mathbf{k}}(\mathbf{r}')$ on the local wave field at $\mathbf{r}'$.

For the two forms (A6) for M^h or (A8) for M^g, we obtain the weighting functions

$$G_{\mathbf{k}}^h(\mathbf{r}') = \exp[-i\mathbf{k} \cdot \boldsymbol{\xi}(\mathbf{r}')]\frac{\sin[(3)^{1/2}k_x|\Delta x|]}{(3)^{1/2}k_x|\Delta x|} \quad (A13)$$

$$G_{\mathbf{k}}^g(\mathbf{r}') = \exp[-i\mathbf{k} \cdot \boldsymbol{\xi}(\mathbf{r}')] \exp\left(\frac{-k_x^2\Delta x^2}{2}\right) \quad (A14)$$

Thus the velocity bunching mapping factor $\exp[-i\mathbf{k} \cdot \boldsymbol{\xi}(\mathbf{r}')]$ in (28) is replaced now in the general case by a mapping factor

$$G_{\mathbf{k}}^{h,g} = \exp[-i\mathbf{k} \cdot \boldsymbol{\xi}(\mathbf{r}')]D^{h,g} \quad (A15)$$

which contains an additional azimuthal acceleration smearing term

$$D^h = \frac{\sin[(3)^{1/2}k_x|\Delta x|]}{(3)^{1/2}k_x|\Delta x|} \quad (A16)$$

or

$$D^g = \exp\left(\frac{-k_x^2\Delta x^2}{2}\right) \quad (A17)$$

The further analysis proceeds as in section 3, with the velocity bunching weighting function $\exp[-i\mathbf{k} \cdot \boldsymbol{\xi}(\mathbf{r}')]$ replaced by the general weighting function $G_{\mathbf{k}}$. The expression (30) for the SAR image variance spectrum becomes

$$P_{\mathbf{k}}^S = A^{-2}\left\langle \iint d\mathbf{r}'\, d\mathbf{r}''\, \exp[-i\mathbf{k}(\mathbf{r}' - \mathbf{r}'')] \right.$$

$$\cdot N_{\mathbf{k}}D(\mathbf{r}')D(\mathbf{r}'')\left\{1 + \sum_{\mathbf{k}'} (T_{\mathbf{k}'}^R\zeta_{\mathbf{k}'} + T_{-\mathbf{k}'}^{R*}\zeta_{-\mathbf{k}'}^*)e^{i\mathbf{k}'\cdot\mathbf{r}'}\right\}$$

$$\left. \cdot \left\{1 + \sum_{\mathbf{k}''} (T_{\mathbf{k}''}^{R*}\zeta_{\mathbf{k}''}^* + T_{-\mathbf{k}''}^R\zeta_{-\mathbf{k}''})e^{-i\mathbf{k}''\cdot\mathbf{r}''}\right\}\right\rangle \quad (A18)$$

The expression differs from (30) only through the inclusion of the acceleration smearing factors $D(\mathbf{r}')D(\mathbf{r}'')$.

10,728   HASSELMANN AND HASSELMANN: MAPPING OF OCEAN WAVE SPECTRUM

To evaluate the expectation value occurring on the right-hand side of (A18), the same technique may be applied as before. The product $D(\mathbf{r}')D(\mathbf{r}'')$ is first expanded in a power series with respect to the wave Fourier amplitudes. For the forms (A16) or (A17), this is straightforward. The expectation value of the product of $N_\mathbf{k}$ with the various wave amplitude products occurring in the rest of the integrand in (A18) is then again computed by expanding $N_\mathbf{k}$ with respect to the particular infinitesimal wave amplitude components appearing in any given product.

The only difference between the extended theory and the previous pure velocity bunching theory is that the wave amplitude products with which $N_\mathbf{k}$ is now correlated are no longer limited to linear and quadratic expressions, as in (32)–(34), but consist of an infinite series. Since it was already found convenient, both theoretically and computationally, to expand the closed integral expression (46) for the pure velocity bunching case in a spectral power series, the present extension involves no significant algebraic or computational overhead. The expansion algebra can be readily incorporated in the (computer coded) algebra used to generate the expansion (50) with respect to velocity bunching and nonlinearity order. The structure of the expansion remains basically unchanged except for the appearance of a third expansion parameter, the ratio of the SAR illumination time τ to the mean wave period.

Acknowledgments. We are grateful for several suggested clarifications to the first version of this paper by the reviewers and for useful discussions with Claus Brüning. This work was partly supported by the U.S. Office of Naval Research through grant N00014-88-J-1025.

REFERENCES

Alpers, W., Monte Carlo simulations for studying the relationship between ocean wave and synthetic aperture radar image spectra, *J. Geophys. Res.*, *88*, 1745–1759, 1983.

Alpers, W., and C. Brüning, On the relative importance of motion-related contributions to the SAR imaging mechanism of ocean surface waves, *IEEE Trans. Geosci. Remote Sensing*, GE-24, 873, 1986.

Alpers, W., and K. Hasselmann, Spectral signal-to-clutter and thermal noise properties of ocean wave imaging synthetic aperture radars, *Int. J. Remote Sensing*, *3*, 423–446, 1982.

Alpers, W., and C. L. Rufenach, The effect of orbital motions on synthetic aperture radar imagery of ocean waves, *IEEE Trans. Antennas Propag.*, AP-27, 685–690, 1979.

Alpers, W., D. B. Ross, and C. L. Rufenach, On the detectability of ocean surface waves by real and synthetic aperture radar, *J. Geophys. Res.*, *86*, 6481–6498, 1981.

Alpers, W., C. Brüning, and K. Richter, Comparison of simulated and measured synthetic aperture radar image spectra with buoy-derived ocean wave spectra during the Shuttle Imaging Radar-B mission, *IEEE Trans. Geosci. Remote Sensing*, SIR-B Spec. Issue, GE-24, 559, 1986.

Beal, R. C., D. G. Tilley, and F. M. Monaldo, Large- and small-scale spatial evolution of digitally processed ocean wave spectra from Seasat Synthetic Aperture Radar, *J. Geophys. Res.*, *88*, 1761–1778, 1983.

Brüning, C., W. Alpers, L. F. Zambresky, and D. G. Tilley, Validation of a synthetic aperture radar ocean wave imaging theory by the Shuttle Imaging Radar-B experiment over the North Sea, *J. Geophys. Res.*, *93*, 15,403–15,425, 1988.

Brüning, K., W. Alpers, and K. Hasselmann, Monte Carlo simulation studies of the nonlinear imaging of a two-dimensional surface wave field by a synthetic aperture radar, *Int. J. Remote Sensing*, *11*, 1695–1727, 1990.

Donelan, M. A., J. Hamilton, and W. H. Hui, Directional spectra of wind-generated waves, *Philos. Trans. R. Soc. London, Ser. A*, *315*, 509–562, 1985.

Feindt, F., Radar-Rückstreuexperimente am Wind-Wellen-Kanal bei sauberer und filmbedeckter Wasseroberfläche im X-Band (9.8 GHz), Ph.D. thesis, Hamburg Univ., Germany, 1985.

Feindt, F., J. Schröter, and W. Alpers, Measurement of the ocean wave–radar modulation transfer function at 35 GHz from a sea-based platform in the North Sea, *J. Geophys. Res.*, *91*, 9701–9708, 1983.

Hasselmann, K., et al., Measurements of wind-wave growth and swell decay during the Joint North Sea Wave Project (JONSWAP), *Ergänzungsh. Dtsch. Hydrogr. Z., Reihe A(8)*, no. 12, 1973.

Hasselmann, K., D. B. Ross, P. Müller, and W. Sell, A parametric wave prediction model, *J. Phys. Oceanogr.*, *6*, 200–228, 1976.

Hasselmann, K., R. K. Raney, W. J. Plant, W. Alpers, R. A. Shuchman, D. R. Lyzenga, C. L. Rufenach, and M. J. Tucker, Theory of synthetic aperture radar ocean imaging: A MARSEN view, *J. Geophys. Res.*, *90*, 4659–4686, 1985.

Hasselmann, K., S. Hasselmann, E. Bauer, C. Brüning, S. Lehner, H. Graber, and P. Lionello, Development of a satellite SAR image spectra and altimeter wave height data assimilation system for ERS-1, *MPI Rep. 19*, Max-Planck-Inst. für Meteorol., Hamburg, Germany, 1988.

Hasselmann, K., S. Hasselmann, C. Brüning, and A. Speidel, Application of a new closed ocean wave–SAR spectral transform relation to LEWEX data, paper presented at Conference on measuring, modeling, predicting and applying directional ocean wave spectra, Johns Hopkins Univ., Baltimore, Md., April 18–20, 1989.

Hasselmann, K., S. Hasselmann, and K. Barthel, Use of a wave model as a validation tool for ERS-1 AMI wave products and as input for the ERS-1 wind retrieval algorithms, *MPI Rep. 55*, 96 pp., Max-Planck-Inst. für Meteorol., Hamburg, Germany, 1990.

Hasselmann, S., K. Hasselmann, J. H. Allender, and T. P. Barnett, Computations and parameterizations of the nonlinear energy transfer in a gravity wave spectrum, II, Parameterizations of the nonlinear energy transfer for application in wave models, *J. Phys. Oceanogr.*, *15*, 1378–1391, 1985.

Holthuijsen, L. H., Observations of the directional distribution of ocean-wave energy in fetch-limited conditions, *J. Phys. Oceanogr.*, *13*, 191–207, 1983.

Jackson, F. C., An analysis of short pulse and dual frequency radar techniques for measuring ocean wave spectra from satellites, *Radio Sci.*, *16*, 1385–1400, 1981.

Jackson, F. C., W. T. Walton, and P. L. Baker, Aircraft and satellite measurement of ocean wave directional spectra using scanning-beam microwave radars, *J. Geophys. Res.*, *90*, 987–1004, 1985a.

Jackson, F. C., W. T. Walton, and C. Y. Peng, A comparison of in situ and air-borne radar observations of ocean wave directionality, *J. Geophys. Res.*, *90*, 1005–1018, 1985b.

Keller, W. C., and J. W. Wright, Microwave scattering and the straining of wind-generated waves, *Radio Sci.*, *10*, 139–147, 1975.

Kenney, J. E., E. A. Uliana, and E. J. Walsh, The surface contour radar: A unique remote sensing instrument, *IEEE Trans. Microwave Theory Techniques*, MTT 27, 1080–1092, 1979.

Komen, G. J., S. Hasselmann, and K. Hasselmann, On the existence of a fully developed wind-sea spectrum, *J. Phys. Oceanogr.*, *14*, 1271–1285, 1984.

Lamb, H., *Hydrodynamics*, 6th ed., 738 pp., Dover, New York, 1932.

Lyzenga, D. R., Numerical simulation of synthetic aperture radar image spectra for ocean waves, *IEEE Trans. Geosci. Remote Sensing*, GE-24(6), 863–871, 1986.

Monaldo, F. M., and D. R. Lyzenga, On the estimation of slope- and height-variance spectra from SAR imagery, *IEEE Trans. Geosci. Remote Sensing*, GE-24, 543–551, 1986.

Monaldo, F. M., and D. R. Lyzenga, Comparison of shuttle imaging radar-B ocean wave spectra with linear model predictions based on aircraft measurements, *J. Geophys. Res.*, *93*, 374–388, 1988.

Pierson, W. J., Jr., and L. Moskowitz, A proposed spectral form for fully developed wind seas based on the similarity theory of S. A. Kitaigorodskii, *J. Geophys. Res.*, *69*, 5181–5190, 1964.

Plant, W. J., The microwave measurement of ocean-wave directional spectra, *Johns Hopkins APL Tech. Dig.*, *8(1)*, 55–59, 1987.

Plant, W. J., W. C. Keller, and A. Cross, Parametric dependence of the ocean wave radar modulation transfer function, *J. Geophys. Res.*, *88*, 9747–9756, 1983.

Raney, R. K., Wave orbital velocity, fade, and SAR response to azimuth waves, *IEEE J. Oceanic Eng.*, *OE-6*, 140–146, 1981.

Rosenthal, W., F. Ziemer, K. Raney, and P. Vachon, Removal of 180° ambiguity in SAR images of ocean waves, paper presented at IGARSS '89, Remote Sensing: An Economic Tool for the Nineties, Vancouver, Canada, July 10–14, 1989.

Schröter, J., F. Feindt, W. Alpers, and W. C. Keller, Measurement of the ocean wave–radar modulation transfer function at 4.3 GHz, *J. Geophys. Res.*, *91*, 932–946, 1986.

Swift, C. F., and L. R. Wilson, Synthetic aperture radar imaging of moving ocean waves, *IEEE Trans. Antennas Propag.*, *AP-27*, 725–729, 1979.

Tucker, M. J., The imaging of waves by satellite synthetic aperture radar: The effects of surface motion, *Int. J. Remote Sensing*, *6*, 1059–1074, 1985.

Vachon, P. W., and R. K. Raney, Resolution of ocean wave propagation direction in single-pass airborne SAR imagery, paper presented at IGARSS '89, Remote Sensing: An Economic Tool for the Nineties, Vancouver, Canada, July 10–14, 1989.

Valenzuela, G. R., An asymptotic formulation for SAR images of the dynamical ocean surface, *Radio Sci.*, *15*, 105–114, 1980.

Walsh, E. J., D. W. Hancock III, D. E. Hines, R. N. Swift, and J. F. Scott, Directional wave spectral measured with the surface contour radar, *J. Phys. Oceanogr.*, *15*, 566–592, 1985.

Walsh, E. J., D. W. Hancock III, D. E. Hines, R. N. Swift, and J. F. Scott, Wave-measurement capabilities of the surface contour radar and the air-borne oceanographic Lidar, *Johns Hopkins APL Tech. Dig.*, *8*(1), 74–81, 1987.

Walsh, E. J., D. W. Hancock III, D. E. Hines, R. N. Swift, and J. F. Scott, Evolution of the directional wave spectrum from shoreline to fully developed, *J. Phys. Oceanogr.*, in press, 1991.

WAMDIG, The WAM model—A third generation ocean wave predition model, *J. Phys. Oceanogr.*, *18*, 1775–1810, 1988.

Wright, J. W., A new model for sea clutter, 1968, *IEEE Trans. Antennas Propag.*, *AP-16*, 217–223, 1968.

K. Hasselmann and S. Hasselmann, Max-Planck-Institut für Meteorologie, Bundesstrasse 55, W-2000 Hamburg 13, Germany.

(Received August 14, 1990;
revised January 28, 1991;
accepted January 28, 1991.)

3.3   Stochastic Climate Model[5]

The difficulty in modelling the climate system is not only due to the variety of physical processes involved, but also to a large extent to the fact that the inter-acting components are characterised by rather different internal timescales: the atmosphere—several days, sea ice and the oceanic surface layer—several months, the deep ocean—several centuries, and the continental ice masses—many millennia. Even if all processes influencing climate variations were completely understood, the fact that the different sub-systems respond over different timescales would still cause considerable problems in numerical modelling.

All models of the individual sub-systems, such as atmosphere, ocean, or ice, with realistic geographical resolutions have been designed for a single timescale range so as to prognostically describe the typical fluctuations of these components. The influence of more rapid processes than the prognostic regime is parameterised by the prognostic variables using the temporal average over the rapid processes. Any components that vary on longer timescales than the prognostic regime are treated as constant boundary values or external parameters.

In many of the climate models used at in the 1960/70 s, the atmo-sphere was not explicitly included and was therefore placed in the model's statistical-diagnostic regime and was represented only through temporally averaged terms. However, in his seminal paper on *Stochastic Climate Models* published in Tellus in 1976 ([18]; see facsimile below) Klaus Hasselmann pointed out that the atmosphere's influence is not limited to these temporally averaged terms and that its variability must also be considered. This results in differential equations for the slow components of the climate system, which include stochastic forcing terms. These short-term atmospheric vari-ations cause (analogous to the Brownian motion) long-term fluctuations in the slow subsystems, which explains the observed red spectrum of the slow climate variables. The theory of Brownian motion has been discussed in many applications since Einstein's paper in 1905 but had not yet been applied to geophysical systems, such as the climate system.

---

[5] Prepared by Peter Lemke.

In his stochastic climate modelling approach, Hasselmann made use of a time-scale separation: a slowly varying dynamic climate variable under the influence of short-term atmospheric variations represented as white noise. Applications of stochastic climate models typically use linearized dynamic equations that describe small fluctuations around an equilibrium state. This approach represents a First Order Markov Process, which is characterised by a memory-term and white noise forcing, and it results in a red spectrum for the slow climate variables.

In two follow-up papers co-published in 1977 with Claude Frankignoul [39] and Peter Lemke,[6] the applicability of the concept was demonstrated through an analysis of sea surface temperatures and thermocline variability and with a global energy balance model. A large variety of different applications of this stochastic approach followed over the subsequent years.

One may of course ask if all this was really new. To some extent it was not, as certain ideas typically float around within the scientific community. In his interview, Hasselmann himself refers to J.M. Mitchell and his 1966-paper[7]: "*the same concept on the generation of different frequency domains of climate variability by the successive forcing of longer timescales by shorter timescales*". Another physicist thinking about such concepts was Chuck Leith, but in hindsight these approaches have not received much attention and did not cause the great epistemological step forward that Hasselmann's physical approach and construction did.

---

[6] Lemke, P., 1977: Stochastic climate models. Part 3. Application to zonally averaged energy models. Tellus 29, 385–392.

[7] Mitchell, J. M., Jr. 1966. Stochastic models of air- sea interaction and climatic fluctuation. (Symp. on the Arctic Heat Budget and Atmospheric Circulation, Lake Arrowhead, Calif., 1966.) Mem. RM-5233-NSF, The Rand Corp., Santa Monica.

Stochastic climate models

Part I. Theory

By K. HASSELMANN, *Max-Planck-Institut für Meteorologie, Hamburg, FRG*

(Manuscript received January 19; in final form April 5, 1976)

ABSTRACT

A stochastic model of climate variability is considered in which slow changes of climate are explained as the integral response to continuous random excitation by short period "weather" disturbances. The coupled ocean–atmosphere–cryosphere–land system is divided into a rapidly varying "weather" system (essentially the atmosphere) and a slowly responding "climate" system (the ocean, cryosphere, land vegetation, etc.). In the usual Statistical Dynamical Model (SDM) only the average transport effects of the rapidly varying weather components are parameterised in the climate system. The resultant prognostic equations are deterministic, and climate variability can normally arise only through variable external conditions. The essential feature of stochastic climate models is that the non-averaged "weather" components are also retained. They appear formally as random forcing terms. The climate system, acting as an integrator of this short-period excitation, exhibits the same random-walk response characteristics as large particles interacting with an ensemble of much smaller particles in the analogous Brownian motion problem. The model predicts "red" variance spectra, in qualitative agreement with observations. The evolution of the climate probability distribution is described by a Fokker-Planck equation, in which the effect of the random weather excitation is represented by diffusion terms. Without stabilising feedback, the model predicts a continuous increase in climate variability, in analogy with the continuous, unbounded dispersion of particles in Brownian motion (or in a homogeneous turbulent fluid). Stabilising feedback yields a statistically stationary climate probability distribution. Feedback also results in a finite degree of climate predictability, but for a stationary climate the predictability is limited to maximal skill parameters of order 0.5.

1. Introduction

A characteristic feature of climatic records is their pronounced variability. The spectral analysis of continuous climatic time series normally reveals a continuous variance distribution encompassing all resolvable frequencies, with higher variance levels at lower frequencies. Combining different data sources of various time scale and resolution (recorded meteorological data, varves, ice and sediment cores, global ice volume) the increase in spectral energy with decreasing frequency can be traced from the high frequency limit of climate variability (approximately 1 cycle per month, following the definitions adopted in GARP Publication 16, 1975) down to frequencies of order 1 cycle per 10^5 years (cf. GARP-US Committee Report (1975), Appendix A). An understanding of the origin of climatic variability, in the entire

spectral range from extreme ice age changes to seasonal anomalies, is a primary goal of climate research. Yet despite the long interest in the ice-age problem and the more recent intensification of climate research there exists today no generally accepted, simple explanation for the observed structure of climate variance spectra.

Various attempts have been made to link climatic changes to variable external factors such as the solar activity, secular changes of the orbital parameters of the earth, or the increased turbidity of the atmosphere following volcanic eruptions (cf. reviews in GARP Publication 16). A persistent difficulty with these investigations is that the postulated input–response relationships, if they exist, are not sufficiently pronounced to be immediately obvious on inspection of the appropriate time series. Thus a detailed statistical analysis is necessary, for which the data base is often only marginally adequate. Summaries of solar-climate relations extracted by statistical techniques may be found in King (1975) and Wilcox (1975); a critical analysis of the statistical significance of some of the claimed correlations has been given by Monin & Vulis (1971).

Climate variations have also often been discussed in terms of internal atmosphere–ocean–cryosphere–land feed-back mechanisms. Positive feedback amplifies the response of the system to changes in the external parameters and, if sufficiently strong, can produce unstable spontaneous transitions from one climate state to another. Feedback mechanisms have generally been formulated in terms of highly simplified energy-budget models containing only a few "climate" variables, such as the zonally averaged surface temperatures, the area of the ice sheets and the albedo of the earth's surface. A basic difficulty of unstable feedback models (apart from— or possibly because of—their high degree of idealization) is that they tend to predict climatic variations as flip-flop transitions and therefore fail to reproduce the observed continuous spectrum of climatic variability.

In this paper an alternative model of climate variability is investigated which predicts the basic structure of climatic spectra without invoking internal instabilities or variable external boundary conditions. The variability of climate is attributed to internal random forcing by the short time scale "weather" components of the system. Slowly reponding components of the system, such as the ice sheets, oceans, or vegetation of the earth's surface, act as integrators of this random input much in the same way as heavy particles imbedded in an ensemble of much lighter particles integrate the forces exerted on them by the light particles. If feedback effects are ignored, the resultant "Brownian motion" of the slowly responding components yields r.m.s. climate variations—relative to a given initial state—which increase as the square root of time. In the frequency domain, the climate variance spectrum is proportional to the inverse frequency squared. The non-integrable singularity of the spectrum at zero frequency is consistent with the non-stationarity of the process. The spectral analysis for a finite-duration record yields a finite peak at zero frequency proportional in energy to the duration of the record.

In order to obtain a statistically stationary response, stabilising negative feedback processes must be invoked. Thus from the viewpoint of the present model, the problem of climate variability is not to discover positive feedback mechanisms which enhance the small variations of external inputs or produce instabilities, but rather to identify the negative feedback processes which must be present to balance the continual generation of climatic fluctuations by the random driving forces associated with the internal "weather" interactions.

Following the derivation of the random-walk characteristics of a stochastically driven climate system in Sections 2 and 3, the basic Fokker-Planck equation governing the evolution of such a system is presented in Section 4. Special solutions for a system with linear feedback are given in Section 5, and the results are then applied to the analysis of climate predictability in Section 6.

Some of the concepts underlying the present stochastic model have been expressed previously by Mitchell (1966) in his investigation of sea-surface temperature (SST) anomalies. An application of the present model to SST data and to temperature fluctuations in the seasonal thermocline is given in Part 2 of this paper (Frankignoul & Hasselmann, 1976). In Part 3, the effect of introducing stochastic forcing into simple statistical dynamical models of the Budyko-Sellers type is investigated (Lemke, 1976).

2. Relationship between GCM's, SDM's and stochastic forcing models

It is useful to introduce a formal notation which is independent of the individual model structure. Let the instantaneous state of the complete system atmosphere–ocean–cryosphere–land be described by a finite set of discrete variables $z = (z_1, z_2, ...)$. The state vector z may be taken to represent the fields of density, velocity, temperature, etc. of the various media, as defined at discrete grid points and levels, or as given by the coefficients of some suitably truncated functional expansion. The evolution of the system will then be described by a series of prognostic equations

$$\frac{dz_i}{dt} = w_i(z) \tag{2.1}$$

where w_i is a known (in general complicated nonlinear) function of z. For the following we ignore the parameterization problems associated with the projection of the complete system on to a finite set of parameters; we assume that for our purposes the prognostic eqs. (2.1) accurately describe the evolution of the system for all times of interest.

A basic assumption of most models is that the complete system z can be divided into two subsystems, $z = (x, y)$, which are characterised by strongly differing response times τ_x, τ_y. Thus writing eq. (2.1) in terms of the two subsystems,

$$\frac{dx_i}{dt} = u_i(x, y) \tag{2.2}$$

$$\frac{dy_i}{dt} = v_i(x, y) \tag{2.3}$$

it is assumed that

$$O\left(x_i\left(\frac{dx_i}{dt}\right)^{-1}\right) = \tau_x \ll \tau_y = O\left(y_i\left(\frac{dy_i}{dt}\right)^{-1}\right) \tag{2.4}$$

The fast responding components x_i may be identified with the normal prognostic "weather" variables used in deterministic numerical weather prediction or General Circulation Models (GCM's), whereas the slowly responding "climate" variables y_i may be associated with variables such as the sea surface temperature, ice coverage, land foliage, etc. which are normally set constant in weather prediction models but represent essential prognostic variables on climatic time scales. τ_x is typically of the order of a few days, whereas most climate variables have response scales τ_y of the order of several months, years or longer. Thus the inequality (2.4) is generally well satisfied.

With presently available computers it is not possible to integrate the complete coupled system (2.2)–(2.3) over periods of climatic time scale $O(\tau_y)$. High resolution GCM's are normally used to integrate the subset of equations (2.2) over an intermediate period τ_i in the range $\tau_x < \tau_i < \tau_y$ for which the "climatic" variables can be regarded as constant, but which is still sufficiently long to define the statistics of the weather variables x for a given climatic state y. Thus although GCM's provide important information for climate studies, they are not suitable for the simulation of climate variability as such.

Dynamical investigations of climate variability have been based in the past largely on Statistical Dynamical Models (SDM's), which address the subset of eqs. (2.3). In the usual approach it is argued that for the time scales τ_y of interest in (2.3), the rapidly fluctuating terms in the prognostic equations can be ignored, so that (2.3) can be averaged over the period τ_i, thereby removing the weather fluctuations while still regarding y in the right hand side of (2.3) as constant,

$$\frac{dy_i}{dt} = \langle v_i(x, y)\rangle \tag{2.5}$$

Formally, it will be more convenient in the following to regard the average $\langle ... \rangle$ as an ensemble average over a set of realisations x for given y. It is assumed that ergodicity holds, so that ensemble averaging and time averaging are equivalent.

Since v_i is in general a nonlinear function of x, the average rate of change $\langle v_i \rangle$ of y_i will depend on the statistical properties of x as well as on y. To close the problem, the statistics of x must therefore be expressed in terms of y through the introduction of some closure hypothesis. For example, in zonally averaged energy budget models of the Budyko (1969)–Sellers (1969) type the meridional heat fluxes by standing and transient eddies must be parameterised in terms of the mean meriodional temperature distributions.

Although this class of model may be termed statistical in the sense that an averaging operation and a statistical closure hypothesis are involved, the reduced eq. (2.5) is in fact deterministic rather than statistical. It is known that the asymptotic solutions of nonlinear deterministic equations containing a relatively small number of degrees of freedom can already exhibit non-periodic, random-type oscillations similar in character to observed weather or climate fluctuations (cf. Lorenz, 1965). However, simple models with these features appear to have been investigated primarily in relation to weather simulation. Most of the better known simple SDM's predict a unique, time-independent asymptotic state for any given initial state. These models appear inherently incapable of generating internally time variable solutions with continuous variance spectra, as required by observation. In the past climate variability

476   K. HASSELMANN

has therefore been explained in the framework
of classical SDM's as the response of the system
(2.5) to variations of external boundary conditions, such as the solar radiation and the turbidity of the atmosphere, rather than through
internal interactions.

By a natural extension of the SDM, however,
one can obtain an alternative climatic model
which yields continuous variance spectra with
the observed "red" distribution directly through
internal interactions. (This, of course, does not
exclude the possible significance of additional
externally induced climatic changes). Returning to eq. (2.3), let $\delta y = y(t) - y_0$ denote the
change of the climate state relative to a given
initial state $y(t=0) = y_0$ in a time $t < \tau_y$ sufficiently small that y can still be regarded as
constant in the forcing term on the right hand
side of the equation. The change may be divided
into mean and fluctuating terms, $\delta y = \langle \delta y \rangle + y'$
where the ensemble average is taken here over
all x states for fixed y_0 (not y). The mean change
$\langle \delta y \rangle$ follows from (2.5),

$$\langle \delta y_i \rangle = \langle v_i \rangle t \qquad (2.6)$$

(for this term it is irrelevant whether the average refers to fixed y or y_0). The rate of change
of the fluctuating term is given by

$$\frac{dy_i'}{dt} = v_i(\mathbf{x}, \mathbf{y}) - \langle v_i \rangle = v_i' \qquad (2.7)$$

where $\langle v_i' \rangle = 0$ and $y_i' = 0$ for $t = 0$.

The statistics of $v_i'(t)$ are defined through the
statistics of the weather variables $\mathbf{x}(t)$ for given
y_0. It is assumed that $\mathbf{x}(t)$, and therefore $\mathbf{v}(t)$,
represents a stationary random process.

Equation (2.7) is identical to the equations
describing the diffusion of a fluid particle in a
turbulent fluid, where y_i' represents the coordinate vector of the particle and v_i' the turbulent
(Lagrangian) velocity It is well known from
this problem (Taylor, 1921, Hinze, 1959) that
for statistically stationary v_i', the integration of
(2.7) yields a non-stationary process y_i', the covariance matrix $\langle y_i' y_j' \rangle$ growing linearly in time
t for $t \gg \tau_x$. Taylor pointed out in his original
paper that this result could be interpreted
physically as the continuum-mechanical analogy to normal molecular diffusion or to Brownian motion. In fact, for $t \gg \tau_x$ it is immaterial
for the (macroscopic) statistical properties of
y_i', involving time scales $\gg \tau_x$, whether the forcing is continuous or discontinuous.

The nonstationary response y_i' to stationary
random forcing v_i' in the stochastic model implies that climate variations would continue to
grow indefinitely if feedback effects were ignored. These, of course, will begin to become
effective as soon as the integration is carried
into the region $t = O(\tau_y)$. The properties of the
random walk model in the ranges $t < \tau_y$ and $t =
O(\tau_y)$ will be discussed in more detail in the fol-
lowing sections.

The relationship between GCM's, SDM's and
stochastic forcing models may be conveniently
summarized in terms of the Brownian motion
analogy. The climate variables y and weather
variables x may be interpreted in the analogous particle picture as the (position and momentum) coordinates of large and small particles, respectively. The analysis of climate
variability in terms of SDM's is then equivalent
to determining the large-particle paths by considering only the interactions between the large
particles themselves and the *mean* pressure and
stress fields set up by the small-particle motions (plus the influence of variable external
forces). Numerical experiments with GCM's
correspond in this picture to the explicit computation of all paths of the small particles for
fixed positions of the large particles. Even if
the large particles were allowed to vary during
the computation, it would normally not be
feasible to carry the integrations sufficiently
far to consider appreciable deviations of the
large particles from their initial positions. Finally, the approach used in the stochastic forcing
model corresponds to the classical statistical
treatment of the Brownian motion problem, in
which the large-particle dispersion is inferred
from the statistics of the small particles with
which they interact. In contrast to the Brownian
motion problem, the variables x in the real
climate–weather system are, of course, not in
thermodynamic equilibrium, so that the statistical properties of x cannot be inferred from
the statistical thermodynamical theory of
energetically closed systems, but must be evaluated from numerical simulations with GCM's
(or from real data). A great reduction of computation is nevertheless achieved through a
statistical treatment, since relatively little statistical information on x is actually needed, and
this can be obtained from GCM experiments of
relatively short duration $\tau_i < \tau_y$.

At first sight it may appear surprising that

a statistical reduction of the complete climate-weather system is possible at all without arbitrary closure hypotheses, since one is accustomed to regarding systems involving turbulent geophysical fluid flows as basically irreducible, strongly nonlinear processes. The reduction in this case is a consequence of the time-scale separation (2.4). This property is lacking in the usual turbulent system. However, the condition is familiar from "weak-turbulence" theories for plasmas (cf. Kadomtsev, 1965) or from similar theories of weakly interacting random wave fields in solid state physics, high energy physics and in various geophysical applications (cf. Hasselmann, 1966, 1967). In essence, the property (2.4) enables statistical closure through the application of the Central Limit Theorem, whereby the response of a system is completely determined statistically by the second moments of the input if the forcing consists of a superposition of a large number of small, statistically independent pulses of time scale short compared with the response time of the system.

3. The local dispersion rate

For times t in the intermediate range $\tau_z < t < \tau_y$ the integration of (2.7) yields linearly increasing covariances in accordance with Taylor's (1921) relation

$$\langle y_i' y_j' \rangle = 2D_{ij}t \qquad (3.1)$$

where

$$D_{ij} = \tfrac{1}{2} \int_{-\infty}^{\infty} P_{ij}(\tau)\, d\tau \qquad (3.2)$$

and $P_{ij}(\tau) = \langle v_i'(t+\tau) v_j'(t) \rangle$ denotes the covariance function.

Physically, the dispersion mechanism may be interpreted as the response to a large number of statistically independent random changes $\Delta y_i = v_i' \cdot \Delta t$ induced in y_i at time increments Δt of the order of the integral correlation time of v_i'.

It is useful to represent the dispersion process also in the Fourier domain. Writing

$$v_i'(t) = \int_{-\infty}^{\infty} V_i(\omega)\, e^{i\omega t}\, d\omega \qquad (3.3)$$

the solution of (2.7) may be expressed as the Fourier integral

$$y_i'(t) = \int_{-\infty}^{\infty} Y_i(\omega)\, e^{i\omega t} d\omega - \int_{-\infty}^{\infty} Y_i(\omega)\, d\omega \qquad (3.4)$$

where

$$Y_i(\omega) = \frac{V_i(\omega)}{i\omega} \qquad (3.5)$$

The second, time independent term on the right hand side of (3.4) arises through the initial condition $y_i' = 0$ for $t = 0$.

For a stationary process, the Fourier components are statistically orthogonal,

$$\langle V_i(\omega)\, V_j^*(\omega') \rangle = \delta(\omega - \omega')\, F_{ij}(\omega)$$

where $F_{ij}(\omega)$ denotes the (two-sided) cross spectrum of v_i. The Fourier components $Y_i(\omega)$ are then also statistically orthogonal, and the cross spectrum of $y_i'(t)$ is given by

$$G_{ij}(\omega) = \frac{F_{ij}(\omega)}{\omega^2} \qquad (\omega \neq 0) \qquad (3.6)$$

The existence of a non-integrable singularity in G_{ij} at $\omega = 0$ is consistent with the non-stationarity of y_i'. The fact that the non-stationary contribution to y_i' is concentrated at zero frequency can be confirmed by evaluating the contribution to the covariance from a narrow band of frequencies $-\Delta\omega < \omega < \Delta\omega$ centered at zero frequency. Noting that the second integral in (3.4) represents a zero-frequency contribution, this is given by

$$\langle y_i' y_j' \rangle_{\Delta\omega} = \int_{-\infty}^{\infty} F_{ij}(\omega)\, \frac{2(1 - \cos \omega t)}{\omega^2}\, d\omega \qquad (3.7)$$

The weighting function $2(1 - \cos \omega t)/\omega^2$ has a maximum value equal to t^2 at $\omega = 0$ and a peak width proportional to $1/t$. Thus its integral is proportional to t, and in the limit of large t, as the peak becomes infinitely sharp, the function can be replaced by the δ-function expression

$$\frac{2(1 - \cos \omega t)}{\omega^2} \approx 2\pi t \delta(\omega) \qquad (\omega t \gg 1) \qquad (3.8)$$

For large t (3.7) therefore becomes

$$\langle y_i' y_j' \rangle = 2\pi t F_{ij}(0) \qquad (3.9)$$

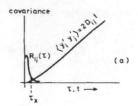

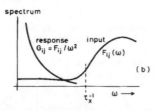

Fig. 1. Input and response functions of stochastically forced climate model without feed-back; (*a*) covariances, (*b*) spectra.

The subscript $\Delta\omega$ has now been dropped, since the contribution to $\langle y_i' y_j' \rangle$ from frequencies $|\omega| > \Delta\omega$ is constant and therefore becomes negligible compared with the nonstationary contribution for large t.

Equation (3.9) represents a special case of the resonant response of an undamped linear system to random external forcing. The general result for such systems states that the energy of the response is concentrated in spectral lines at the eigenfrequencies of the system, and that the energy of each line increases linearly with time at a rate proportional to the spectral density of the input at the eigenfrequency (cf. Hasselmann, 1967). Equation (3.9) corresponds to the case of a system with a single normal mode of frequency $\omega = 0$.

The equivalence of the expressions (3.1), (3.2) and (3.9) can be recognised using the Fourier transform relation

$$F_{ij}(\omega) = \frac{1}{2\pi} \int_{-\infty}^{\infty} P_{ij}(\tau) e^{-i\omega\tau} d\tau \qquad (3.10)$$

It follows from (3.10) that normally, for $F_{ij}(0) \neq 0$, the spectrum of any stationary process v_i becomes white (constant) for sufficiently small frequencies (in other words, one need consider only the first term of the Taylor ex-

pansion of the spectrum). Generally, there exists some cut-off time lag $O(\tau_x)$ such that $P_{ij}(\tau) \approx 0$ for $\tau > \tau_x$. For frequencies $\omega < \tau_x^{-1}$, the exponential in (3.10) can then be set equal to one, so that $F_{ij}(\omega) \approx F_{ij}(0)$. In this range equation (3.6) may then be replaced by

$$G_{ij}(\omega) = \frac{F_{ij}(0)}{\omega^2} \qquad (\tau_y^{-1} \ll \omega \ll \tau_x^{-1}) \qquad (3.11)$$

The left side of the inequality follows from the restriction to integration times $t < \tau_y$, which limits the definition of the spectrum to frequencies large compared with τ_y^{-1}.

The main features of the random walk response in the time and frequency domain are indicated in Fig. 1.

In most climate applications the reponse will lie in the low frequency range $\omega < \tau_x^{-1}$ where the input spectrum can be regarded as white and equation (3.11) is applicable. For the generalization of the theory in the next section it is important to note that the constant level of the input spectrum at low frequencies can be determined from relatively short time series of the input, the record length required being governed by the time scale of the input, rather than the time scale of the response. The length of the time series need only be long enough to evaluate the covariance function for time lags up to the cut-off time lag of order τ_x. For example, in the problem of the generation of SST anomalies by random fluxes at the sea surface (considered in Part 2 of this paper), the statistical structure of the atmospheric input can normally be adequately determined from time series of a few weeks duration (ignoring the seasonal signal). From this the statistical properties of the random walk response according to (3.1), (3.2), and (3.11) can be evaluated for much longer time periods, of the order of several months. The upper limit $t = O(\tau_y)$ of the response time is determined ultimately by the breakdown of the uncoupled random walk model when internal feedback effects begin to come into play.

The dispersion coefficients D_{ij} can be inferred indirectly, without reference to weather data, from the rate of growth of the covariances $\langle y_i' y_j' \rangle$ as evaluated from climatic time series. Alternatively, if the stochastic forcing is known as a function of the weather variables, the zero frequency level of the spectral input can be determined directly from weather data. By either method, application of the random walk model, for example, to ice sheet data or SST anomalies indicates that the r.m.s. rate of divergence of climate from its present state by random weather forcing is considerable: without stabilising feedback the random walk model predicts that changes in the extent of the ice cover comparable with ice-age amplitudes would occur within time periods of the order of a century. The inclusion of feedback is thus essential for a realistic climate model. The generalisation to a model including arbitrary internal coupling is carried out in the next section.

4. The Fokker-Planck equation for a general stochastic climate model

The inequalities $\tau_x < t < \tau_y$ limiting the range of validity of the random walk model without feedback are characteristic of a two-timing theory. With respect to the rapidly varying components of the system the theory represents an asymptotic infinite-time limit, but at the same time the analysis is valid only for infinitesimal changes of the slowly varying components. The standard way of removing the restriction $t < \tau_y$ is to interpret the infinitesimal changes of the slowly varying components as *rates of change*, thereby obtaining a differential equation which is valid for all times, provided the original conditions on which the local theory was based continue to remain valid.

Since y represents a random variable, the appropriate differential equation should be formulated for the probability density distribution $p(y, t)$ of climatic states in the climatic phase space y. For a system in which the mean value and covariance tensor of the infinitesimal changes $\delta y_i = y_i(t) - y_{i,0}$ in an infinitesimal time interval $\delta t < \tau_y$ are both proportional to δt (the effects of the higher moments can be shown to be small on account of the two-timing condition (2.4)) the evolution of the probability distribution $p(y, t)$ is governed by a Fokker-Planck equation (cf. Wang and Uhlenbeck, 1945)

$$\frac{\partial p}{\partial t} + \frac{\partial}{\partial y_i}(\hat{v}_i p) - \frac{\partial}{\partial y_i}\left(D_{ij}\frac{\partial p}{\partial y_j}\right) = 0 \qquad (4.1)$$

where

$$D_{ij} = \frac{\langle y_i' y_j'\rangle}{2\delta t} = \pi F_{ij}(0) \qquad (4.2)$$

with $y_i' = \delta y_i - \langle \delta y_i\rangle$ as before,

and $\hat{v}_i = \langle \delta y_i\rangle/\delta t - \partial D_{ij}/\partial y_j$ or, from (2.6) and (3.1), (3.9)

$$\hat{v}_i = \langle v_i\rangle - \pi \frac{\partial}{\partial y_j} F_{ij}(0) \qquad (4.3)$$

Provided the two-scale approximation remains valid, eq. (4.1) describes the evolution of an ensemble of climatic states with an arbitrary initial distribution for arbitrary large times. The propagation and diffusion coefficients $\hat{v}_i$, D_{ij} will generally be functions of y, both directly and through their dependence on the statistical properties of the weather variables x. The equation includes both direct internal coupling through the propagation term $\hat{v}_i$ and indirect feedback through the dependence of the diffusion coefficients on the climatic state.

In practice, the expectation values and spectra in (4.2) and (4.3), defined as averages over an x-ensemble for fixed y, will normally be determined from time averages, rather than through ensemble averaging. In order that the average values can be regarded as local with respect to the climatic time scale τ_y but still remain adequately defined statistically with respect to the weather variability of time scale τ_x, the averaging time T must satisfy the two-sided inequality $\tau_x < T < \tau_y$. The inequalities imply that the spectral density $F_{ij}(0)$ at "zero frequency" in eqs. (4.2), (4.3) must be interpreted more accurately as the level of the spectrum in the frequency range $\tau_y^{-1} < \omega < \tau_x^{-1}$—as was already pointed out in connection with eq. (3.11). The variance spectra of v_i for lower frequencies $\omega = O(\tau_y^{-1})$ must be attributed, within the framework of the two-timing theory, to the slow variations of the *mean* variables $\langle v_i\rangle$ on the climatic time scale. Since $\langle v_i\rangle$ depends on the local climatic state, the increase of the variance spectra of the climatic variables y_i towards lower frequencies will normally be associated with a corresponding increase of the variance spectra of v_i (and the "weather" variables x_i) in this range. This is not in conflict with the basic premise of a white input spectrum at "low" frequencies. Essential for the application of the two-timing concept is that there exists a spectral gap between the "weather" and "climate" frequency ranges in which the input spectra are flat (cf. Fig. 2).

480   K. HASSELMANN

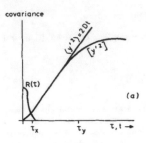

covariance

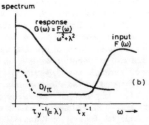

spectrum

Fig. 2. Input and response of stochastically forced (single component) climate model with linear feedback; (a) covariances, (b) spectra. In the ranges $\tau_x \ll \tau \ll \tau_y$ and $\tau_y^{-1} \ll \omega \ll \tau_x^{-1}$ the models with and without feedback are identical. In the range $\omega \lesssim \tau_y^{-1}$ the spectrum $F(\omega)$ cannot be regarded as part of the "weather input", but is coupled to the climate response.

The presence of the diffusion terms in (4.1) implies that climate evolution is necessarily a statistical rather than a deterministic phenomenon. Even if a well defined climate state is prescribed initially in the form of a δ-function distribution for p, the diffusion term immediately leads to a finite spread of the probability distribution p at later times. Without the diffusion term, an initial δ-function distribution would retain its δ-function character and simply propagate along the characteristics $dy_i/dt = \hat{v}_i$ in the climatic phase space.

The analytical integration of eq. (4.1) for an arbitrary nonlinear climate model with several degrees of freedom will normally not be possible. However, solutions can be constructed, for example, by the Monte Carlo method, in which eq. (2.5) and (2.7) are integrated numerically (without the restriction $t \ll \tau_y$) for an ensemble of realisations using an appropriate statistical simulation of v_i'. Within the approximations of

the two-timing theory, v_i' can be represented very simply as a zero'th order Markov process.

For the special case of linear feedback and constant diffusion coefficients, equation (4.1) can be solved explicitly. These solutions are appropriate for climatic systems with small excursions. However, several properties of the linear case discussed in the following two sections may also be expected to apply qualitatively to more general climate models.

Although eq. (4.1) describes the evolution of $p(\mathbf{y}, t)$ in closed form (given the x-statistics for given $\mathbf{y}$), the probability distribution $p(\mathbf{y}, t)$ provides only a partial statistical description of the random process $y(t)$. A complete statistical description would require, for example, the set of joint probability distributions $p(\mathbf{y}_1, ..., \mathbf{y}_p)$ of the climate states for any set of times $t_1, ..., t_p$, or the set of all moments $\langle y_1 ... y_p \rangle$ for all $p > 0$. Generalised Fokker-Planck equations similar to (4.1) can be derived also for multi-time probability distributions, but these will normally be of less immediate interest. In practice, Monte Carlo methods of solving (4.1) actually generate the complete statistics of the process $\mathbf{y}$, as well as yielding $p(\mathbf{y}, t)$, so that the generalised Fokker-Planck equations need not be considered explicitly.

5. Linear feed-back models

(a) *Solution of the Fokker-Planck equation*

For small excursions of the climatic states about an equilibrium state $\mathbf{y} = 0$, say, the diffusion and velocity coefficients in (4.1) can be expanded with respect to $\mathbf{y}$. Since the feedback terms must vanish for the equilibrium state, the coefficients are given to lowest order by

$$D_{ij} = \text{const} \tag{5.1}$$

$$\hat{v}_i = V_{ij} y_j, \quad V_{ij} = \text{const} \tag{5.2}$$

For a stable equilibrium state, the matrix V_{ij} must be negative definite.

The general solution of (4.1) for an arbitrarily prescribed initial distribution $p(\mathbf{y}, t = 0) = p_0(\mathbf{y})$ may be constructed by superposition from the Green-function solution for an initial δ-function distribution $p_0(\mathbf{y}) = \delta(y_1 - y_{10}) ... \delta(y_n - y_{n0})$ at an arbitrary point $\mathbf{y}_0$. This is given by the normal distribution

$$p(\mathbf{y}, t) = (2\pi)^{-n/2} |R|^{-1/2}$$

$$\times \exp\left(-\frac{R_{ij}^{-1}}{2} (y_i - [y_i])(y_j - [y_j]) \right) \quad (5.3)$$

where the mean $[y_i]$ and covariance tensor $R_{ij} = [(y_i - [y]) \cdot (y_j - [y_j])]$ are time dependent functions satisfying the differential equations and initial conditions

$$\frac{d[y_i]}{dt} = V_{ik}[y_k], \quad [y_i] = y_{i0} \text{ for } t = 0 \quad (5.4)$$

$$\frac{dR_{ij}}{dt} = 2D_{ij} + R_{ik}V_{jk} + R_{kj}V_{ik}, \quad R_{ij} = 0 \text{ for } t = 0 \quad (5.5)$$

The square parentheses [ ] denote averages over the ensemble of climatic states $\mathbf{y}$. Equations (5.4), (5.5) can be verified by substitution of (5.3) in (4.1) or can be derived directly from (2.5), (4.1), (4.2) and (4.3). In matrix notation, the solutions may be written

$$[\mathbf{y}] = e^{Vt}\mathbf{y}_0 \quad (5.6)$$

$$R = R_\infty - e^{Vt}R_\infty e^{\overset{+}{Vt}} \quad (5.7)$$

where V^+ denotes the transpose of V and R_∞ is the asymptotic stationary solution of (5.5),

$$2D_{ij} + (R_\infty)_{ik}V_{jk} + (R_\infty)_{kj}V_{ik} = 0 \quad (5.8)$$

R_∞ and the corresponding asymptotic equilibrium distribution p_∞ (with $[\mathbf{y}]_\infty = 0$) are independent of the initial state $\mathbf{y}_0$.

The expressions become particularly simple if the matrix V is diagonal, $V_{ij} = \delta_{ij}\lambda_{(i)}$ (parentheses around the index indicate that the index is excluded from the summation convention). Normally, this can be achieved by a suitable linear transformation of $\mathbf{y}$ to new coordinates. Equations (5.6), (5.7) then become

$$[y_i] = y_{i0}\exp(\lambda_{(i)}t) \quad (5.9)$$

$$R_{ij} = (R_\infty)_{ij}[1 - \exp(\lambda_{(i)} + \lambda_{(j)})t] \quad (5.10)$$

$$(R_\infty)_{ij} = -\frac{2D_{ij}}{\lambda_{(i)} + \lambda_{(j)}} \quad (5.11)$$

(b) *Spectral decomposition of the variance*

The Gaussian form (5.3) of the probability distribution $p(\mathbf{y}, t)$ could have been inferred

directly from the Central Limit Theorem, without invoking the Fokker-Planck equation. The theorem states that, under very general conditions, the response of a linear system driven by a statistically stationary input consisting of a continuous sequence of infinitely short, statistically independent pulses is Gaussian, independent of the detailed statistical structure of the input. This property holds not only for the probability distribution p, but generally for the multi-time joint probability distribution. Thus the statistical structure of the process $\mathbf{y}$ is completely specified if the first moments (given by (5.6)) and the second moments

$$\hat{S}_{ij}(t + \tau) = [(y_i(t + \tau) - [y_i(t + \tau)]) \cdot (y_j(t) - [y_j(t)])] \quad (5.12)$$

are known.

The latter are given by the solution

$$\hat{S}(t, \tau) = e^{V\tau}R(t) \quad (\tau > 0) \quad (5.13)$$

of the differential equation

$$\frac{\partial \hat{S}_{ij}(t, \tau)}{\partial \tau} = V_{ik}\hat{S}_{kj} \quad (\tau > 0) \quad (5.14)$$

under the initial condition $\hat{S}_{ij}(t, \tau = 0) = R_{ij}(t)$, with $R_{ij}(t)$ given by (5.7). Equation (5.14) follows from (2.5), (2.7) and (5.2), noting that in the two-timing limit $v_i'(r + \tau) = (v_i(t + \tau) - \langle v_i(t + \tau)\rangle)$ and $y_j(t)$ are statistically uncorrelated for $\tau > 0$, since the correlation time scale of the random forcing is regarded as infinitely short compared with the correlation time scale of the response. This argument does not hold for $\tau < 0$, since $y_j(t)$ in this case includes the response to v_i' at the earlier time $t + \tau$. However, the solution for $\tau < 0$ can be obtained from (5.13) by interchanging the indices and redefining the time variables.

Of particular interest is the asymptotic stationary solution

$$S(\tau) = \lim_{t \to \infty} S(t, \tau) = e^{V\tau}R_\infty \quad (5.15)$$

which can be compared with the statistical properties of observed, quasi-stationary climatic time series. If the second moments of the input (i.e. D_{ij}) are specified, it is known from linear systems analysis that $S(\tau)$ completely

482   K. HASSELMANN

determines the linear response characteristics (transfer functions) of the system.

The relation corresponding to (5.15) for the climate cross spectrum G_{ij} can best be derived by direct substitution of the Fourier integral representation (3.3) in the basic climate equation

$$\frac{dy_i}{dt} = V_{ij}y_j + v_i'$$

One obtains

$$G_{ij}(\omega) = T_{ik}T_{jl}^* F_{kl}(0) \tag{5.16}$$

where $T = (i\omega I - V)^{-1}$ (I = unit matrix). For diagonal V, eq. (5.16) becomes

$$G_{ij}(\omega) = \frac{F_{ij}(0)}{(\omega - i\lambda_{(i)})(\omega + i\lambda_{(j)})} \tag{5.17}$$

Equations (5.15), (5.16) may be compared with the corresponding relations (3.1), (3.11) for a system without feedback. The deviation covariance $\langle y_i' y_j' \rangle$ considered in section 3 should be compared in the case of a stationary y-process with the expression $[y_i' y_j'] = [(y_i - y_{i,0})(y_j - y_{j,0})]$ (also known as the "structure function", cf. Tatarski (1961)). This can be expressed in terms of the covariance function as

$$[y_i' y_j'] = (S_{ij}(0) - S_{ij}(\tau)) + (S_{ji}(0) - S_{ji}(\tau)) \tag{5.18}$$

The general form of the functions G_{ij}, $\langle y_i' y_j' \rangle$ and $[y_i' y_j']$ for a system with and without linear feed-back is shown in Fig. 2. For $\tau_x < \tau < \tau_y$ and $\tau_y^{-1} < \omega < \tau_x^{-1}$ the behaviour of both systems is identical, but for $\tau \sim O(\tau_y)$ and $\omega = O(\tau_y^{-1})$ the unbounded response of the system without feedback begins to diverge from the bounded response functions of the linearly stabilised system.

6. Climate predictability

The evolution of the probability distribution $p(\mathbf{y}, t)$ as governed by the Fokker-Planck equation (4.1) determines the degree of climate predictability. If the climate state $\mathbf{y}_0$ at time $t = 0$ is known, the initial probability distribution p_0 is a δ-function. For a fully predictable system, $p(\mathbf{y}, t)$ remains a δ-function for all times $t > 0$. As pointed out in Section 4, however, the dif-

fusive term in (4.1) results in a broadening of the probability distribution for $t > 0$, and climate prediction therefore always entails some degree of statistical uncertainty.

A simple quantitative measure of the predictive skill can be defined in terms of the mean climatic state $[y_i]$ and the covariance matrix $R_{ij} = [(y_i - [y_i]) \cdot (y_j - [y_j])]$. The mean may be regarded as the climate "prediction". (In the case of a linear system, this is identical with the most probable climatic state, but in general the most probable state and the mean state will differ.) In order to introduce a measure of skill as a simple number, the distance δ_1 of the predicted climate state from the initial state and the r.m.s. deviation ε from the mean must be defined in terms of some suitable positive definite matrix M_{ij},

$$\delta_1 = \{M_{ij}([y_i] - y_{i,0})([y]_j - y_{j,0})\}^{\frac{1}{2}} \tag{6.1}$$

$$\varepsilon = \{M_{ij}R_{ij}\}^{\frac{1}{2}} \tag{6.2}$$

The usual definition of the skill parameter is then given by the ratio "signal to signal-plus-noise",

$$s_1 = \frac{\delta_1}{(\varepsilon_1^2 + \delta_1^2)^{1/2}} \tag{6.3}$$

For small times $t < \tau_y$, the predicted change δ_1 increases linearly with time

$$\delta_1 \approx (M_{ij}v_{i,0}v_{j,0})^{\frac{1}{2}}t \tag{6.4}$$

whereas the r.m.s. error grows as $t^{\frac{1}{2}}$,

$$\varepsilon \approx (2D_{ij}M_{ij})^{\frac{1}{2}}t^{\frac{1}{2}} \tag{6.5}$$

Thus initially the skill parameter $s_1 \sim t^{\frac{1}{2}}$; the random deviations from the initial state induced by the stochastic forcing dominate over the deterministic changes produced by the internal coupling within the climatic system, and the predictive skill is small.

For very large t, δ_1 and ε will normally approach the limiting values

$$\delta_{1\infty} = \{M_{ij}([y_i]_\infty - y_{i,0})([y_j]_\infty - y_{j,0})\}^{\frac{1}{2}}$$

$$\varepsilon_\infty = (M_{ij}(R_\infty)_{ij})^{\frac{1}{2}}$$

appropriate to the stationary equilibrium distribution $p_\infty(\mathbf{y})$—assuming such a distribution

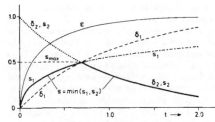

Fig. 3. Predicted climate changes δ_1 relative to initial state and δ_2 relative to asymptotic state, statistical error ε, and skill parameters s_1, s_2 and $s = \min(s_1, s_2)$, for a linear (single component) climate system. The initial value is chosen as $y_0 = (R_\infty)^{\frac{1}{2}} = 1$ (in this case s_2 and δ_2 happen to coincide).

exists—and the skill parameter s_1 will become constant.

The predicted climatic state for large t is simply the stationary climatic mean state $[y]_\infty$. This prediction may be regarded as trivial in the same way as the prediction through persistence for small t is trivial. Since the contribution from straight persistence was subtracted in the definition of s_1, it apears more appropriate to introduce an alternative skill parameter

$$s_2 = \delta_2/(\varepsilon^2 + \delta_2^2)^{1/2} \qquad (6.6)$$

for large t, where

$$\delta_2 = \{M_{ij}([y_i] - [y_i]_\infty)([y_j] - [y_j]_\infty)\}^{\frac{1}{2}} \qquad (6.7)$$

is the deviation of the predicted climatic state from the stationary climatic mean. The net skill parameter may then be defined as $s = \min (s_1, s_2)$.

The behaviour of $s(t)$ in the intermediate range $t = O(\tau_y)$ between the limiting regions in which either s_1 or s_2 is very small depends in detail on the structure of the climate model. The general properties of $s(t)$ to be expected in this range may be inferred, however, from the solution for a linear system, cf. Fig. 3. Provided the initial deviation from the stationary climatic mean is of the same order as the variability of the stationary asymptotic distribution (for each degree of freedom separately), the maximal value of the net skill parameter generally lies in the neighbourhood of 0.5. This is due to the fact that the relaxation times for δ_1 and ε are of the same magnitude, since both are governed

by the same internal feedback processes. Thus both δ_1 and ε increase at approximately the same rate (after the initial period $t < \tau_y$), and the non-trivial (i.e. non-persistent) component of the prediction and the statistical error always remain of comparable magnitude.

These results may be expected to carry over, at least qualitatively, to nonlinear systems, provided there exists a unique stationary equilibrium distribution—i.e. provided the system is transitive in Lorenz' (1968) sense. In fact, the basic properties of the skill parameters s_1, s_2 outlined above are largely independent of the detailed dynamics of the climate system and follow simply from the fact that the evolution of the system corresponds to a first-order Markov process. The prediction problem becomes more complex in the case of intransitive systems, in which more than one stationary distribution may exist (for example, for dynamically disconnected regions of the climate phase space) or for nearly intransitive systems, characterised by two or more quasi-stationary, weakly interacting distributions. However, the discussion of these more complex cases must necessarily remain rather academic without reference to a specific climate model and will not be pursued further here.

7. Conclusions

The principal features of the stochastic climate model discussed in this paper may be summarised as follows:

(1) The time scales of the "weather system" and "climate system" are well separated.

(2) As a consequence of the time-scale separation, the response of the climate system to the random forcing by the weather components can be described as a continuous random walk or diffusion process (first-order Markov process). The response can be completely characterised by a diffusion tensor, which is proportional to the constant spectral density of the random forcing at low frequencies.

(3) The evolution of the climate system is described by a Fokker-Planck equation for the climate probability distribution; the propagation and diffusion coefficients of the equation depend on the instantaneous climate state, both directly and via the weather statistics.

(4) Without stabilising internal feedback

484   K. HASSELMANN

mechanisms, climate variability would grow indefinitely.

(5) Despite the stochastic nature of climate
variability, the internal feedback terms in climate models imply a finite degree of predictability. However, the maximal predicitive skill
for a statistically stationary climate system is
generally no larger than 0.5 and is always
significantly less than unity.

The discussion in this part of the paper has
been restricted to the general structure of
stochastic models, without reference to a specific model. It should be pointed out, however,
that the extension of a typical SDM of, say,

the Budyko-Sellers type to a stochastic model
requires no basic modification of the internal
structure of the model, but simply the addition
of random driving terms. The relevant statistical properties of the stochastic forcing functions
can be obtained directly from numerical experiments with GCM's or from meteorological
data. Thus some of the general properties of
stochastic climate models described in this
paper can be tested rather easily by comparing
observed climatic variability with theoretical
predictions obtained with existing SDM's after
incorporation of appropriate stochastic forcing
terms (Lemke, 1976).

REFERENCES

Budyko, M. I. 1969. The effect of solar radiation
    variations on the climate of the earth. *Tellus 21*,
    611–619.
Frankignoul, C. & Hasselmann, K. 1976. Stochastic
    climate models. Part 2, Application to sea-sur-
    face temperature anomalies and thermocline
    variability (in preparation).
GARP US Committee Report, 1975. Understand-
    ing climate change. A programme for action.
    Nat. Acad. Sciences, Wash.
GARP Publication 16, 1975. The physical basis of
    climate and climate modelling. World Met. Or-
    ganiz., Internat. Council Scient. Unions.
Hasselmann, K. 1966. Feynman diagrams and in-
    teraction rules of wave–wave scattering processes.
    Rev. Geophys. 4, 1–32.
Hasselmann, K. 1967. Non-linear interactions treat-
    ed by the methods of theoretical physics (with
    application to the generation of waves by wind).
    Proc. Roy. Soc. A 299, 77–100.
Hinze, J. O. 1959. *Turbulence*. McGraw-Hill.
Kadomtsev, B. B. 1965. *Plasma turbulence*. Aca-
    demic Press.
King, J. W. 1975. Sun–weather-relationships. *Aero-
    nautics and Astronautics 13*, 10–19.
Lemke, P. 1976. Stochastic climate models. Part 3,
    Application to zonally averaged energy models
    (in preparation).

Lorenz, E. N. 1965. A study of the predictability
    of a 28-variable atmospheric model. *Tellus 17*,
    321–333.
Lorenz, E. N. 1968. Climate determinism. *Meteor.
    Monographs 8*, 1–3.
Mitchell, J. M., Jr. 1966. Stochastic models of air-
    sea interaction and climatic fluctuation. (Symp.
    on the Arctic Heat Budget and Atmospheric
    Circulation, Lake Arrowhead, Calif., 1966.) Mem.
    RM-5233-NSF, The Rand Corp., Santa Monica.
Monin, A. S. & Vulis, I. L. 1971. On the spectra of
    long-period oscillations of geophysical parameters.
    Tellus 23, 337–345.
Sellers, W. D. 1969. A global climate model based
    on the energy balance of the earth–atmosphere
    system. *J. Appl. Met. 8*, 392–400.
Tatarski, V. I. 1961. *Wave propagation in a turbulent
    medium*. McGraw-Hill.
Taylor, G. I. 1921. Diffusion by continuous move-
    ments. *Proc. Lond. Math. Soc. 20*, 196.
Wang, M. C. & Uhlenbek, G. E. 1945. On the theory
    of the Brownian motion. *Rev. Mod. Phys. 17*,
    323–342.
Wilcox, J. M. 1975. Solar activity and the weather.
    J. Atmosph. Terrestr. Phys. 37, 237–256.

СТОХАСТИЧЕСКИЕ МОДЕЛИ КЛИМАТА

Рассматривается стохастическая модель изменчивости климата, в которой медленные
изменения климата объясняются как ин-
тегральная реакция на непрерывное случай-
ное возбуждение короткопериодными «по-
годными» возмущениями. Взаимодействую-
щая система океан–атмосфера–криосфера–
суша разделяется на быстро изменяющуюся
«погодную» систему (атмосфера) и на медленно
откликающуюся «климатическую» систему
(океан, криосфера, растительность суши и
т. д.). В обычной статистически-динамической
модели (СДМ) только средние эффекты пере-

носа быстро меняющихся погодных компо-
нент параметризуются в климатической си-
стеме. Результирующие прогностические
уравнения детерминистичны и климатические
вариации обычно могут возникать только
при изменении внешних условий. Существен-
ной особенностью стохастических климати-
ческих моделей является то, что неосреднен-
ные «погодные» компоненты также сохра-
няются. Формально они появляются как
случайные вынуждающие силы. Климати-
ческая система, действующая как интегратор
этого короткопериодного возбуждения, про-

являет те же самые характеристики реакции случайного блуждания, как крупные частицы, взаимодействующие с ансамблем гораздо более мелких частиц в аналогичной задаче броуновского движения. Модель предсказывает «красные» спектры изменений параметров в качественном согласии с наблюдениями. Эволюция распределения вероятностей климата описывается уравнением Фоккера–Планка, в котором эффект случайного погодного возбуждения описывается диффузионными членами. Без стабилизирую-

щей обратной связи модель предсказывает непрерывное увеличение изменчивости климата по аналогии с непрерывной неограниченной дисперсией частиц при броуновском движении (или в однородном турбулентном потоке). Стабилизирующая обратная связь дает статистически стационарное распределение вероятностей климата. Обратная связь проявляется также в конечной степени предсказуемости климата, но предсказуемость ограничивается максимальной величиной параметра умения предсказывать порядка 0,5.

3.4   Reducing the Phase Space: Signal-to-Noise Analysis and Detection and Attribution[8]

Some would argue that the most significant part of Hasselmann's legacy would be the introduction of the stochastic dimension in the dynamical and analytical concept of the climate system. The first paper that received a great deal of international attention was the one in which he introduced the "stochastic climate model" [38] in 1976 (see Sect. 3.3). Indeed, this first paper[9] (see facsimile in Sect. 3.3) made use of scale separation—a long-term dynamic, given by a climate variable—under the influence of short-term variations summarised as white (or red) noise. One can see this as a separation of two parts of the phase space, one defined by long-term fluctuations, and the remainder as short-term fluctuations. After setting out a few assumptions and discretisation, the prototype of the concept was encapsulated in an autoregressive first order process, with the conclusion that even in the absence of any force acting upon the slow dynamics, the system would show variations on all time scales because of the presence of the white noise of the short-term variability.

In the 1980s, Hasselmann formulated a more general concept[10] of **"Principal Oscillation Patterns"** (POPs) in an—again barely comprehensible, but never published—manuscript and asked Hans von Storch to "bring it to life". He did so, but only after simplifying or "vulgarising" the concept, such that a workable version finally emerged, even if the basic idea was less clear [94]. Hasselmann saved the original concept by introducing a new term: **"Principal Interaction Patterns"** (PIPs), which, however, never became popular—at least so far. His 1988 paper on "Principal Interaction patterns" [86] (see facsimile below) spelled out the idea that it would be possible to divide the phase space into two sets, one with a finite number of dimensions, within which the core of the dynamics would play out. The basis spanning this space, were the PIPs. The rest, spanned by very many if not an infinite number of dimensions would contribute to the core dynamics, but in a kind of slave mode – either independent noise, or noise conditioned by the state of the PIPs (colloquially referred to as "parametrization"). This

---

[8] Prepared by Hans von Storch incorporating comments by Peter Lemke.

[9] As was the case with several of Hasselmann's early papers, this one was written in very complicated manner, and was hardly comprehensible for many people. He would sometimes set out a very clear version of the then matured concept at a later date. But in case of the stochastic climate model, he declined to do so, replying that it would be "too simple," when Hans von Storch suggested that he should do so in the late 1990s. However, developing a complicated but powerful approach, which eventually transforms into something perceived as simple, is the hallmark of a genius.

[10] I believe that he had always had it in mind, although he had hardy spelled it out explicitly.

concept, of a low-dimensional dynamically active part of the phase space, and the high-dimensional part essentially operating as (conditional) noise is at the core of his conceptualization of the stochastic climate system.

Hasselmann had thus introduced a new paradigm,[11] certainly worthy of a Director of a Max-Planck Institute, which enabled an understanding of climate variability based on a split between signals (related to specific causes) and background "noise". But the concept led to another practice in the analysis of climate variability and responses, i.e., the challenge of separating the two components; to find the relevant signal within the sea of noise. To achieve this, Hasselmann introduced the "**detection and attribution**" concept [54], which involved a 2-step process. First, in the detection-step, the relevant change is examined to see if it falls within the range of natural variability. This is done via a conventional statistical hypothesis test. If the result of the first step is the successful rejection of the null hypothesis "consistent with unprovoked variability", then the change is compared to one or several theories derived from numerical experimentation, theoretical arguments, or independent statistical analysis in a second step. If a good fit is found, then the conclusion is drawn that the relevant change can be attributed to the relevant factor(s). This attribution takes the form of a non-rejection of a null-hypothesis and, as such, represents a weaker argument than the successful detection. Hasselmann added another level of complexity by suggesting that one should optimize the potential signal, allowing for an a priori expectation of a favourable signal-to-noise ratio, but this elegant component was hardly ever used.

Again, the original paper posed a challenge for the reader, but Hasselmann wrote an updated version about 15 years later, which was extremely clear and easy to comprehend [110]. Later still he followed this with a version in which he employed Bayesian concepts [138].

The concept was used successfully for investigating whether an external signal, as suggested by climate model simulations, would be detectable in the observational record of global temperatures, and to attribute the change

---

[11] von Storch, H., J.-S. von Storch, and P. Müller, 2001: Noise in the Climate System—Ubiquitous, Constitutive and Concealing. In B. Engquist and W. Schmid (eds.) *Mathematics Unlimited—2001 and beyond*. Part II. Springer Verlag, 1179–1194.

to human emissions [125,135] since the beginning of industrialization. The technique was adopted by the IDAG[12] group at various corners of the academic world and eventually became a corner stone in IPCC reports: **the fingerprint of human activity in changing the global climate**.

The insight that there is internal variability, i.e., variability unrelated to an external force, or sometimes simply called "noise" had already been floating around in scientific circles: in the early 1970s scientists in the USA had already noticed the omnipresence of internal variability, but not the constructive role played by this noise in the formation of gradual variations and signals. The need to discriminate between signal and noise when evaluating the outcome of numerical experiments with global atmospheric models was introduced in the early 1970s.[13] The challenge of detecting a human signal in the observational record had gradually been addressed by others in the 1980s, notably by Jerry North,[14] Ben Santer, and Tom Wigley.[15]

---

[12] Pennell, W., T.P. Barnett, K. Hasselmann, W.R. Holland, T. Karl, G.R. North, M.C. MacCracken, M.E. Moss, G. Pearman, E.M. Rasmusson, B.D. Santer, W.K. Smith, H. von Storch, P. Switzer and F.W. Zwiers, 1993: The detection of anthropogenic climate change. *Proceedings of the Fifth Symposium on Global Change Studies*. Amer. Met. Soc., Jan. 17–22, 1993, Anaheim (California), 21–28 (DOI: https://doi.org/10.13140/2.1.1970.2567).

IDAG, 2005: Detecting and attributing external influences on the climate system. A review of recent advances. J. Climate 18, 1291–1314.

[13] Chervin, R. M., Gates, W. L. and Schneider, S. H. 1974: The effect of time averaging on the noise level of climatological statistics generated by atmospheric general circulation models. *J. Atmos. Sci.* 31, 2216–2219.

[14] North, G., K. Y. Kim, S.S.P. Shen, and J.W. Hardin 1995: Detection of Forced Climate Signals. Part I: Filter Theory. *J. Climate,* **8,** 401–408.

[15] Santer BD, Taylor KE, Penner JE, Wigley TML, Cubasch U, and Jones PD, 1995: Towards the detection and attribution of an anthropogenic effect on climate. Climate Dynamics, 12: 77–100.

JOURNAL OF GEOPHYSICAL RESEARCH, VOL. 93, NO. D9, PAGES 11,015–11,021, SEPTEMBER 20, 1988

PIPs and POPs: The Reduction of Complex Dynamical Systems Using Principal Interaction and Oscillation Patterns

K. Hasselmann

Max-Planck-Institut für Meteorologie, Hamburg, Federal Republic of Germany

A general method is described for constructing simple dynamical models to approximate complex dynamical systems with many degrees of freedom. The technique can be applied to interpret sets of observed time series or numerical simulations with high-resolution models, or to relate observation and simulations. The method is based on a projection of the complete system on to a smaller number of "principal interaction patterns" (PIPs). The coefficients of the PIP expansion are assumed to be governed by a dynamic model containing a small number of adjustable parameters. The optimization of the dynamical model, which in the general case can be both nonlinear and time-dependent, is carried out simultaneously with the construction of the optimal set of interaction patterns. In the linear case the PIPs reduce to the eigenoscillations of a first-order linear vector process with stochastic forcing (principal oscillation patterns, or POPs). POPs are linearly related to the "principal prediction patterns" used in linear forecasting applications. The POP analysis can also be applied as a diagnostic tool to compress the extensive information contained in the high-dimensional cross-spectral covariance matrix representing the complete second-moment structure of the system.

1. Introduction

To gain insight into the behavior of complex dynamical systems with many degrees of freedom, a standard strategy is to devise simpler analog systems which contain only a few degrees of freedom but nevertheless succeed in capturing the principal dynamical properties of the full system. In the case of climate modeling, some form of system reduction is essential if one wishes to model the many interactions between the different climate subsystems which span many orders of magnitude of different time scales. A high-resolution model, such as a general circulation model (GCM), cannot be integrated long enough to cover more than a small fraction of the time scales occurring in natural climate variability, so that a trade-off must be found between the number of degrees of freedom of the model and the spectral bandwidth of the simulation. Apart from the need to remain within finite computational restraints, system reduction is also the standard approach to "understanding" the system. Various reduced systems have been proposed, for example, to deduce the overall response characteristics of the global climate system or to explain particular phenomena, such as the El Niño/Southern Oscillation or atmospheric blocking.

The simplest and most commonly used method of system reduction is scale truncation. The nonresolved components of the system beyond the cutoff scale are normally parameterized in the form of mean interaction terms and a residual stochastic forcing contribution. The latter can often be the dominant term responsible for the time variability of the reduced system [cf. *Hasselmann*, 1976].

In this paper an alternative method of system reduction is considered, based on the observation that the dynamical behavior of complex systems often appears to be dominated by interactions between only a few characteristic "patterns". A number of hypotheses which have been proposed to explain climate fluctuations in terms of internal feedback processes rather than short time scale stochastic forcing have been for-

mulated implicitly in terms of such interaction patterns. However, the identification of the basic interaction structures in observed data or in the simulation data of high-resolution model runs has often proved elusive.

In the following discussion a general method for constructing reduced dynamical models is introduced which addresses this problem. The models combine internal linear or nonlinear interactions within the reduced system with residual stochastic forcing, both of which can contribute to the natural variability of the system. The basic technique is straightforward: the reduced dynamical model is constructed by finding the optimal model, within a given model class, which best fits the data in a generalized least squares sense. In defining the model class for the fitting procedure, the interaction patterns, parameter values of the dynamical model, and statistical structure of the stochastic forcing are not specified. The optimal model fit then yields the set of "principal interaction patterns" (PIPs), the model parameter values, and the (cross) spectra of the stochastic forcing. In the linear case the PIPs reduce to (damped) normal modes (principal oscillation patterns, or POPs).

The approach may be regarded as a combination and extension of standard methods of expanding statistical fields with many degrees of freedom in terms of empirical orthogonal functions (EOFs) or "principal prediction patterns" (PPPs) and the autoregressive moving average (ARMA) technique of constructing dynamical models for systems with a few degrees of freedom.

EOFs yield an optimal representation of the covariance structure of fields at a given time, but they are not designed to reveal the structure of the time evolution or the internal dynamics of the system. Principal prediction patterns provide an optimal representation of the linear prediction of one field in terms of another field [cf. *Davis*, 1976; *Barnett and Preisendorfer*, 1987]. If the two fields represent the same physical field taken at different times, the prediction represents a forecast, and the principal prediction patterns therefore contain some time evolution information. However, this cannot normally be translated into an explicit dynamical model without further assumptions. Although EOFs and PPPs can be used, and occasionally have been used, to construct reduced dynamical models, they are not optimized for this purpose.

Paper number 8D0302.
0148-0227/88/008D-0302$05.00

11,016   HASSELMANN: PIPs AND POPs

ARMA methods, on the other hand, have been developed specifically to construct dynamical models. In its standard form [cf. *Box and Jenkins*, 1976; *Kashyap and Rao*, 1976], the technique is basically linear and applies for systems with relatively few degrees of freedom. The method can be readily generalized to nonlinear systems and to systems with time-dependent model parameters. For many climate modeling applications, the nonlinearity and annual modulation of the system are indeed essential characteristics which need to be included at the outset in the formulation of the reduced dynamical model. The main restriction of such a generalized ARMA approach, however, is still the limitation to a few degrees of freedom. The ARMA technique as such contains no provision for projecting the full system onto a smaller set of dominant patterns. This aspect is addressed in the present study.

Although the primary motivation for the paper is to construct reduced dynamical models, the POPs technique for the linear case may also be applied simply as a diagnostic tool (just as the ARMA technique can be applied to estimate spectra using maximum entropy methods). A complete description of the space-time dependent covariance structure of a statistically stationary field requires the specification of the complete cross-spectral covariance matrix for each frequency band of the spectrum or, equivalently, the complete sets of complex EOFs for each frequency. The POPs essentially identify those regions of the spectrum which can be described by the same patterns for an extended frequency interval. They therefore provide a simultaneous optimization of the representation of the second moments of the field with respect to both the spatial and frequency dependence.

Alternative techniques for combining spatial structure information with reduced time-dependence information have been proposed in which an extended EOF analysis is carried out for an enlarged set of time series, consisting of the original time series argumented by a finite number of time-lagged time series [cf. *Weare and Nasstrom*, 1982] or by the Hilbert transforms of the original time series [cf. *Wallace and Dickinson*, 1972; *Horel*, 1984]. The main difference in the POP technique, compared with these methods, is that it ties the different patterns more closely into the structure of the spectrum (see also section 4).

Another diagnostic application of the POP method is the determination of the perturbation eigenmodes of complex model systems, e.g., an ocean general circulation model (E. Maier-Reimer et al., paper in preparation, 1988). If the model is driven by white noise stochastic forcing, a POP analysis of the response automatically extracts the system's normal modes.

The general nonlinear PIP formalism is developed in section 2, while the simplifications resulting for POPs in the linear case are discussed in section 3. The relation between the complete cross-spectral matrix representation of the second moments of the field and the POPs expansion is considered in section 4. The conclusions are summarized in section 5.

2.   PRINCIPAL INTERACTION PATTERNS: THE GENERAL CASE

Consider a system represented by the state vector $\Phi = (\Phi_1, \Phi_2, \cdots, \Phi_n)$, whose evolution is governed by a set of first-order equations

$$\frac{d\Phi}{dt} = \mathbf{F}(\Phi) \tag{1}$$

where $\mathbf{F}$ is some nonlinear, time-dependent function of Φ (the time dependence is suppressed in the notation). The dimension n of the system is assumed to be high: for numerical high-resolution models, n is typically of order 10^4–10^6, while for observational data, n may be of order 10^2.

attempt now to construct a simplified dynamical model approximating (1) which involves a significantly smaller number of degrees of freedom m, where m is perhaps of order 2–10. The reduction is carried out in two steps.

First, the dimension of the state vector is reduced by approximating Φ as a superposition $\hat{\Phi}$ of m time-independent principal interaction patterns, $\mathbf{p}_v$,

$$\Phi = \hat{\Phi} + \rho \tag{2}$$

where

$$\hat{\Phi} = \sum_v z_v(t)\mathbf{p}_v \tag{3}$$

and ρ is the residual error.

For this section and section 3, it is convenient to introduce a matrix notation in which, in an extension of Dirac's bra-ket notation, the structure of a matrix is depicted by left and right delimiters, indicating the dimensions of the left and right indices of the matrix, respectively. A transposed matrix is represented by a mirror image delimiter pair; a matrix multiplication contains two adjacent mirror image delimiters. Thus the state vector Φ is represented as the one column matrix $\Phi_{i1} \equiv |\Phi\rangle$, the set of patterns $\mathbf{p}_v$ as the rectangular matrix $p_{iv} \equiv |p)$, the set of coefficients z_v as the one-column vector $z_{v1} \equiv (z)$ (or as the transposed row vector $\langle z)$), and the symmetrical $n \times n$ matrix in (4) as $M_{ij} \equiv |M|$ (see Table 1). Following standard practice, we shall replace the delimiter pair represented by double verticals in matrix multiplications by a single vertical delimiter.

The matrix form of (3) is then given by

$$|\hat{\Phi}\rangle = |p)(z) \tag{3'}$$

The coefficients $z_v(t)$ for a given set of patterns $\mathbf{p}_v$ are determined, in the standard manner, by requiring that the square modulus $\langle\rho|M|\rho\rangle$ of the error ρ with respect to some suitably defined metric M is minimized. One obtains

$$(z) = (d|M|\Phi\rangle \tag{4}$$

where $|d)$ is the set of adjoint patterns to $|p)$, defined as the set of vectors within the space spanned by $|p)$, which are orthonormal (with respect to the matrix M) to the set $|p)$,

$$(d|M|p) = (I) \qquad (\equiv \delta_{vp}) \tag{5}$$

or explicitly,

$$|d) = |p)(N^{-1}) \tag{6}$$

where

$$(N) = (p|M|p) \tag{7}$$

In statistical applications, in which Φ is regarded as a particular realization taken from a statistical ensemble of states, the metric M is usually chosen as the inverse of the covariance matrix of Φ. If Φ is Gaussian and is defined such that its expectation value vanishes, surfaces of constant $\langle\Phi|M|\Phi\rangle$ in Φ phase space then correspond to surfaces of constant probability density. This choice of M has the property that it maximizes the statistical significance of patterns extracted from Φ

TABLE 1. Notation

(Right) Delimiter	Associated Dimension)
	1
⟩	n, equal to the dimension of state vector Φ
)	m, equal to the number of patterns $\mathbf{p}_v$, $v = 1, \ldots m$
I	dimension of predictand field Ψ

	Column Vectors			
Vector	Vector Index Notation	Matrix Index Notation	Bra-ket Notation	Definition
Φ	Φ_i	Φ_{i1}	$\lvert\Phi\rangle$	state vector
$\hat{\Phi}$	$\hat{\Phi}_i$	$\hat{\Phi}_{i1}$	$\lvert\hat{\Phi}\rangle$	modeled state vector
$\mathbf{p}$	ρ_i	ρ_{i1}	$\lvert\rho\rangle$	error of modeled state vector
z	z_v	z_{v1}	(z)	pattern expansion coefficients
c	c_v	c_{v1}	(c)	pattern coefficients
Ψ	Ψ_a	Ψ_{a1}	$\lvert\Psi\rangle$	predictand field
p	p_i	p_{i1}	$\lvert p\rangle$	single expansion pattern
q	q_i	q_{i1}	$\lvert q\rangle$	single projection pattern
$\hat{q}$	$\hat{q}_i$	$\hat{q}_{i1}$	$\lvert\hat{q}\rangle$	$\lvert\hat{q}\rangle = \lvert M \lvert q\rangle$

Matrix	Index Notation	Bra-ket Notation	Definition
M	M_{ij}	$\lvert M\rvert$	metric used for pattern expansion
$\bar{M}$	$\bar{M}_{ij}$	$\lvert\bar{M}\rvert$	metric used for PIP model error minimization
M'	$M_{\alpha\beta}$	$\lvert M\rvert$	metric used for PPP model error minimization
$\mathbf{p}_v$	p_{iv}	$\lvert p)$	set of expansion patterns
$\mathbf{d}_v$	d_{iv}	$\lvert d)$	set of adjoint patterns: $(d/M\lvert p) = (I)$
N	$N_{v\mu}$	(N)	$(p/M\lvert p)$
D	$D_{v\mu}$	(D)	linear model matrix
q_v	q_{iv}	$\lvert q)$	set of projection patterns: $\lvert q) = \lvert d) (D^T)$
K	K_{ai}	$\lvert K\rvert$	predictand-predictor covariance matrix
C	C_{ij}	$\lvert C\rvert$	predictor covariance matrix
$\hat{K}$	$\hat{K}_{ia}$	$\lvert\hat{K}\rvert$	$\lvert C^{-1}\rvert K^T\rvert M'\rvert$

which are associated with specific externally generated "signals", as opposed to the internal background noise of the statistical ensemble of states Φ [cf. *Hasselmann, 1979*].

In the second step of the reduction procedure, a set of m (in general nonlinear, time-dependent) evolution equations

$$\frac{dz_v}{dt} = G_v(z; \alpha_1, \alpha_2, \cdots, \alpha_p) + n_v \quad (8)$$

is postulated for the coefficients $z_v(t)$ of the expansion (4). The reduced model (8) is specified a priori only as a member of a model class: the evolution equations contain a number of free parameters $\alpha_1, \alpha_2, \cdots \alpha_p$, which still need to be determined. In addition to the deterministic evolution functions G_v, the evolution equations contain an (unknown stochastic) forcing term n_v, representing the residual errors of the reduced dynamical system. The class of model G_v must be specified a priori in accordance with some preconceived notion or hypothesis regarding the type of dynamical process governing the evolution of the system.

The unknown model parameters α_j and PIPs $\mathbf{p}_v$ are now determined simultaneously by minimizing the error

$$\varepsilon = \{\langle\dot{\hat{\Phi}} - \dot{\Phi}\,\lvert\,\bar{M}\,\lvert\,\dot{\hat{\Phi}} - \dot{\Phi}\rangle\} \quad (9)$$

between the rate of change $\dot{\Phi} \equiv d\Phi/dt$ of the true system and the rate of change $\dot{\hat{\Phi}} \equiv d\hat{\Phi}/dt$ of the approximate system, as determined from (3), (4), (6), and (8), but without inclusion of the unknown noise term n_v. The braces in (9) denote expectation values (or, if the system cannot be regarded as a stochastic process, as the time integral over the period for which the reduced model is applied).

The metric $\bar{M}$ of the scalar product in (9) may be defined differently from the scalar product (4). For example, it may be appropriate to choose the matrix $\bar{M}$ as the inverse of the covariance matrix of $\dot{\Phi}$, rather than Φ.

Substituting (8) into (3), we obtain

$$\varepsilon = \{\langle G(p)\lvert\bar{M}\lvert p)(G) - 2\langle\Phi\lvert M\lvert p)(G) + \langle\Phi\lvert M\lvert\Phi\rangle\} \quad (10)$$

Variation of ε with respect to α_j and $\mathbf{p}_v$, respectively, then yields as the determining equations for our model

$$\frac{1}{2}\frac{\partial\varepsilon}{\partial\alpha_j} = 0 = \left\{\langle G(p\lvert\bar{M}\lvert p)\left(\frac{\partial G}{\partial\alpha_j}\right) - \langle\Phi\lvert M\lvert p)\left(\frac{\partial G}{\partial\alpha_j}\right)\right\} \quad (11)$$

$$\frac{1}{2}\frac{\partial\varepsilon}{\partial\lvert p)}\,\delta\lvert p) = 0 = \{[\langle G)(p\lvert - \langle\Phi[]\lvert\bar{M}\lvert\delta p)(G)\}$$
$$+ \left\{[\langle G)(p\lvert\bar{M}\lvert p) - \langle\Phi\lvert\bar{M}\lvert p)]\frac{\partial(G)}{\partial(z)}\,\delta(z)\right\} \quad (12)$$

where, according to (4), (6)

$$\delta(z) = (N^{-1})(\delta p\lvert M_*\lvert\lvert\Phi\rangle - (N^{-1})$$
$$\cdot[(\delta p\lvert M\lvert p) + (p\lvert M\lvert\delta p)](N^{-1})(p\lvert M\lvert\Phi\rangle \quad (13)$$

Equations (11) and (12) are in general nonlinear and can be solved only by iterative techniques. In practice, the determination of the optimal solution will be less forbidding than the structure of the equations appears to imply. For a prescribed set of patterns, the minimization with respect to α_j can be carried out using minimization routines available in standard libraries (the dimension of the parameter vector α_j is assumed to be small). Conversely, for given α_j the matrix equation (12)

11,018   HASSELMANN: PIPs AND POPs

for the determination of the optimal patterns can also be solved iteratively, for example, by considering small perturbations about a reference set of patterns and then applying linear techniques. For a linear model, (12) reduces to a straightforward matrix eigenvalue problem (compare section 3). The full minimization problem can thus be solved by minimizing iteratively with respect to $\mathbf{p}_v$ and α_j. Alternatively, the full parameter set $\mathbf{p}_v$, α_j can be determined simultaneously, using recently developed adjoint gradient techniques designed for optimizing systems with many degrees of freedom [cf. *Navon and Legler*, 1987].

It has been assumed that the class of models is formulated such that the optimal solution which minimizes ε is uniquely determined. This will generally require some restriction in the form of the dynamical model. For example, an alternative representation equivalent to (3) can be obtained by any linear transformation of the set of patterns within the space spanned by the patterns. If the resulting linear transformation of the coefficient vector z_v yields a model which still belongs to the class of models defined by (8), the optimal model is clearly defined only to within an arbitrary linear transformation. In this particular case the solution can be made unique by requiring, for example, that the linear terms in the Taylor expansion of the dynamical equations are diagonal. For the case of a linear system, the principal interaction patterns reduce then to principal oscillation patterns, which are defined as the normal modes of the system. Alternatively, the representation can be made unique by introducing the patterns successively, keeping the previously defined patterns fixed when the next pattern is determined.

The general technique outlined in this section can clearly be modified in various respects. The set of adjoint vectors $|d\rangle$ need not be defined through (4) and (5), but can be determined simultaneously with the PIPs as part of the minimization condition on ε. Furthermore, the model can be required to satisfy additional side conditions, either rigorously or approximately, which can be included with appropriate weighting in the definition of the error function ε. In most applications the model will be constructed for a particular time scale range. This can be taken into consideration by passing the original data $\Phi(t)$ through an appropriate filter. Alternatively, the minimizing function itself can be defined as a weighted integral over the frequency domain (this procedure lends itself most readily to linear models).

3. PRINCIPAL OSCILLATION PATTERNS: THE LINEAR CASE

The evolution equations (8) for the reduced dynamical system are given in the linear case by

$$\frac{d\langle z\rangle}{dt} = (D)\langle z\rangle + \langle n\rangle \tag{14}$$

or, invoking (4),

$$\frac{d|\hat{\Phi}\rangle}{dt} = |p\rangle(D)\langle d|M|\Phi\rangle + \text{residual forcing} \tag{15}$$

where $(D) \equiv D_{v\mu}$ represents a constant matrix. The elements of (D) correspond to the parameters α_j in the general formulation of section 2. Since the square matrix (D) will normally be nonsymmetrical, the transpose matrix $(D^T) \equiv D_{\mu v}$ is indicated explicitly by an superscript T in the following discussion to avoid ambiguity in the delimiter notation.

Through a (complex) linear transformation, the form (14) can normally be diagonalized (we shall ignore degenerate cases where this is not possible). We denote the linear patterns, $\mathbf{p}_v$, of the optimal model after diagonalization, which represent the eigenmodes of the linear system (14), as principal oscillation patterns. The POPs occur as complex conjugate pairs, if the eigenvalues are complex (damped oscillations), or as single real patterns for real eigenvalues (exponentially damped modes).

Before considering the POPs further, it is helpful to relate the present approach to the alternative, but essentially equivalent, expansion in terms of principal prediction patterns (PPPs). For this purpose, we introduce a set of "projection patterns" $|q\rangle \equiv q_{iv}$, defined by

$$|q\rangle = |d\rangle(D^T) \tag{16}$$

where $|d\rangle$ is the set of adjoint patterns given by (6).

Equation (15) then takes the form

$$\frac{d}{dt}|\hat{\Phi}\rangle = |p\rangle\langle q|M|\hat{\Phi}\rangle \tag{17}$$

Since, for a given set of patterns $|p\rangle$ (and nonsingular (D)), the patterns $|q\rangle$ are uniquely determined through (16) if (D) is given and vice versa, the model can be optimized with respect to the set of parameters $|p\rangle$, (D), as in the original formulation of the model or, alternatively, with respect to the pattern sets $|p\rangle$, $|q\rangle$. The latter approach corresponds to an expansion in terms of PPPs.

PPPs are normally introduced in the more general context of deriving an optimal linear prediction of a predictand field $\Psi \equiv \Psi_\alpha$ from a predictor field $\phi \equiv \phi_i$, using an expansion of only a finite number of predictand and predictor (projection) patterns [cf. *Davis*, 1976; *Barnett and Preisendorfer*, 1987]. The dimensions of Ψ_α and ϕ_i can be different. We denote the field Ψ in matrix notation as $|\Psi\rangle$. In our particular case, however, $\Psi \equiv \Phi$, so that the predictor and predictand fields have identical dimension.

In the general problem one seeks an optimal prediction

$$|\Psi(t)\rangle = |p\rangle\langle c(t)\rangle \tag{18}$$

for the predictand field Ψ in terms of m constant patterns $\mathbf{p}_v \equiv p_{\alpha v} \equiv |p\rangle$, $v = 1, \cdots m$, where the coefficients $c_v(t) \equiv \langle c\rangle$ of the expansion are derived from the predictor field Φ by a projection

$$\langle c\rangle = \langle q|M|\Phi\rangle \tag{19}$$

using m projection patterns $\mathbf{q}_v \equiv q_{iv} \equiv |q\rangle$, $v = 1, \cdots . m$

Thus

$$|\Psi\rangle = |p\rangle\langle q|M|\Phi\rangle \tag{20}$$

The pattern sets $|p\rangle$ and $|q\rangle$ are determined by minimizing the mean square error

$$\varepsilon = \{\langle\Psi - \hat{\Psi}|M'|\Psi - \hat{\Psi}\rangle\} = \min \tag{21}$$

defined with respect to some metric $|M'| \equiv M_{\alpha\beta}'$. (This will generally differ from the metric $|\bar{M}| \equiv \bar{M}_{ij}$ in (9), introduced in section 2, since Ψ_α and ϕ_i have different dimensions. However, in our application we may set $M' = \bar{M}$).

The condition (21) alone clearly does not specify the pattern sets $|p\rangle$, $|q\rangle$ uniquely, since (20) and (21) are invariant with respect to an arbitrary linear transformation $|p'\rangle = |p\rangle(L)$, $|q'\rangle = |q\rangle(L^{-1T})$. However, we may make the solution unique by requiring that the pattern pairs $\mathbf{p}_v$, $\mathbf{q}_v$ are introduced in sequence (and, say, by suitably normalizing the patterns p_v). In this case, we need consider only one pattern pair, $\mathbf{p}_v$, $\mathbf{q}_v \equiv \mathbf{p}$, $\mathbf{q}$, at a time and can thus drop the index v (i.e., we may replace $| p)$ by $| p\rangle$. The predictand field Ψ is redefined at each step as the original field minus the field already predicted from the patterns introduced previously.

The (local) minimization of ε yields the relations

$$\frac{1}{2}\frac{\partial \varepsilon}{\partial | p\rangle} = \langle q | M | [- | K^T | + | C | M | q\rangle\langle p |] | M' | = 0 \quad (22)$$

$$\frac{1}{2}\frac{\partial \varepsilon}{\partial | q\rangle} = \langle p | M' | [- | K | + | p\rangle\langle q | M | C |] | M | = 0 \quad (23)$$

where

$$| K | \equiv K_{ai} = \{\Psi_a \phi_i\} \quad (24)$$

and

$$| C | \equiv C_{ij} = \{\phi_i \phi_j\} \quad (25)$$

Multiplying (22) and (23) from the right with $| M' |^{-1}$, and $(| C | M |)^{-1}$, respectively, one obtains two coupled eigenvalue equations for $| p\rangle$ and

$$| \bar{q}\rangle = | M | q\rangle \quad (26)$$

namely,

$$| K | \bar{q}\rangle - \lambda' | p\rangle = 0 \quad (27)$$

$$| \bar{K} | p\rangle - \lambda'' | \bar{q}\rangle = 0 \quad (28)$$

where

$$\lambda' = \langle \bar{q} | C | \bar{q}\rangle \quad (29)$$

$$\lambda'' = \langle p | M' | p\rangle \quad (30)$$

and

$$| \bar{K} | = | C^{-1} | K^T | M' | \quad (31)$$

Separation of $| p\rangle$ and $| \bar{q}\rangle$ then yields the pair of eigenvalue equations

$$| K | \bar{K} | p\rangle - \lambda | p\rangle = 0 \quad (32)$$

$$| \bar{K} | K | \bar{q}\rangle - \lambda | \bar{q}\rangle = 0 \quad (33)$$

with the same eigenvalues

$$\lambda = \lambda' \lambda'' \quad (34)$$

for $| p\rangle$ and $| \bar{q}\rangle$. The values λ', λ'' follow from λ after introduction of a suitable normalization for $| p\rangle$, say, for example, $\lambda'' = 1$ (equation (30)).

The matrices $| K | \bar{K} | = [| K | C^{-1} | K^T |] [| M |]$ and $| \bar{K} | K | = [| C^{-1} |] [| K^T | M' | K |]$ in the eigenvalue equations (32) and (33) consist of quadratic products of symmetrical, positive definite matrices, indicated here by the square parentheses. It follows that the eigenvalues are real and positive and that the eigenvectors are also real. The eigenvectors themselves are in general not orthogonal but are related to an equivalent orthogonal set by a linear transformation.

The absolute minimum of ε is given by the eigenvector pair with the largest eigenvalue λ, and the sequence of predictor patterns is accordingly given by the ordered sequence of eigensolutions of (32) and (33).

To derive the POPs for a linear model, the most straightfor-

ward procedure is accordingly the following: (1) determine the sequence of PPPs by solving (32) and (33); (2) evaluate the linear model coefficient matrix D by inverting (16); and (3) diagonalize (14) by transforming to the (complex) eigenoscillations of the linear system (14).

The inversion of (16) in the second step is readily carried out by making use of the orthogonality of the vector sets $| p)$ and $| d)$. One obtains, applying (5)-(7),

$$(D) = (q | M | p) \quad (35)$$

Although this approach is presumably the simplest when dealing with purely linear systems, it may be more convenient to resort to the general nonlinear formalism when the linear model is to be investigated within a hierarchy of nonlinear models in which the linear model occurs as a limiting case. The dynamical model would then be formulated at the outset in terms of a model which is diagonalized in the first-order, linear approximation.

In the diagonalized reference frame the POP expansion of the field Φ is given by

$$\Phi = \sum_v A_v(t)\mathbf{p}_v(t) + \text{complex conjugate} \quad (36)$$

where the complex amplitudes A_v satisfy the standard damped harmonic oscillator equation

$$\frac{dA_v}{dt} - i\Omega_v A_v = N_v(t) \quad (37)$$

$$\Omega_v = \omega_v + i\mu_v$$

The residual forcing term $N_v(t)$, frequencies Ω_v, and patterns

$$\mathbf{p}_v = \mathbf{p}_v^{(1)} + i\mathbf{p}_v^{(2)} \quad (38)$$

are generally complex. For eigenmodes with zero frequency ω_v, however, the complex conjugate pattern pair $\mathbf{p}_v$, $\mathbf{p}_v^*$ reduces to a single, real pattern $\mathbf{p}_v^{(1)}$ representing an exponentially decaying mode $\sim\exp(-\mu_v t)$.

In general, both real and imaginary patterns $\mathbf{p}_v^{(1)}$, $\mathbf{p}_v^{(2)}$ will be continually excited by the residual complex stochastic forcing $N_v(t)$. Each excitation pulse gives rise to a damped oscillation, in which the originally excited pattern $\mathbf{p}_v^{(1)}$, say, is transformed into the pattern $\mathbf{p}_v^{(2)}$ after a quarter period $\pi/2\omega_v$, returning back to the original pattern with opposite sign another quarter period, and so on. Depending on the form of the patterns $\mathbf{p}_v^{(1)}$ $\mathbf{p}_v^{(2)}$, the oscillation can appear as a standing wave, a traveling wave, a local pulsation, or various combinations of these. The general oscillation represents a damped amphodromic wave of the form considered extensively in the harmonic analysis of tides or in the eigenoscillation theory for ocean basins.

In contrast to the nondynamical expansion in terms of EOFs, the amplitudes of different POPs are generally correlated. This applies also for the amplitudes of a PPP expansion, and it must be expected to hold generally for any expansion procedure based on dynamical models. Even when the different system components are dynamically decoupled by transforming to normal-mode coordinates, as in the POP representation, correlations between the different modes are introduced by the residual forcing $N_v(t)$ in (38), which cannot be assumed to be uncorrelated.

A more detailed discussion of POPs, in the context of a meteorological application, is given by von Storch et al. [this issue].

11,020   HASSELMANN: PIPs AND POPs

4. RELATION BETWEEN POPs AND CROSS-SPECTRAL
EOF ANALYSIS

In applications to a multivariate, stationary, stochastic process $\Phi_i(t)$, the POPs analysis is not only useful as method of constructing optimal linear dynamical models, but also provides a simple diagnostic tool for compressing the extensive information required for a complete characterization of the second-moment statistics of the process into a manageable set of numbers and patterns.

The second moments of a statistically stationary process $\Phi_i(t)$ can be characterized by the covariance function

$$R(t) \equiv R_{ij}(t) = \{\Phi_i(t+\tau)\Phi_j(t)\}$$

or equivalently, by its Fourier transform, the complex cross spectrum $F_{ij}(\omega)$.

The cross spectrum may also be formed directly from the Fourier representation of the process itself,

$$\Phi_i(t) = \int \phi_i(\omega) \exp(i\omega t)\, d\omega \qquad (39)$$

through the relation

$$\{\phi_i^*(\omega)\phi_j(\omega')\} = F_{ij}(\omega)\delta(\omega-\omega') \qquad (40)$$

where $\phi_j^*(\omega)$ denotes the complex conjugate, and the reality of Φ_i requires

$$\phi_j^*(\omega) = \phi_j(-\omega) \qquad (41)$$

$$F_{ij}(\omega) = F_{ji}(-\omega) = F_{ij}^*(-\omega) \qquad (42)$$

At each frequency ω, the cross spectrum F_{ij} may be diagonalized by transforming from $\phi_i(\omega)$ to the coefficients $c_\alpha(\omega)$ of the expansion with respect to the complex EOFs, $e_{i\alpha}(\omega)$ [cf. Brillinger, 1981]:

$$\phi_i(\omega) = \sum_\alpha e_{i\alpha}c_\alpha(\omega) \qquad (43)$$

where the EOFs satisfy the eigenvalue equation

$$\sum_j F_{ij}e_{j\alpha} = \lambda_\alpha e_{i\alpha} \qquad (44)$$

The set of variables $c_\alpha(\omega)$ provides a completely orthogonalized representation of $\Phi_i(t)$,

$$\{c_\alpha^*(\omega)c_\beta(\omega')\} = \delta_{\alpha\beta}\delta(\omega-\omega')\lambda_\alpha(\omega) \qquad (45)$$

Thus the spectrum $\lambda_\alpha(\omega)$, together with the set of EOFs $e_{i\alpha}(\omega)$, may be regarded as the most compact complete representation of the second-momentum structure of $\Phi_{ij}(t)$. However, the information content is still formidable, as a different set of EOFs and eigenvalues is required for each frequency band.

The POP analysis may be regarded as an attempt to interpolate both the EOF pattern structure and the EOF energy levels across the frequency spectrum. Translating (37) into the frequency domain and again using lower-case symbols for the Fourier transforms, the POP amplitude cross spectrum for a given noise cross spectrum is given by

$$\{a_\nu(-\omega)a_\mu(\omega)\} = \frac{\{n_\nu(-\omega)n_\mu(\omega)\}}{(\Omega_\nu + \omega)(\Omega_\mu - \omega)} \qquad (46)$$

(the quadratic products must be formed here using the negative frequency amplitudes rather than complex conjugate amplitudes, since (41) and the second part of (42) do not hold for the complex processes $A_\nu(t)$ and $N(t)$).

Up to this point the second moments of the POP amplitudes have simply been expressed in terms of the second moments of the forcing. If we require the same number of POPs to describe the process as spectral EOFs, and if the forcing cross spectrum is as complex as the response, clearly nothing has been gained. However, it is the basic premise of the present dynamic model expansion, as in the general ARMA approach, that the dominant structures in the spectrum may be attributed to a relatively small number of dynamical processes represented explicitly in the model, rather than to the external stochastic forcing which is left as a residual after one has identified the dominant internal processes. Thus it is assumed that the residual forcing is white in the frequency domain, or is at least smoothly varying, exhibiting no marked resonances of the type represented by the denominator in (46). In this case, the structure of the cross spectrum can be characterized by the (relatively few) POP patterns, the positions of the quasi-resonance POP frequencies Ω_ν in the complex frequency plane, and the strengths and cross correlations of the effective forcing at these frequencies.

The advantage of the POP technique as a method of compressing the detailed information contained in the complete cross-spectral matrix clearly comes to bear only for relatively broad spectra containing many frequency bands. In the limit of a single-frequency, very narrow band spectrum, the set of POPs reduces to the set of complex EOFs at that frequency.

An alternative technique for compressing the information content of a broadband cross spectrum is to apply a complex EOF analysis in the time domain to a set of time series consisting of the original time series and their Hilbert transforms [Brillinger, 1981; Horel, 1984]. This is equivalent to treating the entire cross spectrum formally as a single-frequency band, i.e., to averaging over the (one-sided) cross spectrum. Both techniques yield similar sets of reduced patterns, but the POP analysis, in addition, provides information on the frequency structure of the spectrum by identifying the spectral peaks (the POP eigenfrequencies) and peak widths (the inverse damping time scales) associated with different patterns. In the limit of the single-frequency, narrow-band spectrum, the Hilbert transform method again becomes equivalent to the POP and the spectral EOF analysis techniques.

5. CONCLUSIONS

A general method has been described for constructing optimal reduced dynamical models for systems with many degrees of freedom. The technique combines the approach used in empirical orthogonal function or principal prediction pattern analyses, in which systems with a large number of degrees of freedom are reduced to a few dominant patterns, with ARMA methods for constructing simple dynamical models from data. Both methods have been generalized to nonlinear and time-dependent systems.

The simultaneous determination of the optimal set of principal interaction patterns and the optimal dynamical model describing the evolution of the PIP amplitudes yields a coupled nonlinear eigenvalue problem for the PIPs and a standard minimization problem for the dynamical model parameters. The full nonlinear problem can generally be solved numerically by a Newton method, i.e., by iterating a local linear minimization problem.

In the linear case the PIPs reduce to principal oscillation patterns, which represent the eigenoscillations of the reduced linear dynamical system.

As a diagnostic tool, the POPs provide a smoothed representation of the cross-spectral covariance matrix of the full system in the frequency domain. The extensive information contained in the complete cross spectrum is reduced to a finite set of patterns characterized by a finite set of complex resonant frequencies.

Examples and applications are given in the papers of *von Storch et al.* [this issue] and E. Maier-Reimer et al. (manuscript in preparation, 1988).

Acknowledgments. The author is grateful for critical reviews and helpful discussions of the concepts presented in this paper by H. von Storch, T. Bruns, I. Fischer-Bruns, and U. Weese, and to useful comments by a reviewer.

REFERENCES

Barnett, T. P., and R. Preisendorfer, Origins and levels of monthly and seasonal forecast skill for United States surface air temperatures determined by canonical correlation analysis, *Mon. Weather Rev., 115,* 1825–1850, 1987.

Box, G. E. P., and G. M. Jenkins, *Time Series Analysis, Forecasting and Control,* Holden-Day, Oakland, Calif., 1976.

Brillinger, D. R., *Time Series—Data Analysis and Theory,* 540 pp., Holden-Day, Oakland, Calif., 1981.

Davis, R. E., Predictability of sea-surface temperature and sea level pressure anomalies over the North Pacific Ocean, *J. Phys. Ocean., 6,* 249–266, 1976.

Hasselmann, K., Stochastic climate models, I, Theory, *Tellus, 28,* 473–485, 1976.

Hasselmann, K., On the signal-to-noise problem in atmospheric response studies, in *Meteorology of Tropical Oceans,* pp. 251–259, Royal Meteorological Society, London, 1979.

Horel, J. D., Complex principal component analysis: Theory and examples, *J. Clim. Appl. Meteorol., 23,* 1660–1673, 1984.

Kashyap, R. L., and A. R. Rao, *Dynamic Stochastic Models From Empirical Data,* Academic, San Diego, Calif., 1976.

Navon, I. M., and D. M. Legler, Conjugate gradient methods for large-scale minimization in meteorology, *Mon. Weather Rev., 115,* 1479–1502, 1987.

von Storch, H., T. Bruns, I. Fischer-Bruns, and K. Hasselmann, Principal oscillation pattern analysis of the 30- to 60-day oscillation in general climate model equatorial troposphere, *J. Geophys. Res.,* this issue.

Wallace, J. M., and R. E. Dickinson, Empirical orthogonal representation of time series in the frequency domain, I, Theoretical consideration, *J. Appl. Meteorol., 11,* 887–892, 1972.

Weare, B. C., and J. S. Nasstrom, Examples of extended empirical orthogonal function analysis, *Mon. Weather Rev., 110,* 481–485, 1982.

K. Hasselmann, Max-Planck-Institut für Meteorologie, Bundstrasse 55, D-2000 Hamburg 13, Federal Republic of Germany.

(Received September 18, 1987;
revised April 12, 1988;
accepted April 13, 1988.)

3.5   Climate and Society[16]

The insight that climate change is not merely a natural science issue, and its link to policy making and coordinated climate action shaped Klaus Hasselmann's interest in socioeconomic modelling and modelling human decision-making, both in the context of climate action. These modelling activities resulted in several publications, and also in his active contributions to recent major research projects.[17]

Hasselmann was quite skeptical about certain dominant approaches and paradigms of mainstream economics, arguing that their basic assumptions do not adequately reflect the ways in which economic stakeholders interact and shape decision making, and seeing these conceptual shortcomings as a reason for their limited ability to describe and predict real-world economic processes—especially when things go worse than usual. For instance, during the 2008 global economic crisis Hasselmann was highlighting the need for rethinking certain paradigms of economic modelling in his talks and papers, as no mainstream models had been able to foresee and predict the coming crisis.

In particular, Hasselmann was skeptical about computable general equilibrium (CGE) models. His criticism of this was primarily based on two objections: on one hand, at the conceptual level, he argued that real-world economic processes were fundamentally out of equilibrium; on the other, he referred to the technical difficulties involved in calibrating and validating the model, and questioned the extent to which economic data could effectively support overly complex multi-regional, multi-sector applied CGE.

Hasselmann also adopted a critical position towards another cornerstone of mainstream economic modelling, intertemporal optimisation, which is common to the majority of economic growth models and, accordingly, broadly used in relation to the economics of climate change and in integrated assessment models (IAM). From a conceptual perspective, in his opinion,

---

16 Prepared by Dmitry V. Kovalevsky, with some additions by Hans von Storch.

17 This includes EU FP7 COMPLEX Project "Knowledge Based Climate Mitigation Systems for a Low Carbon Economy" (2012–2016, Grant Agreement No. 308601), where models developed by Klaus Hasselmann and his colleagues following the actor-based system dynamics approach proposed by Klaus Hasselmann were an essential element of a project model suite for assessing climate mitigation options.

neither individual nor collective decision making follows the mechanism that has been translated to the mathematics of intertemporal optimisation schemes. In addition, following the publication of the influential Stern Review in 2006, a lively discussion emerged concerning the economics of climate change particularly in relation to the high sensitivity of IAM to the value of discount rate, a fundamental parameter of intertemporal optimisation models, and therefore about the 'correct' value of this parameter. Hasselmann also participated in this debate, and it made him even more concerned about the extent to which the intertemporal optimisation approach could serve as a solid basis for informing climate action.

Concerns about excessive complexity and the related difficulties of calibration and validation mentioned above with respect to CGE, were also the reasons for Hasselmann's mixed feelings towards certain innovative modelling approaches, such as agent-based modelling (ABM), that have been very popular since the 1990s. Whilst acknowledging that ABM is conceptually much more satisfactory than, say, CGE, in the attempt to describe the decision-making processes of various stakeholders, he questioned whether a high level of disaggregation of ABMs with their extremely large populations of individual agents is really justified, and whether such strongly disaggregated models can be reliably supported by the available data.

Given his creativity and independence of mind, Hasselmann followed his own, original way when it came to modelling the dynamics of coupled climate-socioeconomic systems for the reasons outlined above. He used the term *actor-based system dynamics modelling* to describe his approach to the construction of socioeconomic models. In essence, his approach involves describing a socioeconomic system via a dynamic model that includes a few interacting aggregate actors, pursuing their own, often conflicting, goals. Mathematically, the system is described by ordinary differential equations, and stakeholder decision making is also parameterised within this mathematical scheme using actor control strategies. Unlike in CGE, the socioeconomic dynamics are described as fundamentally lacking equilibrium. Unlike intertemporal optimisation models, this one is mathematically a dynamic system, and the maximisation of any goal functions is avoided. Unlike in ABM, there are only a few aggregate stakeholders included in the model, rather than a very large population of individual actors, which in turn reduces the dimensionality of the model.

Some of actor-based system dynamics models developed by Hasselmann are of moderate complexity, with relatively few variables and parameters. Some of the simplest models of this kind could, in principle, be designed to partially allow analytical treatment. Despite the fact that analytical work carried out using paper and pencil is still very much respected, for example, in the realm of mainstream economics, Hasselmann sees actor-based system dynamics modelling as a substantially numerical approach. He developed these models for simulations and numerical experiments, not for the elegance of abstract thinking. For Klaus Hasselmann, the narrative told by a model and the results of 'what-if' simulations are ultimately important, as opposed to rigorous propositions and their proofs so popular in the realm of mathematical economics.

Another important element of the actor-based system dynamics approach, to which Hasselmann continuously draws attention in his papers, is a strategy for developing a hierarchy of model families. A hierarchy should start with designing the simplest possible root model and this model should be thoroughly explored via simulations to determine its strengths and limitations. Based on this experience, the complexity of the model can then be increased by adding new actors and processes. This yields one or several models at the next level of the hierarchy, after which the model building process is reiterated. However, as Hasselmann stresses, there is no need for such a model tree to grow infinitely high: the complexity of models should not go beyond the level at which they are no longer supported by the available data.

Hasselmann's early thoughts on the topic of coupled climate-socioeconomic modelling are reflected in his 1991 conference paper "*How Well Can We Predict the Climate Crisis?*" [99], a facsimile of which is reproduced below. The term "climate crisis", which appears in the title, was already in use when the paper was published but was not at all as widespread as it currently is: its permanent inclusion in the climate change related lexicon came at a much later date. Whilst most of the paper is devoted to the natural science-related aspects of climate change, the first and the last sections are particularly relevant to the current topic.

The introductory section calls for the development of a comprehensive Global Environment and Man (GEM) model and sketches out its conceptual design. He later renamed it "Global Environment and Society" (GES). It includes a review of the building blocks from which a GES model could be assembled which are, paradoxically, already available and the yet missing. In the subsequent parts of the paper, Hasselmann also discusses the required improvements in some of these building blocks, which are yet to be made. Another remarkable point in the Introduction is the stress he places on the interplay between climate and environmental change problems.

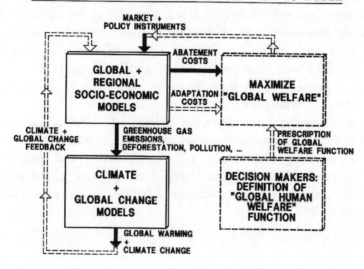

GLOBAL ENVIRONMENT AND SOCIETY (GES) MODEL

Klaus Hasselmann's GES model

In the concluding section, Hasselmann makes several important points relating to the development of GEM-type models that have later been explored in more detail in his own socioeconomic modelling studies, and that one can now see were (and still are) important points for many other researchers in this field. Hasselmann highlights the dynamical and multi-time-scale nature of the GEM system, both in its natural science and human parts. He argues that the inherent uncertainty of all GEM model components calls for the development of GEM-like models as statistical optimisation models. Hasselmann reminds us that the development of GEM should not be seen as "curiosity-driven science", but rather that its ultimate objective is to inform climate-related policy making. Finally, he warns that the limits of our knowledge and the uncertainties inherent in the model that are alluded to above are not an excuse for postponing co-ordinated climate action or waiting for a "perfect" model instead.

The GES approach has met with some critical reactions, one of which involves the question as to whether it would even be possible to define a "global welfare function". Another was that the system reduced society, and the variety of cultures to the choice of a global welfare function, whilst the determination of policy and measures, conditional upon the welfare function, would be a matter only for experts. It was argued that the objective determination of the adaptation and of the abatement costs would not be possible, but that these costs would go through a filter of—possibly interest-led—experts, modified by a variety of different social constructs, so that society would not respond to the state of the environment but to the perception of the state of the environment.[18]

These comments did not target the concept or mathematical implementation of actor-based system dynamics approach proposed by Hasselmann; rather, they broadly apply to the overall architecture and design of such models of the economics of climate change and integrated assessment models, and therefore, so far remain unanswered by the majority of mainstream models used in this area.

[18] von Storch, H., and N. Stehr, 1997: The case for the social sciences in climate research.—*Ambio* 26, 66–71.

Hasselmann, K., 1990: How well can we predict the climate crisis?
In: H. Siebert (ed) Environmental Scarcity - the International Dimension.
JCB Mohr, Tübingen, 165 - 183

Klaus Hasselmann

How Well Can We Predict the Climate Crisis?

1. Interrelation between Climate Models, Economic Models and Comprehensive Global Environment and Man (GEM) Models

In our attempt to model the system earth and human interactions within this system, climate models may be regarded as a simple black box. The input to the black box is man's impact (primarily the emission of greenhouse gases), the output is global warming (Figure 1, panel a). Most of the economic models discussed at this conference may be represented as another black box, with one input, market or policy instruments, and two outputs, greenhouse gas emissions and the costs incurred by greenhouse gas abatement measures (Figure 1, panel b). Some of the models presented here have also addressed the decision-making process. This may be represented by a third black box, with economic costs as input and market and policy instrument strategy as output (Figure 1, panel c).

All of these models may be regarded as sub-models of a comprehensive Global Environment and Man (GEM) model, which represents the ultimate goal of our joint modelling efforts (Figure 1, panel d). The full lines in panel d denote sub-components which have been largely developed. The broken lines indicate missing feedback links or inadequately developed sub-system models. The task of the decision-makers (bottom right box) in such a GEM model is to agree upon the definition of a global "cost-to-mankind" function. The task of the technical and political administration (top right box) is then to minimize the given cost function through the implementation of an optimal mix of market and policy instruments. For this purpose, the interactions between the climate/environment system and human activities (left bottom and top boxes) must be determined. This will require a close collaboration between the climate and environment research community and economic analysts.

The construction of an integrated GEM model requires not only the improvement of climate models — discussed in this paper — and economic and decision-making models — discussed in other contributions to this conference volume — and their ultimate coupling, but also a significant extension of the models. Economic models, for example, need to be extended to include the socio-economic impacts of a climate change. These can be properly modelled only through the inclusion of an additional

Figure 1 —   Integration of Climate Models (Panel a), Economic Models (Panel b) and Models of the Decision-Making Process (Panel c) in a Comprehensive Global Environment and Man (GEM) Model (Panel d)

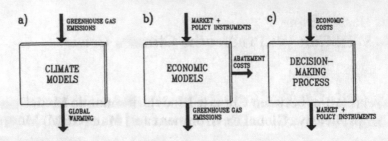

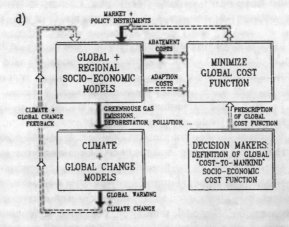

feedback loop, the climate and global change loop, and an additional output term, the socio-economic costs ensuing from climate and global change (compare broken feedback lines in panel d with panels a, b). The latter is required because the cost minimization sector (top right, panel d) requires two inputs, the costs due to greenhouse gas abatement policies and also the costs arising from the climate change resulting from a business-as-usual scenario. Only if both inputs are provided is it possible to meaningfully define an optimal political response strategy consisting of a mix of prevention and adaption strategies. These additional linkages will need to receive more attention if we wish to improve our understanding of the operation of the GEM model as an integrated system.

Another necessary model extension is the generalization of climate models to complete earth system models. This goal is being actively pursued in the international Global Change Programme. The merging of climate and general environmental research in this programme represents in many respects a natural development: it has never been possible to strictly separate the problems of climate and environmental change. Atmospheric chemistry models, for example, are needed for climate research because ozone, chlorofluorocarbons (CFCs), methane, nitrous oxide and other atmospheric trace gases represent important greenhouse gases. But the same models are also needed independently of the climate problem because stratospheric ozone shields the biosphere and human beings from dangerous ultraviolet radiation, because high tropospheric ozone levels are harmful to health, and because most of the gases mentioned interact with other trace gases in complex biogeochemical cycles that directly determine the concentration of many atmospheric pollutants. A similar linkage between applications to climate and environment in general is found in models of the hydrological cycle, atmospheric and ocean transports, changes in the biosphere, soil erosion and many other processes.

In summary, it may be anticipated that many of the specific aspects of climate modelling that I shall attempt to review in this paper will be merged in the course of the developing Global Change Programme with other disciplines. It is hoped that within this general context this review will contribute to a closer interaction between climate modellers and the socio-economic modelling community.

2. The Greenhouse Effect

The greenhouse effect is an important component of the earth's radiation balance. Natural trace gases in the atmosphere (mainly water vapour, carbon dioxide, ozone, methane and nitrous oxide) absorb — in the same way as the glass in a greenhouse — most of the thermal infrared (IR) radiation emitted from the earth's surface, reemitting the absorbed radiation both out to space and back to earth. The back radiation results in an increase of the earth's global mean surface temperature relative to an atmosphere without greenhouse gases by about 35° C. Natural greenhouse gases are therefore essential for the maintenance of life on our planet. However, the accelerating increase in greenhouse gas concentrations since the beginning of the industrial era due to man-made emissions is expected to lead to a further temperature increase of 2–5° C in the next 50–100 years. This is of the same order as the temperature increase since the last ice age and represents a major change in the climatic conditions on which present human society depends.

There exists a general scientific consensus on the strength of the additional man-made greenhouse radiative heating and on the order of magnitude of the expected climate change. However, quantitative estimates of the detailed impact of this additional radiative forcing on the climate system are more uncertain. I shall try to summarize in the following what aspects of the response of the climate system

168

to man's activities are believed to be fairly well understood, and where our principal uncertainties still lie.

3. The Climate Heat Engine

To assess the reliability of climate model predictions, let me first briefly review the structure of the climate system that these models attempt to simulate. The climate system is driven by the sun's radiation. If the earth were a perfectly conducting, perfect black body, which absorbed all of the sun's radiation, it would heat to a uniform temperature of 4° C. At this temperature, the thermal radiation emitted by the earth would exactly balance the absorbed solar radiation. Fortunately, the earth is not a black body (Figure 2):

— First, the earth absorbs only 70 per cent of the incident solar radiation; the remaining 30 per cent is reflected, principally by clouds, but also by the earth's surface itself. The equilibrium black body radiation temperature corresponding to this reduced absorbed radiation is −20° C.
— Second, the occurrence of natural greenhouse gases in the atmosphere raises this temperature again by 35° C, to a habitable mean global temperature of 15° C.
— Third, the earth is not a perfect heat conductor, nor a perfect heat isolator. The temperature distribution on the earth is governed not only by radiative processes but also by the horizontal heat transport via the atmospheric and oceanic circulation systems. Without this redistribution of heat from the tropical belt to polar regions through winds and ocean currents, the tropics — which receive significantly more solar radiation than the polar regions — would be approximately 20° C warmer than observed today, and the polar regions would be 30–40° C colder. The atmospheric and oceanic circulation systems therefore play a major role in maintaining the present habitable temperature distribution over most of the earth. The poleward heat transports by the atmosphere and the ocean are of comparable magnitude. Thus both systems need to be included in a realistic climate model.

Figure 3 shows schematically the major processes and sub-systems which need to be considered in a complete description of the climatic system. In addition to circulation models of the atmosphere and the ocean, a complete climate model must include also models of the cryosphere, the biosphere and the biogeochemical cycles. The cryosphere (consisting of the components sea ice, ice shelves, continental ice sheets and snow) affects the earth's climate through the storage of water (principally in the form of ice sheets), through the high reflectivity (albedo) of snow and ice, and through the shielding of the heat and moisture flux from the ocean to the

Figure 2 — Principal Energy Fluxes Determining the Energy Balance of the Global Climate System

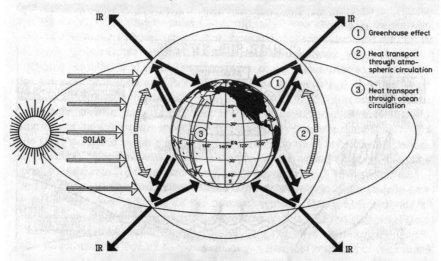

atmosphere by sea ice. The biosphere has a strong influence on the concentration of CO_2 in the atmosphere — the second most important natural greenhouse gas (following water vapour) and the most important anthropogenic greenhouse gas — and also affects the climate through the land vegetation cover, which modifies the albedo and the heat and moisture transfer. The biogeochemical cycles, finally, control the atmospheric concentrations of other important greenhouse gases such as ozone, methane and the CFCs.

With the exception of the biogeochemical cycles, reasonably realistic three-dimensional models of all of the climate sub-systems have now been developed. The typical horizontal resolution of atmospheric and oceanic general circulation models used for climate research is of the order of 500–1,000 km, while the vertical structure is generally described by 10–20 separate layers [Washington, Parkinson, 1986]. Three-dimensional global carbon cycle models based on atmospheric and ocean circulation models [Maier-Reimer, Hasselmann, 1987; Bacastow, Maier-Reimer, 1990] and ice sheet models have a similar resolution.

Modelling biogeochemical cycles is an area of active current research. However, the complexities of the biogeochemical system, involving several hundred different chemical species interacting through a wide variety of radiation-dependent photochemical processes, are so great that realistic models that can be incorporated into existing three-dimensional global climate models will probably not be available for several years.

170

Figure 3 —   The Principal Processes Governing the Dynamics and Interactions
between the Five Major Climate Sub-Systems: Atmosphere, Ocean,
Cryosphere, Biosphere and Land Surface, and Biogeochemical Cycles

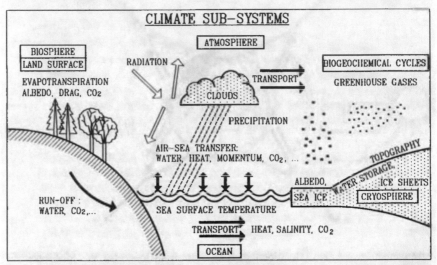

Despite significant progress in the modelling of most climate sub-systems, the
various sub-system models have nevertheless not yet been systematically coupled
together to produce integrated climate models. It is only quite recently that a few
modelling groups (National Center of Atmospheric Research, Boulder; Geophysical Fluid Dynamics Laboratory, Princeton; Max Planck Institute of Meteorology,
Hamburg; and the Meteorological Institute of the University of Hamburg) have
begun to carry out climate simulations with realistic coupled ocean-atmosphere
models.

Nearly all published predictions on future global warming have been based on
simulations with atmospheric general circulation models alone, without considering
the remaining climate sub-systems. The usual approach is to first estimate the future
CO_2 concentrations in the atmosphere for a given future CO_2 emission scenario
using a carbon cycle model, and then to use the computed CO_2 concentration as
input for a subsequent simulation with an atmospheric general circulation model.
An attempt is then sometimes made to determine the possible effects of the
computed changes in the atmospheric climate on the ocean circulation, the biosphere and the cryosphere — for example, to determine the rise in the global sea
level (estimated as 1 m ± 0.5 m for a CO_2 doubling). However, these estimates are
normally not derived from realistic three-dimensional models, and the effects of
these changes on the original computations of the atmospheric CO_2 concentration
and the resulting atmospheric climate change are not considered.

The main reason for the sparsity of coupled model simulations lies in various computational difficulties arising from the disparity of the time scales of the different sub-systems. Whereas the atmosphere responds to a change in external forcing within a few days or weeks — maximally a few months — the response time of the ocean (whose density is a thousand times greater than that of the atmosphere) lies in the range of hundreds to a few thousand years. A separate numerical simulation of either the atmosphere or the ocean over a time period of the order of the natural equilibration time of the sub-system is computationally feasible. Comparable computation times are required for both systems. The shorter response time of the atmosphere is accompanied by a smaller numerical integration time step of the atmosphere model, so that essentially the same number of integration steps is needed to compute the equilibrium response of either the atmosphere or the ocean. When the atmosphere is coupled to the ocean, however, the atmospheric model must be integrated — with its inherently much smaller time step — over the longer time periods characteristic of the ocean response time scale in order to follow the slow evolution of the coupled system. This greatly exceeds the available capacities of even present-day supercomputers. The few numerical simulations with coupled ocean-atmosphere models which have been carried out so far have therefore been limited to the study of the transient response of the climate system over a relatively short integration time period of the order of a few decades to a century and were unable to investigate the transition to the final equilibrium state.

The same difficulty arises when the atmosphere is coupled to the biosphere, whose time scales are comparable with those of the ocean, or the continental ice sheets, which have still longer time scales of the order of several thousand years. Various techniques have been proposed to overcome the basic time-scale mismatch problem — such as iterative integrations, or intermittent ("burst") coupling techniques. However, these are still in the development stage and have not yet been systematically tested. Further work in this area is needed.

4. Some Results of Climate Model Simulations

Extrapolations of the measured increase in CO_2 and other greenhouse gas concentrations in the atmosphere (Figures 4 and 5) indicate that the "equivalent CO_2 concentration" (the net effect of all greenhouse gases expressed in terms of an equivalent CO_2 increase) will double within the next 50–100 years if greenhouse gas emissions continue to grow at present rates. Approximately half of the greenhouse gas forcing is seen to be due to CO_2 (predominantly from fossil fuel use by the industrialized nations, with a small contribution of about 20 per cent from

Figure 4 —   Measured Increase in the Concentration of Carbon Dioxide, Methane,
              Nitrous Oxide and CFC 11 since the Beginning of Industrialization

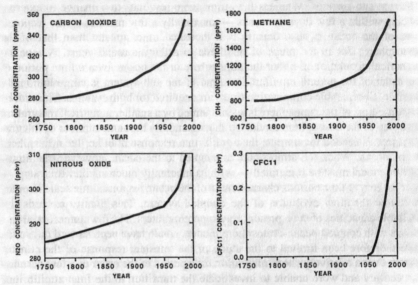

Source: IPCC [1990].

tropical deforestation), while the CFCs and methane constitute most of the remainder.

Figure 6 shows the equilibrium global warming for summer and winter (Northern hemisphere) computed with three different atmospheric general circulation
models (A–GCMs) for a doubling of present CO_2 levels. The global mean warming
for all three models is about 3° C. All three models show higher temperatures in
higher latitudes than in the tropics, particularly during the winter.

The atmospheric models used for such computations are essentially the same as
those used for numerical weather prediction, but are run at a lower resolution
(500–1,000 km, rather than 100 km for weather forecast models, which are integrated for only a few days). They simulate the full time-dependent dynamics of the
atmosphere, including clouds, precipitation, monsoons and all transient tropical and
mid-latitude weather phenomena. However, such simulations must still be regarded
only as order-of-magnitude estimates in view of the significant differences between
different models, particularly on the regional scale. The discrepancies can be
attributed to different representations of poorly known physical processes.

One of the main sources of uncertainty is the treatment of clouds. Clouds strongly
affect the climate through two properties: the high reflectivity (albedo) of clouds,

Figure 5 —   Estimated Contributions of Different Greenhouse Gases to Climate
Warming in the 1980s

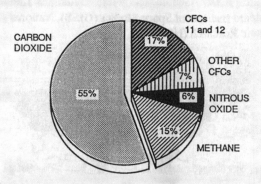

Source: IPCC [1990].

which cools the earth's surface by shielding it from the sun's radiation, and the
greenhouse effect of the water vapour and liquid water in clouds, which increases
the surface temperature. The net change in surface temperature can be of either sign,
depending on the physical properties, latitude and height of the clouds. For low
clouds the albedo effect is normally dominant, while for high clouds the greenhouse
effect is more important. In general, changes in cloud cover in the order of 10 per
cent can produce changes in the global mean temperature which are comparable to
the changes induced by a CO_2 doubling. Unfortunately, we are not yet able to predict
changes in cloud cover to this accuracy.

Figure 7 illustrates the sensitivity of climate models to cloud parameterizations
determined in a model intercomparison study of 14 A–GCMs of comparable
resolution [Cess et al., 1989]. The ordinate represents a measure of the change in
global temperature induced by a given change in the external forcing of the
atmosphere — for example a change of the CO_2 concentration, the solar constant
or the sea surface temperature. The abscissa denotes a parameter characterizing the
"climate effectiveness" of clouds. Without entering into the details of the definitions, the essential point illustrated by the figure is that one and the same forcing
(in the particular experiment shown, a change in the sea-surface temperature) can
produce changes in the global mean temperature which differ by factors of more
than three for models using different cloud parameterizations but otherwise comparable physics and numerics.

The second major source of uncertainty in climate simulations using only
atmospheric general circulation models is the effect of changes in the ocean
circulation. Since the poleward ocean heat transport is comparable to that of the
atmosphere, the neglect of changes in such a first-order process necessarily invali-

174

Figure 6 —   Global Warming (Surface Air Temperature) for the Northern Hemisphere for a Doubling of CO_2, Computed with Three Different Atmosphere GCMs (Geophysical Fluid Dynamics Laboratory (GFDL),
Goddard Institute of Space Studies (GISS), National Center of Atmospheric Research (NCAR))

a) Summer

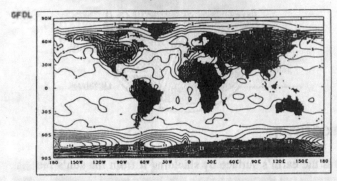

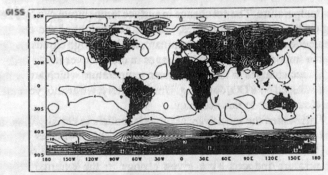

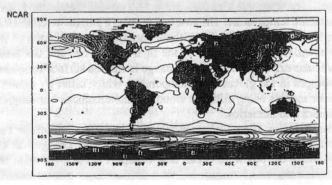

Figure 6 continued

b) Winter

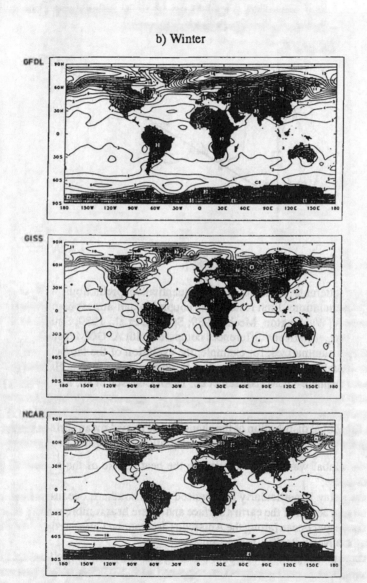

Source: Adapted from Schlesinger and Mitchell [1987].

176

Figure 7 — The Global Sensitivity Parameter λ Plotted against the Cloud Feedback Parameter Δ CRF/G for 14 GCM Simulations (The solid line
represents the best-fit linear regression)

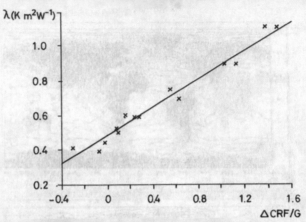

Source: Adapted from Cess et al. [1989].

dates any attempts to make detailed quantitative predictions with equilibrium
A–GCM simulations. First results with coupled ocean-atmosphere general circulation models [Washington, Meehl, 1989; Stouffer et al., 1989; Sausen et al., 1990]
show, in fact, significant differences compared with A–GCM results.

Figure 8 summarizes, for example, the evolution of the temperature field for a
sudden doubling of CO_2 at time t = 0, computed with the Hamburg coupled
ocean-atmosphere-GCM [Sausen et al., 1990; Cubasch et al., forthcoming] over a
100-year period. The changes of the zonally averaged annual mean surface temperatures are shown as a function of latitude and time (Hovmöller diagram). Figure 9
shows the global distribution of the change in the annual mean surface temperatures
averaged over the last ten years of the integration. The figures demonstrate that

— the global warming is delayed by the heat uptake of the oceans by several
decades;
— the delay is particularly pronounced in the southern oceans, which cover a
large fraction of the earth's surface and where heat is more readily mixed into
the deeper ocean (along the Antarctic Circumpolar Current) than in the tropics
or the North Pacific, for example;
— the regional patterns of global warming generally differ significantly from the
equilibrium response patterns computed with A–GCMs alone; however,
— some features of the regional response, such as the enhanced warming in the
mid-continents, are common to both A–GCMs and coupled O–A–GCM
simulations.

Figure 8 — Evolution of Zonally Averaged Surface Temperature as a Function of Time Computed with a Coupled Ocean-Atmosphere Model for a Sudden Doubling of CO_2 at Time t = 0

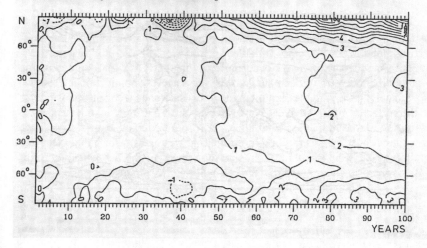

Figure 9 — Global Warming of Surface Temperature Computed with a Coupled Ocean-Atmosphere Model Averaged over Integration Years 91–100 after a Sudden Doubling of CO_2 at Time t = 0

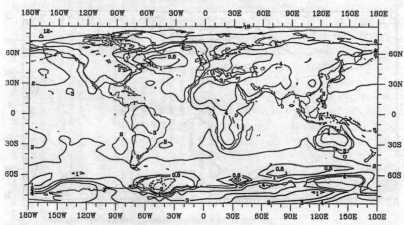

Source: Cubasch et al. [forthcoming].

Although surface temperature is the most commonly used variable for summarizing global warming, other variables such as precipitation, soil moisture or a drought index characterizing the occurrence of long periods without precipitation are generally more useful climate indices for socio-economic impact studies. Unfortunately, these variables exhibit a still higher scatter between different models than the surface temperature does. Nevertheless, certain general features appear to be common to most models, including A–GCMs and coupled O–A–GCMs, such as a tendency for dry areas to become dryer and wet areas to become wetter, an overall increase in the global mean precipitation due to the higher water content of the atmosphere and a more rapid overturning of the hydrological cycle.

5. Greenhouse Gas Predictions

Until we are able to carry out simulations with integrated climate-chemical cycle models, model computations of the climate change due to a given increase in atmospheric greenhouse gas concentrations require as input estimates of the future greenhouse gas concentrations derived from known or assumed anthropogenic emission sources. The future atmospheric concentrations of non-CO_2 greenhouse gases cannot yet be reliably derived from chemical cycle models and are therefore normally estimated by the straightforward extrapolation of present trends. However, for CO_2 fairly realistic carbon cycle models have been developed which can be used to compute future atmospheric concentrations for given emission scenarios.

Figure 10 shows, as an example, four predictions of future atmospheric CO_2 concentrations for four different scenarios of future CO_2 emissions, computed with the Hamburg carbon cycle model [Maier-Reimer, Hasselmann, 1987]. An important conclusion from these simulations is that the future atmospheric CO_2 concentration depends not only on the total amount of CO_2 emitted into the atmosphere but also on the rate at which this emission occurs. This is due to the time-dependent absorption of CO_2 in the ocean (the largest CO_2 reservoir in the carbon cycle system). Up to the present, about half of the CO_2 which has been emitted into the atmosphere has remained in the atmosphere, while the remainder has been absorbed by the oceans. This airborne fraction of approximately 50 per cent is reproduced by the two scenarios (a) and (b), which represent 4 and 2 per cent exponential-logistic growth extrapolations of the past CO_2 emission curve. The zero growth rate scenario (c) and the decreasing emission scenario (d), however, yield significantly lower airborne fractions. The lower emission rates lead therefore not only to a smaller total amount of CO_2 released into the atmosphere but also to a significantly smaller fraction of the total input that remains in the atmosphere.

This bonus is due to the slow transfer of CO_2 into the main CO_2 reservoir in the deep ocean. If a small increment of CO_2 is introduced into the atmosphere, the

Figure 10 — Annual CO_2 Emissions and Atmospheric CO_2 Concentration Computed with the Hamburg Carbon Cycle Model for the Four Emission Scenarios a–d

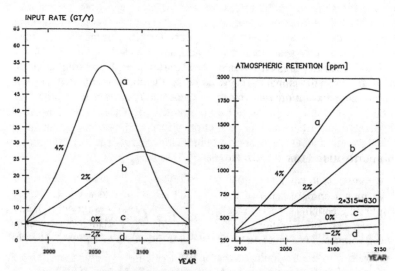

Source: Maier-Reimer, Hasselmann [1987].

carbon system adjusts to a new equilibrium in which the CO_2 increment is ultimately partitioned between the atmosphere and the ocean in the ratio 0.15 : 0.85. However, this asymptotic chemical equilibrium is attained only in the course of several hundred to a thousand years after the additional CO_2 has penetrated into the deep ocean. For the zero growth rate and decreasing emission rate scenarios (c) and (d), the equilibrium airborne fraction of 15 per cent can nevertheless be approached asymptotically. However, for the exponential growth rate scenarios (a) and (b), the large deep-ocean reservoir is unable to keep up with the continually increasing input, and a significantly higher airborne fraction is maintained.

The time constant for CO_2 storage in the deep ocean, which controls the airborne fraction, depends rather sensitively on the rate of deep-water formation. This occurs in today's oceans in the high-latitude regions of the North Atlantic and the southern oceans (particularly in the Weddell Sea). Numerical simulations with ocean general circulation models have demonstrated that relatively small changes in the air-sea heat transfer or surface evaporation and precipitation distributions, which together with the wind drive the global ocean circulation, can significantly modify the rate of deep-water formation in these areas. Maier-Reimer and Mikolajewicz [1989], for example, were able to explain the sudden resurgence of the ice sheets during the brief Younger Dryas interruption of the last post-glacial warming approximately

11,000 years ago by a rapid change in the Atlantic circulation that triggered a reduction of deep-water formation in the North Atlantic through the influx of light surface melt-water from the ice sheets. Simulations with coupled ocean-atmosphere models suggest that a similar reduction in the rate of deep-water formation could be induced by the surface ocean warming in high latitudes accompanying a general global warming. This would seriously impede the ability of the ocean to transport CO_2-saturated surface water into the deep ocean.

This feedback process within the coupled ocean-atmosphere carbon cycle system could lead to a potentially serious global warming amplification that has not been considered in past non-interactive carbon cycle and climate change simulations. It underscores that the prediction of future atmospheric CO_2 concentrations should not be treated separately from the prediction of future climate change: an integrated coupled ocean-atmosphere carbon cycle model is clearly needed. Simulations with such a model are currently being planned in Hamburg.

6. Impact of Climate Models on the Construction of Global Change and Global Environment and Man (GEM) Models

Let me return to the beginning. What inferences can be drawn from the general structure of climate models that I have tried to briefly review for the construction of extended global change models and the integrated GEM models outlined in Figure 1?

Perhaps the most important inference is that the climate system encompasses a broad spectrum of natural time scales. The response of climate to anthropogenic forcing cannot be treated, even in a first approximation, as an equilibrium response problem. The anthropogenic input is inherently time dependent, and the climatic response characteristics are strongly dependent on the form of this time dependence. Since the socio-economic system also contains a broad spectrum of natural time scales, a GEM model must from the outset be conceived as a time-dependent, dynamic system.

A second important result is that the predictions of climate model simulations are always characterized by a finite band of uncertainty. This is the case even if the input (for example in the form of an assumed greenhouse gas emission scenario) is regarded as precisely defined. Thus the input into a socio-economic model provided by a climate (or global change) model, and the resulting costs computed as output by the socio-economic model and used further downstream again as input to the cost-minimizing sector of the GEM model (see Figure 1), must be regarded as probabilistic rather than deterministic. Since the socio-economic and cost-minimization model sectors will also contribute their own uncertainty bands, the problem of minimizing the "total cost-to-mankind function" must necessarily be formulated as a statistical optimization problem. From the point of view of formal mathematics the statistical cost minimization problem is fortunately not less well-defined than a deterministic optimization problem.

However, significant research work is clearly still needed before a reasonably realistic GEM model can be formulated which can be used as the basis for such an optimization procedure. We should nevertheless not lose sight of the ultimate goal of applying climate models together with socio-economic impact models in a general decision analysis and policy definition framework, even while struggling with the complex intricacies of global climate models. This framework should provide the guiding principle for our long-term climate modelling strategy.

In conclusion, I should like to emphasize that despite the many uncertainties in the quantitative details of present climate predictions — on which this summary has naturally focused — and despite the present lack of adequate models with which to assess the socio-economic impacts of the predicted climate change, there exists no serious doubt within the scientific community that the predicted global warming is real; that the estimated orders of magnitude of the predicted climate change are reliable; and that if no corrective measures are adopted, we may expect within the next 100 years the warmest climate ever experienced in the history of mankind. It would be unwise to delay remedial action only because our predictions today — and in the foreseeable future — must be qualified with finite uncertainty bounds. We cannot afford to wait for the perfect GEM model before making political decisions. Policy and model development should be pursued as parallel, interactive, iterative processes.

Bibliography

BACASTOW, R., E. MAIER-REIMER, "Circulation Model of the Oceanic Carbon Cycle". Climate Dynamics, Vol. 4, 1990, pp. 95–125.

CESS, R.D., G.L. POTTER, J.P. BLANCHET, G.J. BOER, S.J. GHAN, J.T. KIEHL, H. Le TREUT, Z.-X. LI, X.-Z. LIANG, J.F.B. MITCHELL, J.-J. MORCRETTE, D.A. RANDALL, M.R. RICHES, E. ROECKNER, U. SCHLESE, A. SLINGO, K.E. TAYLOR, W.M. WASHINGTON, R.T. WETHERALD, I. YAGAI, "Interpretation of Cloud-Climate Feedback as Produced by 14 Atmospheric General Circulation Models". Science, Vol. 245, 1989, pp. 513–516.

CUBASCH, U., K. HASSELMANN, H. HÖCK, E. MAIER-REIMER, U. MIKOLAJEWICZ, B. SANTER, R. SAUSEN, "Transient Greenhouse Warming Computations with a Coupled Ocean-Atmosphere Model". Paper submitted to Nature. Forthcoming 1991.

INTERGOVERNMENTAL PANEL OF CLIMATE CHANGE (IPCC), Working Group 1, Climate Change. The IPCC Scientific Assessment. Edited by J.T. Houghton, G.J. Jenkins, and J.J. Ephraums. Cambridge 1990.

MAIER-REIMER, E., K. HASSELMANN, "Transport and Storage of CO_2 in the Ocean: An Inorganic Ocean-Circulation Carbon Cycle Model". Climate Dynamics, Vol. 2, 1987, pp. 63–90.

—, U. MIKOLAJEWICZ, "Experiments with an O-GCM on the Cause of the Younger Dryas". In: A. AYALA-CASTANARES, W. WOOSTER, A. YANEZ-ARANCIBIA (Eds.), Oceanography. Mexico D.F., 1989, pp. 8–100.

MIKOLAJEWICZ, U., B.D. SANTER, E. MAIER-REIMER, "Ocean Response to Greenhouse Warming". Nature, Vol. 345, 1990, pp. 589–593.

SAUSEN, R., U. CUBASCH,M. BÖTTINGER, K. HASSELMANN, F. LUNKEIT, G. LÜDGENS, E. MAIER-REIMER, U. MIKOLAJEWICZ, J.M. OBERHUBER, R. RODZUN, E. ROECKNER, B. SANTER, D. SCHRIEVER, Simulation des transienten CO_2-Treibhauseffektes mit gekoppelten Atmosphäre-Ozean-Modellen. Vorläufiger Bericht des Max-Planck-Instituts für Meteorologie und des Meteorologischen Instituts der Universität Hamburg. Hamburg 1990.

SCHLESINGER, M.E., J.F.B. MITCHELL, "Climate Model Simulations of the Equilibrium Climate Response to Increased Carbon Dioxide". Review Geophysics, Vol. 25, 1987, pp. 760-798.

STOUFFER, R.J., S. MANABE, K. BRYAN, "Interhemispheric Asymmetry in Climate Response to a Gradual Increase of Atmospheric CO_2". Nature, Vol. 324, 1989, pp. 660–662.

WASHINGTON, W.M., C.L. PARKINSON, An Introduction to Three-dimensional Climate Modeling. Mill Valley, CA, 1986.

—, G.A. MEEHL, "Climate Sensitivity due to Increased CO_2: Experiments with a Coupled Atmosphere and Ocean General Circulation Model". Climate Dynamics, Vol. 4, 1989, pp. 1–38.

3.6   Strategy in Climate Modelling at MPI[19]

As computers became more powerful, comprehensive modelling of the Earth's climate system developed rapidly and increasingly came to be considered as a key tool for gaining a better understanding of the climate system. Progress had been made in the use of global weather prediction models in particular at ECMWF, which was established by a number of European member states in 1979. The parameterisation of many physical and fine scale dynamical processes requires systematic experimentation in operational daily predictions to define a well-functioning forecasting model. To use such a well-tested global model for climate simulation and forecasting and in much longer integration was found to be a most useful and a practical strategy, as many aspects of the weather forecast model could naturally be adjusted to climate research.

Following discussions between Klaus Hasselmann and ECMWF it was agreed that this was the approach to be taken. The ECHAM model was than combined with different ocean models developed at MPI and the University of Hamburg in such a way as the exchange of energy and momentum fluxes between atmosphere and oceans could be handled consistently. Different subsystems, such as that for atmospheric transport and the full carbon cycle were subsequently added and integrated into the coupled atmospheric-ocean model.

Thus, from the 1990s onward, the MPI had a comprehensive set of climate models that constituted a numerical laboratory for all kinds of climate studies including climate change simulation studies made available for all IPCC assessments. The system was set up in a systematic and flexible way, which made all sorts of climate studies possible. It was not only used by scientists at the MPI and the University of Hamburg but also by a large number of research groups in Germany as well by associated European groups and by visiting scientists from all over the world. Important studies to understand and predict the ENSO phenomenon as well as tropical and extra-tropical cyclone were carried out successfully. We were happy to learn that the ECHAM model was found to be one of the most realistic ones in several evaluation studies, as its results came closest to the observed climate.

---

[19] By Lennart Bengtsson.

Of particular importance were the diagnostic systems developed by Klaus Hasselmann and his group, the aim of which was to identify anthropogenic climate change through a multi-dimensional search for a climate change "fingerprint" in the modelled data sets. This turned out to be a powerful tool for enabling the detection of climate change as early as the late 1980s. It played a very important role for the IPCC in its bid to convince the world that anthropogenic climate change is really happening. Today, 30-years later, the signal of climate change is obvious for everybody.

One crucial factor in the successful modelling work was the positive, unbureaucratic, and open atmosphere that was due in large part to Klaus Hasselmann's clear mind and stimulating personality.

Summary of the Report

DKRZ 2000 +

Proposal for the Development of DKRZ into the next Century

K. Hasselmann and W. Sell

September 1995

Summary

I. <u>Responsibilities of the German Climate Computing Centre</u>

Humanity is confronted today with the threat of a drastic change in global climate due to increasing emissions of greenhouse gases and other human activities. In order to develop effective climate-protection strategies, possible future climate changes and their consequences for society must be understood and quantified. This can be achieved only by means of complex model simulations using high-performance computer systems.

Reliable climate models must be based on detailed knowledge of the physical, chemical and biological processes which control the climate system. This requires also an understanding of natural climate variability, which is superimposed on the anthropogenic climate change signal over a broad range of time scales.

Natural and anthropogenic climate change, climate impact studies and climate protection strategies are closely interrelated and can be addressed successfully only through an integrated systems analysis approach.

The establishment of the German Climate Computing Centre (Deutsches Klimarechenzentrum, DKRZ) in 1987 as one of the few dedicated high-performance climate computing centres world-wide (together with the National Center for Atmospheric Research - NCAR - and the Geophysical Fluid Dynamics Laboratory - GFDL - in the USA, the Hadley Centre in Britain, and the French Centre Meteo-France) laid the foundation for effective climate research in Germany. The international standing of German climate research today can be attributed in large part to the computational work carried out at the DKRZ.

However, the German Climate Computing Centre faces considerably more difficult challenges in the last decade of this century than was the case at the time of its establishment eight years ago.

These result from

- **recent advances of climate modelling together with the increasing public awareness of the climate problem; these have created a strong demand for reliable predictions of future climate,**

which are urgently needed as a basis for policymaking, in particular with regard to the negotiation of international commitments for the reduction of CO_2 emissions within the framework of the Climate Convention (Berlin Mandate);

- **the related need to provide climate input data for climate impact studies,**

which require greatly increased spatial resolution of climate model simulations down to regional scales;

- 2 -

- the substantial additional computing time required for climate impact research,

resulting in particular from the Berlin Mandate and the anticipated high costs of an effective climate protection strategy (the Enquête-Commission "Protection of the Atmosphere" of the Bundestag estimates Germany's share alone at 300 billion DM); this requires a major expansion of climate research into new fields;

- the requirements of the international "Global Change" Programme,

of which climate research represents an essential element, and which will demand an extension of the present climate model hierarchy to the significantly more complex and computer-intensive areas of atmospheric chemistry, biogeochemical cycles, hydrology and ecology;

- the increasing role of the DKRZ as a centre of expertise,

which follows from the evolving networking of climate research and the growing demand of external users for active participation in complex climate simulations carried out at DKRZ;

- the concomitant increasing demand for high-performance computing resources by European research partners within the framework of the ECCN (European Climate Computer Network),

which arises through the reputation of the DKRZ as a European centre of expertise and which can be projected to expand if the DKRZ, with support from Hamburg's climate research institutes, continues to play a leading role in the European climate research community;

- the rapid developments in the computer market,

particularly the forthcoming introduction of considerably more powerful moderately, highly or massively parallel computers, which will require support of a new programming paradigm and a stronger commitment of the DKRZ in the area of scientific computing,

- the need for effective access to climate data,

particularly in support of the increasing number of users seeking simple and efficient access to the enormous quantities of data being produced by ever more powerful computer systems,

- the need for efficient data communication lines

in order to connect an increasing number of clients more effectively to the DKRZ.

- 3 -

II.   Consequences

Computer Performance

In order to meet these challenges, the computing power of DKRZ will need to be increased by at least two orders of magnitude within the next decade. This will become technically feasible through the development of new highly and massively parallel computers. The annual investments for the new computer generation are estimated to be of the order of 8-10 million DM (at today's costs). The high-performance compute-server used for large-scale simulations will need to be complemented by computer systems for data services, model development, model diagnostics and visualization, together with a high-speed communications network connecting the individual systems. These additional components, which are essential for an efficient utilization of the high-performance compute-server, will require an additional annual investment cost of 3-4 million DM. The operating costs of the computer system and the present level of services should require only a moderate increase compared to the present level of expenditures, which is of the order of 13 million DM. The enhancement of the data and model support services, together with the planned new scientific computing activities, could cost an additional 2-3 million DM per year. The total costs are high, but it must be realized that high-performance computers represent an indispensable tool for modern climate research - comparable to accelerators in particle physics or satellites in space research.

Although a trend towards a decentralization of computer resources can be observed in several research fields, in climate research the need for high-performance computers for carrying out complex simulations with high-resolution global climate models will, if any thing, increase in the foreseeable future. The future computer requirements in climate research will need to be met at three levels. The DKRZ facilities serve on the first level mainly as a powerful resource for elaborate large-scale simulations with high-resolution climate models, planned and carried out cooperatively by several institutes. A second level of computer systems, located in various climate research institutes, should be used for advanced simulations carried out at the individual institute level, while a third level of workstations should be used for the local processing of simulation results and the development of model components, algorithms, diagnostic software etc.

Global Change and Climate Impact Research

The incorporation of problems related to global change and climate impact within the scope of DKRZ responsibilities will automatically enlarge the field of DKRZ users. Because of the close interrelationship between research in climate and climate impact, the DKRZ would welcome the Potsdam Institute of Climate Impact Research (PIK) as a fifth shareholder. In order to enable PIK and other external research groups involved in these broadened research activities to use the DKRZ resources effectively, they should be connected to DKRZ via high-speed data links. The investments which have already been made in the national and European communication infrastructure could yield a far greater return for science through an appropriate financing policy for a scientific data network.

- 4 -

Evolution to a Service and Expertise Centre

By the turn of the century, DKRZ will have evolved from a computing centre with a primarily technical and operational orientation into a combined service and expertise centre. For the successful application of sophisticated climate models, consisting in general of a number of coupled complex subsystem models, climate researchers not directly engaged in model development must be advised by experts at DKRZ who are thoroughly acquainted with the models. The advisory role will gain in importance as DKRZ becomes more involved in the field of global change and with the increasing use of DKRZ by external research groups. In addition to providing general modelling support, the DKRZ user support group should also carry out routine modelling tasks, both to relieve basic research groups from these activities and to maintain the expertise of the group. These could include, for example, scenario computations of global greenhouse warming or model intercomparisons.

Scientific Computing Group at DKRZ

In the course of the evolution into a centre of expertise, a Scientific Computing Group should be established at DKRZ. Its principal task would be to test and implement algorithms and computing methods suitable for the coming generation of parallel computers and to assist users in the optimization of their models for the new parallel systems. Such an advisory service by a Scientific Computing Group familiar with new algorithms will be required also for future procurements of parallel computer systems selected to optimally satisfy the DKRZ needs.

Scientific Computing Network

The new DKRZ Scientific Computing Group will need to cooperate closely with similar groups at the Potsdam Institute for Climate Impact Research (PIK) and the Alfred-Wegener-Institute for Polar and Marine Research (AWI) in Bremerhaven. Work in this field should be coordinated through a Scientific Computing Network, in which, in addition to DKRZ, PIK and AWI, all institutes involved in climate and global change research would participate. Joint projects carried out within this network would be directed by scientific project coordinators reporting to the institutes involved in the project. Responsibilities would be divided between institutes and project coordinators in accordance with established matrix structures for such cooperative projects.

Joint Large-Scale Simulations

New forms of coordination need to be introduced at DKRZ in order to carry out extensive simulation experiments in the form of joint undertakings by a larger number of institutes. The increasing importance of such large joint simulations - also within the European context - requires organizational structures similar to those developed for experiments at high-energy particle-accelerator installations. The European Climate Computer Network (ECCN) and a proposed European Concerted Action Project "Cooperative Climate Simulation Experiments" may be regarded as first steps in this direction.

- 5 -

Institutional Structure

These developments demand reconsideration of the terms of reference, structure, and financing of DKRZ. The experience in the past has been good both with respect to the present institutionalized close interaction between the DKRZ and its supporting institutions, i.e. covering basic research and applications, and with respect to the legal status of the DKRZ as a limited corporation, jointly financed by the German Ministry of Education, Science, Research and Technology (BMBF) and the DKRZ shareholders.

This organizational structure was designed to meet the original responsibilities of DKRZ. It was largely oriented towards establishing an optimal balance between the diverse interests of the participating partners in basic and applied research. It had the great advantage during the initial evolution of the DKRZ of offering a high degree of flexibility. This would almost certainly not have been possible with other conventional institutional structures (e.g. "Blue List" Institute or Large Research Facility).

In considering the future organizational structure and financial security of the DKRZ, the positive experience gained with the present organizational and legal status should be kept in mind, especially with regard to the ability of reacting flexibly to new situations and the evolving, not necessarily parallel interests of the various DKRZ partners.

- 6 -

III. <u>Relation to Problems of Global Climate and Environment</u>

The costs for the DKRZ should be viewed in relation to the considerably greater national expenditures for modern climate and environmental research, including the international systems of earth-observing satellites and global observational networks. Without high- performance computers and complex models, the enormous flood of data provided by these systems cannot be effectively evaluated and interpreted.

The total expenditures for climate and environmental research are in turn infinitesimal compared with the price of adapting to global warming or of preventing or ameliorating climate change. This is of the order of hundreds of billions of DM. The far smaller expenditures for reliable climate and climate impact studies as input for the development of optimized climate-change prevention and adaptation strategies promise an exceptionally high return for investment. But a cost-benefit analysis hardly seems appropriate in this context. Humanity is confronted today with far-reaching decisions which will have an irreversible effect on the lives of all people. We cannot experiment with the earth, and it is too complex to be reproduced in the laboratory. The only available prognostic tool is the computer model. Climate model simulations are therefore indispensable if the impending far-reaching political decisions are to be founded on a rational scientific basis.

Although present-day climate models provide invaluable information regarding the expected future change in climate, the spatial resolution and physical-chemical-biological foundations of the models must be improved considerably in order to provide fully relevant data for detailed political planning purposes. In addition, extensive studies are required to clarify the interrelationships between climate change, climate change-impact and climate-protection strategies. As a leading industrial nation with a proclaimed vanguard role in climate-protection policy, and as host to the secretariat of the Berlin Mandate for establishing a climate-protection protocol, Germany should continue to provide the best available computational and technical conditions for climate model simulations on which political decisions will necessarily strongly depend.

3.7   Metrons—Particle Physics[20]

Inspired by his work on interactions between ocean surface waves and other wave phenomena, which he approached using perturbation theory with the aid of Feynman diagrams in the 1960s, Klaus Hasselmann went on to develop a unified deterministic theory of fields and particles thereby realising Einstein's dream of a deterministic description of all elementary particles and their interactions [121, 122, 131, 132]. Quantum theory is regarded only as a first approximation.

The theory is based on solutions to an extension of Einstein's vacuum equation with an additional attenuation tensor using solitons or solitary waves, which he set out in papers published in 1996 and 1997. These solutions were then verified in computational models.

In his twelve-dimensional theory (four space–time and eight extra dimensions, representing interacting non gravitational wave lengths as well as electromagnetic, strong, and weak forces.) he takes a classical view of real particles and their guiding force free waves (de Broglie waves). He sees solutions in which solitons are trapped by waveguides, in accordance with the theory proposed by Theodor Kaluza and Oskar Klein, as being exemplary.

His theory was rejected by particle physicists. Reactions ranged from polite smiles to pronounced aggression, as already stated above in the interview between Hasselmann, Olbers, and von Storch (Sect. 2.1). So, he started writing a book: *The Metron Model: A Classical Unified Theory of Fields and Particles*. We present the first chapter below, which contains the basic theory and a short description of the following chapters in an overview section.

Chapters 5–8 have not yet been completed. The manuscript can be downloaded in its present state. The model programmes, which are written in Fortran V, are also available from Susanne Hasselmann.

Maybe someday, when the careers of young physicists no longer depend on the sheer volume of their publications but rather on their originality, some young physicist may feel inspired to flesh out and complete the metron theory: this is the hope of Klaus and Susanne.

---

[20] Prepared by Susanne Hasselmann.

The Metron Model: Elements of a Deterministic Unified Theory of Fields and Particles

Klaus Hasselmann

Max Planck Institute for Meteorology

file: Metron-book-18thNov-2016-Integrated version

December 13, 2016

Contents

Chapter 1

The Metron Concept

1.1 Introduction

The basic paradoxes faced by physicists in the early part of the last century produced two fundamental revolutions in our understanding of physical reality. The paradoxes were resolved, however, through two diametrically opposite conceptual innovations that have yet to be reconciled within a unified theory. This book is an attempt to overcome the long-standing dichotomy by introducing a new perspective on the interrelation between particles and fields.

The paradox of the Galilean invariance of the speed of light was resolved by Einstein through the recognition that the concepts of space and time could not be accepted as self-evident *a priori*, but must be defined explicitly, together with the properties of the other physical objects with which they were to be combined in a physical theory. The approach was then generalized further by Einstein to explain gravitation as a geometric property of spacetime. In contrast, the second paradox, the simultaneous observation of both wave-like and corpuscular features in microphysical phenomena, was resolved by Born, Heisenberg, Jordan, Bohr, Pauli and other proponents of quantum theory by the opposite premise that microphysical objects could not be precisely defined. Meaningful was not the concept of a precisely defined real existing particle or field, but only the ability to statistically predict

1

the outcome of experiments. The experiments themselves, as emphasized by Bohr, must nevertheless still be described in terms of the microphysical objects that had been declared before to defy precise definition. The search for an escape from this apparent logical circularity has spurred innumerable ontological and epistemological investigations and a continual debate among philosophers of science.[1]

Despite their diametrically opposite ontological foundations, both general relativity and quantum theory have been extremely successful in explaining a wide array of physical phenomena. Quantum theory, and its relativistic generalization, quantum field theory, in particular, have reproduced an extremely diverse ensemble of experimental data, many with impressive accuracy. And it has withstood for more than eight decades all attempts to demonstrate that the theory is axiomatically inconsistent. Indeed, the much discussed critical thought experiment of Einstein, Podolsky and Rosen [25], suggesting that quantum theory was incomplete, is widely (mis)interpreted today, in conjunction with the experimental verification of Bell's [4] inequality theorem on the EPR experiment, as demonstrating that classical theories are basically incapable of explaining the phenomena of entanglement exemplified by the EPR experiment. It will be shown later, however, that entanglement is in fact a necessary consequence of any microphysical theory, whether classical or quantum theoretical, that satisfies the general accepted requirement of micro-physical time-reversal symmetry.

It is therefore understandable that most unification approaches, such as supergravity,[29],[58], string theory [35],[63],[69],[44],[71], quantum loop gravity [70],[11] or the AdS-CFT correspondence [19], seek to incorporate gravity in a broader mathematical framework that accepts the basic tenets of quantum field theory. Nevertheless, in the following the opposite route is explored: the explanation of quantum phenomena in terms of real existing particles and fields representing solutions of a higher-dimensional generalization of the Einstein equations.

The difference between the standard quantum theoretical and alternative classical picture developed in the following can be illustrated by the standard example of the double-slit particle diffraction experiment, which, in Feynman's words,[26], "is impossible [....] to explain in any classical way,

---

[1]See, for example, [5],[10],[20],[22],[28], [48],[53],[60]-[62] ,[72],[76].

and which has in it the heart of quantum mechanics. In reality, it contains the only mystery [of quantum mechanics]." A monochromatic beam of particles incident on a double-slit diffraction screen appears on the one hand (for sufficiently low beam intensities) as an ensemble of individual particles: each individual particle can be localized, with acceptable accuracy, both before the particle passes through the diffraction screen and after striking a detector screen some distance behind the screen. On the other hand, the statistical distribution of a large ensemble of diffracted particles is found to correspond to that of an incident periodic wave field, rather than that of an ensemble of localized particles. How can this contradiction be resolved?

Quantum theory asserts simply that it cannot be resolved. At least, not within a classical theory in which both fields and particles are defined as simultaneously existing real objects. Feasible is only a probabilistic description of particles and fields in terms of quantum states, in which the simultaneous exact specification of complementary properties establishing precisely both the corpuscular and the wave-field properties of an individual quantum state is explicitly excluded.

It is generally accepted that this interpretation of physical reality does not lead to logical contradictions. In particular, the much discussed question as to which of the two slits a particular particle has actually passed through is simply rejected as meaningless. For it cannot be answered without modifying the experimental set-up (for example, by closing one of the two slits). On the other hand, Einstein was not the only physicist who was uncomfortable with the implication that it is meaningful to speak of the physical state of a system only if it is directly amenable to measurement [60] - as evidenced by the never-ending discussions over the fate of Schrödinger's cat [67] or Einstein's question "When is a moon not a moon?".

The explanation of the double-slit experiment proposed in the following is more straightforward. The existence of objects exhibiting both corpuscular and wave-like features is simply accepted as a fact. If physics is to remain within its classical explanatory foundations, there must therefore exist a theory that explains the existence and structure of real objects exhibiting such dual properties. It is postulated that the sought-for objects represent soliton solutions of some appropriate set of nonlinear field equations. A natural candidate for the proposed field equations, following Kaluza [49] and Klein [51],

3

is Einstein's theory of gravity, suitably generalized to the metric of a higher-dimensional space to include the electromagnetic, weak and strong forces in addition to gravity. The dependence of the metric on the higher dimensional coordinates is severely restricted, however, following again Klein, to be either constant or limited to a small number of prescribed periodicities. Thus the model world remains fundamentally four dimensional. Consistent with this restriction, "extra space" will normally be referred to in the following as a *fibre*, and full space, composed of four-dimensional spacetime plus the fibre, as a *bundle* [73], [34],[47],[2]. Nevertheless, the extra-space picture will also occasionally be invoked, for example in the derivation of inter-particle forces as higher-dimensional geodetic accelerations.

A plausible structure of the hypothesized *metric* solit*ons* (or *metrons*) is not difficult to conceive. Consider, for example, the soliton representation of an electron. Assume the soliton contains a strongly nonlinear spherically symmetric core with a spatial scale of order κ^{-1}. From this there emanate two fields, corresponding to the classical gravitational and electromagnetic fields of a point-like particle, decreasing asymptotically as $1/r$ with the distance r from the particle core. Assume further that the gravitational and electromagnetic fields are produced by nonlinear interactions of a further soliton field that is periodic with respect to time and the fibre variables and is trapped within the soliton core. The periodic core field falls off exponentially as $\exp(-\kappa r)/(\kappa r)$ beyond the core region. It represents a time-symmetrical standing wave; there is no energy loss through radiation to infinity.

Finally, let the periodicity ω of the core field with respect to time be given by de Broglie's basic wave-particle duality relation, $\hbar\omega = mc^2$, where m is the particle mass. The periodic core field and the gravitational and electromagnetic fields can then be identified as the *fermion, graviton* and *boson* components, respectively, of a composite particle exhibiting both corpuscular properties, in the form of a localized core, and field properties, composed of the non-periodic gravitational and electromagnetic fields and the periodic fermion field.

This simple soliton picture can then be generalized to describe more complex objects, leading, finally, to a detailed representation of the observed elementary particle spectrum. Common to all metron particle models is the existence of mutually supporting periodic core fields, representing fermions,

4

and non-periodic (or, in some cases, periodic) fields, representing bosons and gravitons.

Once the existence of such composite objects is accepted, the wave-particle duality paradoxes of micro-physics can be readily resolved. In interactions between well separated particles, the exponentially decreasing periodic fermion field is too small to be observed. Moreover, for non-relativistic particles, the frequency of the field is too high to have a significant direct impact on the path of a particle. However, in interactions of a particle with some other object at smaller separations $d \leq \kappa^{-1}$, the particle's periodic field scattered by the other object interacts back on the original particle field. Since the velocities of the particle and scattering object differ, the frequencies of the primary and scattered fields also differ, resulting in the generation of a low-frequency difference-interaction field. This creates the typical wave-like interference patterns observed in particle diffraction experiments.

Applied to the double-slit diffraction experiment, the path of a diffracted soliton is well defined, but depends on the initial random scattering impulse received by the particle (the "hidden variable") on its passage through the screen. The subsequent interaction with the diffracted field then creates the diffraction pattern. Each individual particle does indeed pass through only one of the two slits, but the path taken before and after passing through a particular slit depends on the interactions of the particle's periodic core field with the net diffracted field produced by both slits.

A similar classical picture can be invoked to explain also the double-slit diffraction experiment for photons. However, in contrast to quantum theory, in which the material-particle and photon diffraction cases are treated as essentially identical, requiring only an interchange of the concepts "material particle" and "photon", the simple metric soliton picture sketched above cannot be applied directly also to the photon diffraction case. The model contained a periodic core component, identified as a fermion field, together with non-periodic gravitational and electromagnetic far fields, but no massless periodic electromagnetic field that could be identified with a photon. In the metron picture, photons are not represented as massless solitons, but rather as the far-field interactions between separated particles undergoing atomic transitions or other changes of state, following Einstein [24]. To explain the photon double-slit experiment, the metron model therefore needs to be extended first to atomic theory. The wave-particle duality paradox can then be readily resolved also for the photon diffraction case by interpreting the photon as the time-symmetrical electromagnetic field connecting the atomic state that emits the photon, before the screen, with the atomic state that absorbs the photon, behind the screen. Apart from the time-symmetrical rather than the usual time-asymetrical treatment of the electromagnetic field, the analysis is then identical to the standard photon diffraction case.

According to the classical picture, an atom consists of a discrete set of electrons orbiting a central kernel. Considering only the Coulomb forces that balance the individual electron's centrifugal forces, one would anticipate a continuum of possible orbits, which should furthermore gradually collapse due to radiation damping. How can one then explain the observed discrete structure of atomic spectra? This results in the proposed soliton model from the additional interaction of the orbiting electron with the low-frequency field produced by the quadratic interaction between the electron's primary periodic field and the secondary periodic field scattered at the atomic nucleus. The resultant low-frequency difference-interaction field depends in general on the orbit. There exist, however, special orbits – or, more precisely, discrete stationary positions of the electron – for which the force resulting from the difference-interaction of the back-scattered fermion field with the original fermion field exactly balances the electromagnetic attraction between the opposite electric charges of the electron and atomic kernel. This difference-interaction force replaces the centrifugal force of an orbiting electron. The electron is at rest; there exists no centrifugal force, and therefore no radiation damping. Starting from an arbitrary initial orbit, the electron drifts through radiation damping into these stable equilibrium positions.

The thereby resurrected modified version of the original Bohr orbital model is at the same time closely related to modern quantum field theory: the field equations of the difference-interaction field are found to be identical to the field equations of quantum electrodynamics.

Applied to the photon version of the double-slit diffraction experiment, the analysis of the interactions between the incident and diffracted photon fields then follows closely the corresponding analysis for finite-mass particles. The coupling between the initial photon-emitting and final photon-absorbing states is represented by a time symmetrical Lagrangian.

6

Similar interactions involving the fermion fields of the proposed composite particle model also yield straightforward classical explanations of Compton scattering, the photo-electric effect, and Planck's law.

The representation of elementary particles as solutions of the higher dimensional Einstein equations, finally, yields a simple geometrical explanation of the basic $U(1) \times SU(2) \times SU(3)$ symmetries of the Standard Model of quantum field theory: the symmetries follow from the invariance of the Einstein equations with respect to coordinate transformations.

1.2   Historical development

The metron model may be regarded as a synthesis of a soliton model with the pilot-wave concept of de Broglie and Bohm ([17], [18],[7], [45]). Soliton-particle models were popular at the turn of the 19th century (see, for example, [1], or Lorentz's [55], [56] model of the electron). However, interest in solitons was lost when the mounting paradoxes of microphysics undermined the belief that particles could be represented by something as simple as a classical soliton, leading instead to the creation of quantum theory. (The soliton solutions developed in the following will, indeed, no longer be simple, but will exhibit successive levels of complexity comparable, finally, to the Standard Model of quantum field theory – as, of course, must be expected, if quantum field theory is to be translated into an analogous classical picture including gravity).

The de Broglie-Bohm model, despite its attractive classical foundation, suffered from the basic shortcoming that the guiding pilot wave that produced the wave-like interference patterns was introduced *ad hoc* as an additional field, independent of the particle representation. In the present soliton model, the relevant periodic fermion field is an integral component of the soliton particle itself, fully integrated within the basic particle dynamics. The interactions between the primary and scattered fermion fields that produce the interference patterns also differ fundamentally from the relevant interactions in the de Broglie-Bohm model.

Essential for the metron model is that fermions, bosons and gravitons are not

7

regarded as independent particles, but as different components of a single particle. A photon emitted from an atom, for example, is represented, in the spirit of Einstein [24], as the electromagnetic far-field expression of an electron undergoing a transition from one atomic orbital state to another, rather than as an independent particle. The distinction is immaterial in many applications, but is important, for example, for the explanation of entanglement (see Chapter 3).

The proposed classical soliton model presents a picture of physical reality that clearly differs in its very core from the established quantum theoretical picture that evolved from the fundamental "Dreimäennerarbeit" of Born, Heisenberg and Jordan [8] and the equivalent alternative formalism proposed independently by Schrödinger [66]. The subsequent generalization of these non-relativistic approaches to a relativistic theory by Dirac [21] required then the re-interpretation of the fields as operators, leading to a higher level of abstraction and the need to remove the resultant divergences in the perturbation expansions through formal renormalization methods. Quantum field theory has since been extended and elaborated through a host of further developments, including parity-violating weak interactions, strong interactions, the generation of fermion and weak-interaction boson masses by the Higgs mechanism, and the representation, finally, of the elementary particle spectrum in terms of the Standard Model.

Nevertheless, despite these impressive achievements, quantum field theory suffers from two fundamental shortcomings: the need to remove divergences through rather artificial renormalization techniques, and the exclusion of gravity. From the viewpoint of the present approach, both problems have the same origin: quantum field theory represents an ingenious theory of incomplete information[2]. Its successes derive from its ability to capture the field properties of quantum phenomena. Its difficulties stem from its inability to represent the associated corpuscular features through its rejection of the concept of real existing particles.

An alternative classical theory that strives to overcome the shortcomings of quantum field theory faces two challenges: first, the derivation of the spectrum of elementary particles from the soliton solutions of the proposed field

---

[2]In the more explicit words of Holland[45]: "Quantum mechanics is the subject where we never know what we are talking about."

equations; and second, the quantitative explanation in terms of this alternative particle picture of the basic microphysical phenomena that motivated the creation of quantum theory. The first task focuses on the individual soliton solutions of the field equations, the second on the interactions between the computed soliton solutions. The conclusion of the exercise should be the translation of the accepted picture of quantum field theory into a classical, divergence-free representation of real existing particles and fields, including gravity.

This is clearly an ambitious program that one cannot expect to complete and present in a single publication. The basic concepts were developed in an earlier four-part paper ([37] - [40]; see also [41], [54] and progress reports [43], [42]). However, in these first papers, the postulated soliton solutions were not explicitly computed, their structure being inferred only from an inverse analysis; numerical computations were limited to a simpler scalar analogue of the proposed field equations illustrating the basic soliton trapping mechanism. In the following, these earlier investigations are extended (and in important points revised) by providing numerical solutions of the fully specified field equations and establishing more clearly the relation of the proposed metron model to standard quantum field theory.

The development of the metron model in the following chapters alternates between the above two goals: the representation of elementary particles, and the application of the elementary particle models to explain quantum phenomena. The complexity of the models increases as the theory evolves, reflecting the analogous historical development of quantum theory and quantum field theory. However, all models are based on the generalized Einstein equations and therefore exhibit general-relativistic invariance from the start.

1.3   Metron properties

Before entering into the detailed development of the model, it is useful as orientation to summarize briefly the principal properties of the model:

P1. Elementary articles represent soliton solutions of the generalized higher-dimensional Einstein vacuum equations

$$R_{LM} + A_{LM} = 0 \tag{1.1}$$

where R_{LM} and A_{LM} denote, respectively, the Ricci tensor and a generalized "cosmological" or – in the more appropriate terminology of the later definition, eq.(2.12) – *attenuation* tensor.

The solitons represent spherically symmetric perturbations relative to a constant background metric $\eta_{LM} = \mathrm{diag}(1, -1, -1, -1; -1, -1, \ldots, -1, 1)$, in which the extra-space signature $(-1, -1, \ldots, -1, 1)$ mirrors the signature $(1, -1, -1, -1)$ of spacetime.

The solitons are composed of periodic and non-periodic components. Following Kaluza [49] and Klein [51], all components are assumed to be independent of or periodic with respect to the extra-space coordinates. The extra-space periodicities are prescribed, with separate wavenumber components k_l representing the charges of the electromagnetic, weak and strong forces.

Formally, the restriction of the extra-space dependence to a few prescribed periodicities reduces extra-space to a template for deriving field equations in four-dimensional spacetime. The extra-space coordinates no longer appear in the final field equations. As mentioned previously, it is therefore more consistent to refer to the restricted extra-space as a *fibre* defined over physical spacetime as *base space*, and to the full higher dimensional space as a *bundle* It will nevertheless remain useful to refer occasionally to the concept of an - albeit restricted - higher-dimensional space. The symmetries of the Standard Model, for example, follow from the coordinate invariance of the higher-dimensional Einstein equations, and the far-field forces between well-separated particles will be derived later from the geodetic particle accelerations in extra-space.

The periodic soliton components exhibit periodicity also with respect to time, the wavenumber k_0 (in the particle restframe) corresponding to the particle mass. Thus the four forces of nature are represented by four separate wavenumber constants. In contrast to the fibre periodicities, the mass wavenumber k_0 is not prescribed, but is determined by interactions with an homogeneous fibre-wave field, which plays an analogous role to the Higgs field of quantum field theory.

10

The periodic fields fall off exponentially $\sim \exp(-\kappa r)/(\kappa r)$ with distance r from the particle origin, where κ^{-1} represents the particle scale. In addition to the periodic fields, the solitons contain non-periodic fields generated by quadratic interactions of the periodic fields. The non-periodic fields consist of two classes: fields that decrease asymptotically as $1/r$ with distance r from the origin, in accordance with the classical gravitational and electromagnetic fields of a point-like particle, and fields that (due to the attenuation term in eq.(1.1)) decrease exponentially at twice the rate of the periodic fields.

Thus the metron soliton is composed of three classes of fields: periodic components, identified as *fermions*, non-periodic fields decreasing asympototically as $1/r$, identified as *gravitational* and *boson* fields, and exponentially decreasing non-periodic components, termed *kaluzons*. The kaluzon fields, first discovered by Kaluza [49], have no counterpart in quantum field theory. However, it is shown in the next chapter that, although not observed as far fields, they are the key fields enabling soliton solutions[3].

Each fermion, boson or kaluzon component of each particle is identified with a different metric component characterized by a different pair of metric indices. The dimension of the full metron space accordingly increases with the complexity of the successive models introduced in the following chapters, reaching 25 finally for the metron equivalent of the three-family Standard Model.

P2. The soliton structure of metrons is governed by the interactions between the fermion and kaluzon components. The fermion field represents an eigenmode trapped in a wave-guide created by the kaluzon field. The two fields are mutually supporting: the eigenmode generates the kaluzon field, that in turn traps the eigenmode.[4] The attenuation term in eq.(1.1) confines the kaluzon fields to the relatively small region of the particle core, beyond which the kaluzons are in effect non-observable.

---

[3]It is an irony of history that the kaluzon fields presumably contributed to the difficulties in publishing Kaluza's seminal paper. As noted by Einstein, the fields – if unattenuated, as assumed by Kaluza – would greatly exceed all other fields.)

[4]In the metron rest-frame, both the fermion and kaluzon fields represent local, spherically symmetric fields. The term "wave-guide" is nevertheless used, since in a moving frame the fermion represents a propagating wave group of finite extent that is trapped in a guiding "world tube".

Analagous difference-frequency interactions of the fermion component generate also the electromagnetic and gravitational fields of the metron solution. However, the electromagnetic field has the wrong sign to trap the eigenmode, while the gravitational field has the right sign but is too small.

P3. If follows that in the metron model particles, in the traditional sense of localized corpuscular objects, and fields, as extended, distributed entities, represent different features of the same real existing objects. These are referred to in the following simply as particles, or, where misunderstandings could arise, as metron particles.

All particles of quantum field theory have counterparts in the metron model. However, in contrast to the fermion and boson particles of quantum field theory, which are treated as separate fields represented by operators acting in an abstract Hilbert space, their metron counterparts represent the different field components of a single real particle.

This implies, in particular, that photons, weak interaction gauge bosons and gluons (as well as gravitons) represent classical fields originating in the core regions of particles, rather than independent particles.

P4. The observed disrete nature of the photon follows, as argued by Einstein [24], from the discrete structure of the atomic spectra or particle states from which the photon was emitted - i.e. from the transition of an electron from one discrete orbit to another.

P5. Interactions between particles satisfy time-reversal symmetry. The time-symmetrical coupling between past and future states of separated particles is the origin of "entanglement", as observed in the EPR experiment. Entanglement is a universal phenomenon common to all microphysical theories, whether quantum theoretical or classical, that satisfy time-reversal symmetry, including in particular the metron model.

P6. The metron equivalent of Heisenberg's uncertainty principle follows from the structure of the basic core components of the metron soliton. These exhibit both localized and periodic properties with respect to spacetime (in the particle restframe: localized with respect to $\mathbf{x}$, periodic with respect to t). Thus all components of the four-dimensional wavenumber and spacetime-location of a particle cannot, by definition, be simultaneously specified with arbitrary accuracy. The indeterminacy product δ (location) $\times \delta$ (wavenumber) is proportional to the Planck constant, as in the Heisenberg relation.(*Mean* locations and wavenumbers of individual metron particles remain, of course, well defined, but in practice cannot be determined accurately due to the unavoidable interaction of the measurement process with the object of measurement. Relevant is that a particle is always completely specified conceptually as a well defined composite object, including its inherent coupled spacetime location and momentum widths, independent of the measurement process).

P7. An important assumption of the metron model is the existence of a spacetime-independent background fibre-wave field. This is the origin of the particle mass, the chirality of the weak interactions, the coupling between left and right chiral leptons, and the mass of the weak interaction gauge bosons. It plays a role similar to the Higgs field of the Standard Model.

The generation of the particle mass through interactions with the fibre-wave field explains the exceptional role of the very much weaker gravitational force in relation to the other three fundamental forces. While the electromagnetic, weak and strong force are defined directly via their prescribed fibre wavenumbers, the relevant wavenumber k_0 characterizing the particle mass and gravitational force is a derived constant proportional to the prescribed (weak) fibre-wave field.

P8. The different structures of the fermion and boson components of the metron solution are consistent with the Pauli exclusion principle of quantum theory. Fermions are nonlinear eigensolutions, concentrated in the metron core and falling off exponentially outside the core region. Since two particles cannot occupy the same position, the fermion fields of different particles cannot be superimposed in the regions in which the fields are non-negligible – as expressed by the Pauli exclusion principle. Boson fields, in contrast, represent wave-guide fields that fall off as $1/r$ beyond the particle core; the fields from different particles can therefore be meaningfully superimposed at positions distant from the individual sources of the fields (cf. Figure 1.1).

P9. The symmetries of the Standard Model follow in the metron model from the invariance of the Einstein equations with respect to coordinate transformations, the $SU(3) \times SU(2) \times U(1)$ symmetries corresponding to particular classes of coordinate transformation that respect the invariance of the prescribed background metric and fibre wavenumbers.

P10. The existence of three generations of elementary particles is reproduced in the metron model, in analogy with the Standard Model, by simply introducing three generations of metric components, each generation being characterized by different sets of tensor indices. (The anticipation, expressed in earlier publications, that the second and third generations correspond to higher nonlinear eigenmodes, could not be confirmed.)

P11. The metron model distinguishes between three types of particle interactions: internal interactions, far-field interactions and collisions. Internal interactions describe the basic soliton dynamics of an individual particle. These have no counterpart in quantum field theory, which postulates the existence of various types of particles *a priori*. Far-field interactions represent interactions between particles that are sufficiently separated that their core regions do not intersect. The metron model reproduces in this case the classical gravitational and electromagnetic far-field interactions between point-like particles, while explaining also the interference phenomena characteristic of microphysical processes (see next item).

Collisions, finally, correspond to the case in which the interacting particles penetrate each other's core regions. These are computed in the metron model using wave-wave interaction diagrams rather than Feynman diagrams. The latter reduce the higher products of creation and annihilation operators appearing in the representation of collision processes to products of number densities by applying the relevant commutation or anti-commutation relations of the associated field operators. In contrast, the analogous higher-order interaction diagrams that arise in the classical metron representation are reduced to products of number densities by invoking the statistical independence of the interacting field components ([36]). In contrast to quantum field theory, the perturbation expansions for the metron model encounter no divergences.

P12. Quantum phenomena at lower energies are explained in the metron picture by the interaction of the particle's periodic fermion field with other matter. The interference patterns of the single or double slit particle diffraction experiment referred to above, for example, are explained in the metron model by the quadratic interaction of the primary fermion fields of the particles with their secondary fields scattered at the screen slit(s). This produces

14

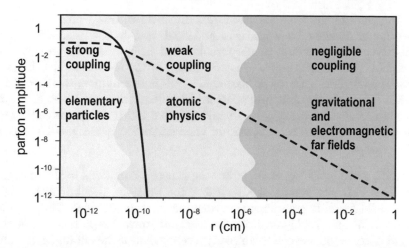

Figure 1.1: *Near- and far-field components of metron solutions. Full line: eigenmodes (fermions), decreasing exponentially for large r; dashed line: wave-guide fields (massless bosons), decreasing asymptotically as $1/r$.*

a field with a low difference wavenumber that modifies the particles' trajectories, creating the observed interference patterns. The discrete structure of atomic spectra can be similarly explained by the interactions of the primary periodic core fields of orbiting electrons with their secondary fields scattered at the atomic nucleus. Similar explanations in terms of particle-field interactions can be given for the photo-electric effect, Compton scattering and Planck's law.

Not investigated in the present analysis are the cosmological implications of the metron concept. This restriction is imposed by the representation of particles as perturbations relative to a constant background metric and the assumption of a background fibre wave field that is independent of spacetime. The embedding of the present analysis in more general coordinate systems relevant for cosmological investigations is left for future investigations.

In summary, the metron model translates the basic results of quantum field theory, ranging from atomic scales to elementary-particles, into a classical picture of real existing particles, encompassing all four forces. Its starting point of combining corpuscular and field properties in a single real object is orthogonal to the foundations of quantum field theory, which negates the objective existence of real particles and fields due to the assumed irreconcilability of the existence of corpuscular and field properties in the same physical object. The following representation of the metron model therefore pursues an inverse path to the historical development of quantum theory and quantum field theory. Starting point is the generalization of Einstein's equations to derive elementary particle models, after which the discrete structure of atomic spectra and other quantum phenomena are investigated as second step.

However, a straightforward leap from the Einstein equations to the Standard Model of elementary particles is clearly not possible. Beginning from the generalized Einstein equations it is therefore first shown that these support soliton solutions displaying both particle and wave properties. This first simple particle picture enables already a resolution of the entanglement phenomenon, which has been widely (incorrectly) argued to rule out classical explanations of microphysical phenomena. It also permits a first simple representation of the basic single and double-slit particle diffraction experiments. From this starting point the metron model is then successively generalized, yielding first a representation of atomic spectra and other quantum phenomena, and culminating finally in the level of detail of the Standard Model of elementary particles.

The relation between the iterative development of the metron model in the following chapters and the historical development of quantum field theory and quantum field theory is summarized in the next section.

1.4   Overview

In Chapter 2, numerical computations are first presented for some simple scalar particle models. Section 2.3 describes the general mode trapping mechanism. The simplest example of a soliton with mass and electric charge, but without attenuation of the kaluzon fields, is presented in Section 2.4. In Section 2.5, the far-field forces acting between separated particles are then derived from the geodetic acceleration of a wave group, identified with the periodic soliton, propagating in a slowly varying gravitational field in higher-dimensional space. In addition to the correct expressions for the gravitational and electromagnetic far fields, the model yields unrealistic super-strong kaluzon fields. The kaluzon far fields are removed in Section 2.6 through the inclusion of an attenuation tensor, eq.(2.12), yielding an acceptable model of a scalar charged particle. After a comparison of the model with data for the electron in Section 2.7, the model is generalized in Section 2.8 to a simplified scalar model of a nucleon composed of three scalar quarks coupled through gluon fields.

Although lacking important further properties such as the particle spin and the parity breaking weak interactions, the scalar particle models are already sufficiently realistic to demonstrate in the following two chapters that a classical theory of real existing particles incorporating both corpusucular and wave-like field properties is able to explain quantum phenomena that are normally cited as defying a classical explanation.

In Chapter 3 it is shown that entanglement, contrary to the prevalent view, is not a defining feature of quantum theory, but follows simply from the relativistic generalization of Newton's third law: actio = reactio. Entanglement is a necessary consequence of any microphysical theory, whether classical or quantum theoretical, that satisfies the basic requirement of time-reversal symmetry. This is demonstrated in Section 3.2 for the simplest case of the photon realization of the EPR experiment and is then generalized in Section 3.3 to the original particle version of the experiment.

Applying the same simplified scalar particle models, it is shown in Chapter 4 that in the standard single or double-slit particle diffraction experiments, the interaction of the particles' period fermion field with the period field scattered at the single or double slit reproduces the observed diffraction patterns to the same level of accuracy as quantum theory.

In Chapter 5 the particle models are then developed in full detail, reflecting the complexity of the elementary particle spectrum as summarized in the Standard Model. This requires the introduction of particle spin (Section 5.1), weak interactions (Section 5.2), strong interactions (Section 5.3), and a background wave field corresponding to the Higgs field (Section 5.4). The Higgs field is not only the origin of the particle mass, but also determines the particle's stability or instability. In Section 5.6 these results are generalized to three particle families, while the last Section 5.7, finally, derives the symmetries of the Standard Model from the coordinate invariance of the higher dimensional Einstein equations.

With the availability of a real-particle interpretation of the Standard Model, the metron model can be applied in Chapter 6 to provide alternative interpretations of a number of basic quantum phenomena. Examples chosen are the computation of atomic spectra (Section 6.1), the photo-electric effect (Section 6.2), Compton scattering (Section 6.3) and Planck's law (Section 6.4).

Chapter 7 returns again to the dynamics of elementary particles. The investigation of far-field interactions of Chapter 2 is broadened to the more complex case of near-field interactions, or collision processes, in which the interacting particles penetrate each other's core regions. The interactions can be divided into elastic processes (Section 7.1), in which the same particles are found before and after the collision, and inelastic processes (Section 7.2), in which particles are created or destroyed. In both cases, the far-field approximation (Section 2.5) of an undisturbed periodic soliton propagating in a slowly varying higher dimensional space is no longer applicable. However, to the extent that a perturbative approach is permissible, the interactions can be computed using a classical version [36] of the Feynman diagram approach.

The interacting fields (solitons) are represented as a statistical ensemble of quasi-periodic wave components. The wave spectra represent the probabilities of observing a given wave component (soliton) at a particular time and location. The evolution of the wave spectra is determined by a transport equation derived from the higher-dimensional Einstein-Hilbert Lagrangian. In both the Feynman diagram and the analogous classical formulation, one is faced with a statistical closure problem: the reduction of the expectation values of the higher-order products of field amplitudes that arise in the perturbation expansions of the particle interactions to the expectation values of the quadratic products of field amplitudes that characterize the initial input fields. In the quantum theoretical case, the closure is achieved through the commutation and anti-commutation relations of the creation and annihilation operators. In the classical case, the closure follows from the Gaussian hypothesis: the assumption that fields exhibiting different spacetime periodicities are statistically independent. The hypotheses apply in both cases only for the evolution forwards in time, not for the reconstruction of the past, in accordance with the Boltzmann hypothesis.

18

The classical interaction diagrams summarizing the structure of the resulting transfer integrals are similar to the quantum theoretical Feynman diagrams. However, they differ in a key feature: they contain no closed-loop terms with associated divergences. There is therefore no need for renormalization.

The last Chapter 8 summarizes the present status of the metron model and ventures an outlook on future developments.

4

Personal Accounts by Colleagues and Co-workers

During his long career, Klaus Hasselmann has been a boss and teacher but also a colleague to many people. Therefore, we have asked quite a few of these people about how they remember their time with him. Specifically, we asked:

- How did you meet Klaus?
- What is the legacy of Klaus' scientific work in your field?
- Is there a personal advice from Klaus that helped you in your career?
- How did Klaus' thinking influence your scientific work?

We left it open to the addressees of our survey as to whether they would prefer to answer these questions or if they would like to discuss their experience in a different way. The people whom we approached and who gave us a wealth of answers were:

- **Susanne Hasselmann**, Klaus' wife and research partner during his scientific work on ocean waves and particles,
- **Dirk Olbers, Jürgen Willebrand, Peter Müller,** and **Peter Lemke**—first generation of co-workers during the early years of Hasselmann's move into the field of climate science,
- A second generation of co-workers at the Max Planck Institute that included **Martin Heimann, Christoph Heinze, Mojib Latif, Hans Graf,**

Compiled by Martin Heimann and Hans von Storch

© The Author(s) 2022
H. von Storch, *From Decoding Turbulence to Unveiling the Fingerprint of Climate Change*,
https://doi.org/10.1007/978-3-030-91716-6_4

Gabriele Hegerl, Jin-Song von Storch, Hans von Storch, Patrick Heimbach Jörg Wolff, Ben Santer, Ulrich Cubasch, Achim Stössel, Robert Sausen, Dmitry V. Kovalevsky, Carola Kauhs,

- Colleagues who shared his interest in ocean waves and remote sensing: **Gerbrand Komen, Luigi Cavaleri, Kristina Katsaros, Peter Janssen,** and **Ola M. Johannessen,**
- Colleagues, who assisted Hasselmann in constructing the network of competence: **Lennart Bengtsson, Jürgen Sündermann, Klaus Fraedrich, Udo Simonis, and Hartmut Graßl**.

Placing the various characters into these categories is not always perfect; indeed, in many cases, people would fit in several categories rather than just one. However, this placing them in these categories is sufficient to provide a rough overview.

We allow these people to speak their minds in the following sections and, as the reader will soon learn, discussions with Hasselmann could sometimes be stormy, but were always honest, and constructive, so that the overarching conclusion is: **respect for a great scientist and a great person**.

4.1   Susanne Hasselmann: Klaus—Scientist, Husband, Father, Grandfather, Great-Grandfather

We met in 1955 in Hamburg. Klaus had just finished his diploma in physics and started his Ph.D. work in Göttingen. I was a student of mathematics and physics in Hamburg. I was fascinated by the intensity with which his mind constantly worked. Any problem was trivial for him and could be solved in two or three lines of formulas. He was full of humor and very fond of sports. All in all, a very attractive young man.

We married in 1957, because a little apartment had been offered to us. One has to keep in mind that it was only 10 years after the war and Hamburg had been bombed immensely. So a two room (14 m^2 and 16 m^2) was divine for us. Within the span of one month, Klaus finished his Ph.D., started a position as an assistant at the Institute for Shipbuilding in Hamburg, and got married. Our plan was that I would finish my diploma. However, times were different then. Only three girls from my school started university after the High School Exam. Women got married and had children.

Therefore, when our daughter was born, I stayed home. However, I could take part in Klaus' work. We were happy. For instance, when he thought he had solved the Turbulence Problem, even if the next day showed an error in the computations. Or after long walks in the park, he announced that he would have to go one order higher in the computations. And out came the wave-wave interaction theory.

He was invited to a conference in Easton, Maryland, on wave dynamics in 1961. There he was able to offer the link that scientists had been looking for years. So he was invited to several places in the US and was offered jobs. Meanwhile, I was at home with two little kids, one newborn and the other with a very bad case of the measles. However, he was so happy on the phone about the sun and the blooming bougainvillea in California and the lively science there that I could only prepare myself and the family for years of packing and travelling. One has to keep in mind that science in Germany at that time was underdeveloped and the scientific community here was generally old and stuffy. For a young man at that time, gaining entry into a lively scientific atmosphere was just wonderful.

However, for the children it was not easy, especially for our oldest daughter. Three and a half years of California, back to Hamburg, then six months in Cambridge, England, followed by two years in Woods Hole, where the JONSWAP data were worked on, because there were no efficient computer facilities in Hamburg.

Much later, we were invited to a party at a friend's house in Hamburg. Every guest was asked to introduce him/herself with a picture on a black board. Klaus drew himself sitting on a rocking chair, smoking a pipe and flying over the globe. I added myself to the same picture, gripping with one hand the rocking chair and holding suitcases and three children in the other.

However, to see the first curve of the wave-wave interaction in the JONSWAP spectrum and seeing Klaus' theory verified was an experience that we enjoyed immensely. And taking part in all this was worth the inconveniences for the family.

His work on the stochastic nonlinear interaction on ocean waves and other wave phenomena in geophysics in the 1960s, for which he used Feynman diagrams, led to ideas of a new Elementary Particle theory, which he followed up with deep interest on the side. However, I saw how much this theory worked in him. Therefore, when the directorship of the Max Planck Institute for Meteorology (really for Climate Research) was offered to him, I was against accepting the offer. However, he knew that this would give him complete freedom for research and he accepted the position. For the inauguration he quickly developed a stochastic climate model, which he was able to present.

In the mean time I had finished my Mathematics Diploma and was thinking of my future career. For Klaus it was clear: I would work with him. I could follow up the wave-wave interaction and develop a global wave model. However, working at his institute would mean seeing more of him, which was, of course, a good thing. Money for my salary came from ONR. It certainly was an experiment for a woman to work in the institute that her husband was directing. However, the colleagues were very friendly and even found advantages to this arrangement. For example, if you had any problem that needed to be conveyed directly to the top, just mention it if Mrs Hasselmann happens to be in the room. Or, people would call me to say that they had sent Klaus a message weeks earlier and that they needed a response. Etc.

Another question was, how does that work to be his wife and his coworker? Is she only his programmer? We are different. We complemented one another. He presented me with a new theory and I did the untying of the knots, which means that I corrected his mathematics and formed it into something that could be programmed for the computer. For example, the eight-fold integral of the nonlinear interactions. To compute one spectrum cost lots of computer time. The coupling coefficients had to be separated from the integration. Then the integration had to be reduced to the main contributions, etc.

If he had a new idea, he asked me to try it out. After that he followed it up with other coworkers or myself.

The longest and most difficult job was the Metron Theory. It took almost 20 years of my retirement time. It is disappointing, that physicists refused to even think about it.

The title above was: Klaus, Scientist, husband, father, grandfather and great grandfather. Therefore, I have to say something about the family. Most people live a life period first for the family, then profession, then grandchildren and if they live long enough great grandchildren. When we lived in Hamburg in the 1950s, Klaus was very close to his daughter. She adored him (today daddy goes to the institute, tomorrow Meike goes to the institute). In the evenings, he played a puppet show for her. She had admired him all her life and became a very successful scientist herself. She was three years old, when we moved to California in 1961 and was losing him. This was hard, however understandable from both sides and I had to make the best of it. When we had almost lost her, Klaus finally made the decision for the family to move back to Germany. The years to come were travelling years. He tried his best besides Science to be a father and bravely chauffeured the family every Saturday from Woods Hole to the New England Conservatory in Boston. He cuddled with his youngest daughter, enjoyed his son's musical talent. Best was when he could have long discussions with his oldest daughter. When she reached puberty, he managed many occasions with his humor. She was a little talking waterfall. At one dinner, she asked, "What would you do if I would not talk to you anymore?" And he answered, "We would take you to the psychiatrist and ask how we could keep this status." Everyone laughed and the situation again was under control. His humor spread in the family and his fondness of discussion was transferred to his children, too. The older they became the more he could take part in their lives. And he was happy when his son, who never was interested in school much, later after becoming a professional musician, taught himself science to perfection.

It was fun later on to also have professional contact with our children. Our older daughter told me about her research into gene manipulation in the fight against AIDS, and I told her we could put this problem in a system of linear differential equations and compute her free parameters on the computer. We published two papers together on the topic [163, 164].

And with our younger daughter, who creates exhibits about nature and the environment, we could work on climate change or on ocean wave development.

With music we had a problem. Klaus played the flute. However, he thought he did not have to practice. The better the children became on their instruments, the more this became a problem: so they sent him off to practice.

When we had grandchildren, he became a storyteller. He created the character "Little Joe," an angel, after a Christmas show one year. Little Joe always wanted to help but somehow managed to completely mess everything he got involved in. The kids loved it and remembered every subject. Klaus had to create new stories every time.

In 2021, his real family time is now as a great grandfather. He enjoys those little ones enormously, and they adore him. They play together for hours. "Where is grandpa," are their first words when they come to visit.

It is now 64 years that we have been married. It was not always easy, but with a husband, a father, a grandfather and great grandfather like Klaus, it was the richest life one could possibly have dreamt of.

4.2   Dirk Olbers: How to Cook an Ostrich Egg

My first contact with Klaus was in 1968/69 when he dropped into Wolfgang Kundt's seminar on statistical physics at the University of Hamburg, which we attended to find topics for our respective diploma theses. Peter Müller was in that group as well as Hans Juranek, Hajo Leschke and Arne Richter. As I remember it, we had all been searching for a couple of years and we were all well educated in the techniques and concepts of statistical physics but had no idea what to do with it. Klaus hijacked almost all of Wolfgang's students, we all had research topics immediately, and writing our theses took a matter of months. My topic was on plasma physics because Klaus had a proposal on interplanetary space physics which, however, was not approved, and so we (Peter and myself) were suddenly oceanographers (not real ones, this took probably more than a decade) working on JONSWAP and all kind of waves.

Our internal wave research began in 1971 at the Sonderforschungsbereich 94, which Klaus had created and of which he was the head spokesman. Of course, he spoke but our supervisor also vanished immediately to Woods Hole for two years and the best we could do was to follow him. My time at WHOI was full of new experiences, work, learning, and enjoyment. Wednesday dinners at Susanne and Klaus' home, where we worked on weak-interaction theory and the JONSWAP data, were outstanding (Thursdays at Bob Long's). I don't remember what I actually did for the latter except for carrying magnetic tapes, punching cards and fitting the spectral shape to the data measured at Sylt and the profile of Klaus' nose. The JONSWAP paper—which was co-authored by 16 researchers (I am number 13)—is my most cited paper (having been read over 50,000 times on ResearchGate). Another vivid memory of that time is the MODE workshop in Boulder. Klaus was invited, but didn't go himself, instead sending Peter and myself "to tell the people what to do". The people in question were the top theoreticians and observers in the field of US American ocean science with whom we now shared student housing for 6 weeks. So, every day one could find two innocent German diploma physicists (my contract with WHOI referred to me as a "diplomatic physicist") sitting by the pool with Walter Munk, Henk Stommel, Pierre Welander, Carl Wunsch, Francis Bretherton, Kirk Bryan, Peter Rhines, Jim McWilliams, and a dozen famous others. I think that we didn't contribute much to MODE but started learning oceanography instead.

Another experiment to which we made a major contribution was IWEX, the internal wave experiment in the Sargasso Sea, which was originated by Klaus during his stay at WHOI, and performed by Mel Briscoe and Terry Joyce of the WHOI in 1973, and then evaluated by Jürgen Willebrand, Peter and me in Kiel and Hamburg over the following years. Another matter of note for my career is that the Garrett-Munk model of the internal wave spectrum was first introduced at the WHOI in 1971 in the form of a preprint and a lecture from Walter, the result of which for me was that I found in it a foundation and question for my Ph.D. thesis: what is the role of the wave field in the ocean interior for dynamics and mixing? This a problem that still keeps me busy even today.

Peter and I had a joint Ph.D. viva in Klaus' office in Hamburg in 1973 and one of Klaus' questions was how deep the temperature signal of daily insolation would go? No idea! We certainly could write down the solution of the diffusion equation with a delta-function initial condition, but a number, and from what? The simple dimensional argument later led me to my most popular exam question (lectures at Bremen University): how long would you cook an ostrich egg if you knew to cook a hen's egg?

One of the most influential meetings Klaus took me to was the conference on oceans and climate in Helsinki in 1975. Manabe 's talk on CO_2-doubling was disturbing, and we thought it was clear to do, we thought. It was my first contact with the climate problem. The MPI was founded that same year.

In 1979 I went to Kiel to follow Fritz Schott as lecturer in physical oceanography; it was a move in which Klaus played no part (I think; other than my later move to AWI in 1985). I was not happy in Kiel; my friend Jürgen was still in Princeton, and I continued to live in Hamburg for personal reasons. It was a relief when Klaus called me shortly before I had to take the train and offered me a position at the MPI. The negotiation took 3 min, I had to catch the train. I spent another few years working in Klaus' sphere of influence and could follow the early development of the MPI. Klaus also advised me during my habilitation (1981) telling me that "you must be convinced that you're right, not the committee".

Tim Barnett was visiting the MPI around 1984 and brought 14 years of wind field data over the equatorial Pacific with him. Mojib Latif and I had the idea of inputting this data into an existing ocean model to see whether El Niño would pop up. We took the idea to Klaus and were harshly dismissed. But Ernst reached into his desk drawer and pulled out a couple of punching cards, an equatorial circulation model—and El Niño did appear, which launched the career of a promising young scientist.

Most of what I learned from Klaus was communicated in seminars. Parallel to the statistical physics seminar, we attended the plasma physics seminar with Gerd Wibberenz in Kiel. Later, in 1970s, when interest had shifted to oceanography, we had the 'Hamburg-Kiel-Seminar' (which our Kiel colleagues called 'Kiel-Hamburg-Seminar'). I remember that one time Klaus was supposed to give a lecture in Kiel but did not appear. He had forgotten to change trains in Hamburg and ended up at the end of the line terminal—"Abstellgleis". Except for this occasion, Klaus dominated the discussion in the seminars, so much so that we invented the '2 min-seminar' at the MPI: Klaus was forbidden to say a word during the first two minutes.

An outstanding event for me was the meeting in Rissen to mark Klaus' 60th birthday in 1991 at which he presented his metron model for the first time. We all saw a glimpse of the great unified theory of physics and the next Nobel prize. I remember many later boring administrative meetings on computer resources where Klaus sat scribbling metron equations under the desk. And then there was Klaus' 80 birthday celebration in 2011. I tried to give an overview of the first Hamburg ocean model, a multiregional construction that Klaus had created back in 1981 during a summer school in Alpbach. Jürgen and I backed up his lectures (Jochem Marotzke and Robert Sausen were there as students). The idea of the ocean model was to couple the different ocean regions together (which differed in terms of their physical properties) to form one dynamical system. Jürgen was to do the western boundary currents, Peter Lemke the mixed layer, Ernst Maier-Reimer the ocean interior (boring) and I the equatorial currents (complicated). I went to Hawaii for a year to carry out local studies and when I returned, Ernst had already done the whole thing and Klaus had published his work on oceans and climate [68]—the foundation of the celebrated Hamburg LSG model—back in 1982, which was an important reference paper for my own work.

(8)

(*not* crossed out!)

O.K.

Linear operator approach

$$\mathcal{L} = \begin{pmatrix} u_1 \\ u_2 \\ u_3 \\ b \\ s \end{pmatrix} \Longleftrightarrow \begin{pmatrix} u_1 \\ u_2 \\ p \end{pmatrix} \quad \text{, because} \quad \boxed{\nabla^2 p = \frac{\partial b}{\partial x_3} - \nabla \cdot (f \times u)}$$

with $\begin{cases} p = gs & \text{at } x_3 = 0 \\ \frac{\partial p}{\partial x_3} = b & \text{at } x_3 = -h \end{cases}$

and $\boxed{u_3 = -\int_{-h}^{x_3} \nabla \cdot u \, dx_3 = I_a \cdot \nabla \cdot u}$

conversely, $\begin{pmatrix} u_1 \\ u_2 \end{pmatrix} \xrightarrow{s \ (bc)} u_3 \text{ (already know)} \xrightarrow{} b \text{ via eqn. of motion for } \dot{u}_1, \dot{u}_2, \dot{u}_3, \text{ eliminating elevation.}$

Now general form of p is $\quad p = G\left[\frac{\partial b}{\partial x_3} - \nabla \cdot (f \times u) \right] + L_1 (gs) + L_2 [b$

Hence $\dot{p} = G\left[-\frac{\partial}{\partial x_3}(N^2 I_a \nabla \cdot u) - \nabla (f \times (f \times u)) \right] + L_1 g I_a \nabla \cdot u$

Now $\nabla(f \times (f \times u)) = +\nabla \cdot (f (f \cdot u) - u f^2) = -f^2 (\nabla \cdot u) + u_2 \frac{\partial f}{\partial x_2}$

O.K., the $\dot{p}$ ... Normally,

O.K., so $\boxed{\dot{p} = \left\{ G\left[-\frac{\partial}{\partial x_3}(N^2 I_a) + f^2 \right] + L_1 I_a g \right\} \nabla \cdot u}$

$$\equiv I \nabla \cdot u .$$

Hence $\begin{pmatrix} \dot{u}_1 \\ \dot{u}_2 \\ \dot{p} \end{pmatrix} = \begin{pmatrix} 0 & f & -\partial_1 \\ -f & 0 & -\partial_2 \\ I\partial_1 & I\partial_2 & 0 \end{pmatrix} \begin{pmatrix} u_1 \\ u_2 \\ p \end{pmatrix}$

Now do the same thing with $f = f(x_2)$, and neglect term if $N^2 \gg f^2$.

Klaus' lecture from 1970. Not crossed out!

Jürgen Willebrand, Klaus, and Dirk pondering about ocean dynamics in Alpbach in 1981

We (Jürgen, Carsten Eden, and I) also finished our book on ocean dynamics in 2011 and I asked Klaus to write a foreword. "I can't do that," he said, "I don't know anything about ocean physics". I had thought I had learned everything I knew from Klaus! He never liked and rarely gave student lectures. I think that a counter example (from September 1970) explained all I know about internal gravity and other waves. The foreword to the book was very favourable and I well remember celebrating the book's publishing in 2012 with Susanne and Klaus in our garden in Fischerhude.

The temperature signal has reached 18 m deep (by molecular heat diffusion) since 1973. Who cares? The ostrich egg must be cooked $(6 \text{ cm}/1.5 \text{ cm})^2 = 16$ times as long as the hen's egg.

From left: Carsten Eden, Christoph Völker, Klaus, Dirk, Susanne, Peter Lemke, Christine Klaas, Dieter Wolf-Gladrow and Jürgen 2012 in Fischerhude on occasion of the publication of 'Ocean Dynamics', the book by Olbers, Willebrand and Eden

4.3   Peter Müller

I met Klaus back in 1968/69 when I and Dirk Olbers were working on our 'diplom' thesis in physics at the University of Hamburg. Klaus joined a weekly seminar on statistical mechanics that his friend and our thesis advisor Wolfgang Kundt had organized. Klaus shook up the orderly conduct of the seminar quite a bit with his distinctive interpretation of a scientific discussion, lots of questions, lots of diversions, but in the end usually some profound insights. On completion of our diplom thesis Klaus offered us to do a Ph.D. thesis with him, gladly accepted; he also offered us a well-paid position as a 'scientific employee' that allowed me to rent a one-bedroom apartment and marry my long-time girl friend. Life in this apartment was cut short because we joined Klaus and many others at the Woods Hole Oceanographic Institution on Cape Cod to analyze the JONSWAP data. My task was to parametrize the spectra, a simple curve-fitting exercise. The year at WHOI was a transformative and happy time in my life: immersed in an exciting research program, being exposed to a new culture and the stimulating intellectual life in Woods Hole and Cambridge, being nurtured by weekly dinners at Klaus and Susanne's house; the only problem was Klaus' German shepherd Shiva who stubbornly occupied the front passenger seat in the car, but who would argue with a German shepherd, of his thesis adviser. Despite being hard pressed to complete the JONSWAP analysis Klaus found the time and energy to conceive and secure funding for the Internal Wave Experiment IWEX. Funding for the experiment required a trip to the Applied Physics Lab at the Johns Hopkins University where it took all of Klaus' persuasive skills to get two German grad students (Dirk and me) without passports and other legal documentation past the security guards. It should be mentioned that IWEX resulted in many publications, none of them carried Klaus' name. This was his gift to us, providing the basic idea (and funding) and let us run with it. My Ph.D. thesis worked the same way. I got two hand-written pages from him with some formulas and arrows and some crossed-out parts and was then on my own. Klaus also managed to made me a co-principal investigator of the then emerging MODE project. I still remember the expression on Dennis Moore's face when he realized that this young German scientist sitting at the table with all the esteemed East-Coast oceanographers had not gotten his Ph.D. yet.

During this year in Woods Hole I made my way into the scientific world, with Klaus guidance, help and patronage. I realized that not every scientist is as gifted as Klaus and not every curve-fitting exercise is a contribution to a seminal paper. It took me much longer, 25 years to be exact, to realize that after all that Klaus had done for me, I could do something, whatever so slightly, for him. So I invited Klaus and Susanne to one of my Hawaiian Winter workshops, with some 'relaxation and recreation' added.

In summary, my rewarding personal and professional life would not have been possible without Klaus.

4.4   Jürgen-Willebrand: Kiel-Hamburg Oscillations

The way I came into contact with Klaus was perhaps a bit unusual. In 1970 I joined the theoretical oceanography department at IfM Kiel headed by Wolfgang Krauss, with a diploma in physics and some initial exposure to ocean surface waves, a field in which Klaus already was recognized as the leading authority. Sometime earlier, Klaus and Wolfgang had agreed to intensify the exchange of information between their departments, by having a scientist from Hamburg working for half a year in Kiel and then one from Kiel working in Hamburg. At the time when I arrived, Heinz-Hermann Essen from Hamburg was just completing his term in Kiel. However, the scientist who had been designated to work in Hamburg meanwhile had left the Institute. Now someone from Kiel was "owed" to Klaus' group, and instead of strolling to the Institute in Kiel each day I spent the next half year commuting to the Institute for Theoretical Geophysics at Schlüterstraße, Hamburg. It was a time when I learned a lot without contributing much.

When in Woods Hole, Klaus had conceived the tri-moored Internal Wave Experiment which was carried out in 1973 by Briscoe and Joyce at the WHOI. Together with Fritz Schott, I had just completed an analysis of internal wave spectra from another experiment and felt well prepared to join Dirk Olbers and Peter Müller in analysing the IWEX data. We collaborated very closely over the following two years. Frequent trips between Kiel and Hamburg were necessary to keep up the communications. At times, the travel frequency was so high that my director in Kiel concluded that I must have a girlfriend in Hamburg. In 1976, I was looking for a postdoc position in the USA and received an offer from the Oceanographic Institution in Woods Hole. That offer was appealing, because I had maintained close contacts with the WHOI colleagues during my time at the IWEX, and I knew roughly what to expect there. Klaus, on the other hand, suggested that working at GFDL Princeton—a place about which I knew nothing—would be more attractive. In the end, I took his advice and never regretted it. I came in contact with large-scale ocean circulation modelling in Princeton and had the opportunity to work with George Philander and Kirk Bryan. This was also where Bryan and Manabe developed the first coupled atmosphere–ocean GCMs, which, among other things, enabled the first 3-d simulations of the effect of a rising CO_2 concentration on climate.

I was again back in Kiel in 1980, holding a tenured position and working on my habilitation. The process of building a home for my family had just begun when, quite unexpectedly, Klaus called to ask if I wanted to work in his group at the MPI? There really could be no question, and after a brief consultation with my family, it was decided that the house plan had to be stopped. A few truly exciting years at the MPI followed before I finally oscillated back to Kiel. My interactions with Klaus on many further occasions, such as the memorable workshop in Alpbach which resulted in Jochem Marotzke getting involved in oceanography, are nicely described in Dirk Olbers' contribution to this volume, and there is no need to add to this here.

When Klaus became the founding director of the MPI in Hamburg, I recall that in the morning of the opening ceremony (which was attended by officials and dignitaries), he highlighted the importance of climate change prediction. In the afternoon (when only scientists were present) he discussed his new linear statistical climate model from which it follows that climate variations are not predictable. Initially, Klaus had been somewhat skeptical regarding GCMs, which at that time were indeed rather far from representing ocean–atmosphere dynamics in a valid manner. Over time, of course, GCM-based climate modelling at the MPI has achieved an international level of excellence. Klaus has been a leading light within our field of science for many decades. The most amazing thing is that he has also been able to contribute to a completely different field with his metron model. And last but not least, the community (and I personally) have also benefitted from his leadership in national and international climate research programmes.

4.5   Peter Lemke: A Stochastic Decision?

My first encounter with Klaus was indeed of a stochastic nature. It was in late June 1975, after I had submitted my Diploma Thesis in Theoretical Solid-State Physics about plasmons in quasi-one-dimensional metals and the question, why a disordering of the atoms in the metal chain destroyed the beginning of superconductivity. One evening, I rushed down the stairs of the Physics building at the University of Hamburg and Wolfgang Kundt was standing halfway down the stairs talking to a colleague. While passing them, I overheard one sentence of their conversation: "Klaus Hasselmann is looking for physicists for climate research." This sentence stuck in my head the whole evening, even though I had settled on the idea of becoming a high-school teacher in mathematics and physics in Hamburg, which I was going to start on the 1st of August. Klaus was looking for physicists doing climate research in the newly established Max-Plank-Institut für Meteorologie (MPI-M). This sounded so interesting that I suddenly could no longer envisage the "secure" job as a high-school teacher, which I already had in hand. I first went to see Wolfgang Kundt, then I called Klaus, got an appointment and, following an interesting discussion on what was expected, i.e., "the application of stochastic methods to the climate system", I expressed my strong interest. I called Klaus again after my final oral exam, and he offered me a job whereby I could work on my Ph.D. whilst I was also expected to help him to write research proposals and reports. This took up some of my working time, but for me it was an excellent learning period on how to write research proposals with a high probability of obtaining funding.

Yet, the first test in my new job was not of a scientific nature. The official opening of the Institute was scheduled for the 5th of December 1975. I was responsible for the technical appliances during the opening ceremony in the big lecture hall in the basement of the Geomatikum, where the Institute was located. This high-rise university building had just been built, but it was far from being finished. Nothing really worked perfectly. Each one of us got stuck in the elevator several times per month. After learning a bit about stochastic models, the Geomatikum—in my mind—was already displaying many similarities with some sort of stochastic creature with on–off functions following a random process. Consequently, as a theoretical physicist with responsibility for the technical infrastructure, I was really worried about the potential malfunction of all the devices. Fortunately, the lecture hall was in a deterministic phase during the time of the ceremony. After a deep sigh of relief, my heart slowed down and I enjoyed the reception afterwards.

Klaus' attitude with respect to his Ph.D. students was to give them a long leash provided that they made successful progress by themselves. I handed a draft of my first paper "Stochastic climate models, part 3. Application to zonally averaged energy models"[1] to Klaus after a year. A few days later he gave it back to me "with a few editorial remarks". Needless to say, being the optimistic student, these remarks were certainly more than just editorial; instead, they provided clear guidance on how the paper should present the basics, the logic of the model and the results. This was another excellent learning process for me.

The Institute's Science Advisory Board met in 1978 and, following my presentation, Joe Smagorinsky, Director of the Geophysical Fluid Dynamics Laboratory in Princeton, invited me to apply for a Postdoc position in the Atmosphere–Ocean Programme at Princeton University after finishing my Ph.D.. Klaus supported my application, which was eventually approved. He also supported my application for the Woods Hole Summer Study Programme on Polar Oceanography in 1979. This was a marvellous opportunity to talk to eminent scientists next door in Walsh Cottage, where the summer programme took place. Knut Aagaard, Kirk Bryan, Adrian Gill, Peter Killworth, Peter Rhines, Melvin Stern, George Veronis, and Pierre Welander were sitting in the various rooms ready for an intense discussion. During this time, I started my work on "A model for the seasonal variation of the mixed layer in the Arctic Ocean", supported by many valuable suggestions by Ken Hunkins, Peter Killworth, and Adrian Gill.

Very early on, Klaus sent me to several international meetings on his behalf, all of which presented wonderful opportunities to meet with established climate scientists—and for earning a bit of recognition for myself.

---

[1] Lemke, P., 1977: Stochastic climate models. Part 3. Application to zonally averaged energy models. Tellus 29, 385–392.

This was the manner in which Klaus provided me with an excellent spring-board for my international work in World Climate Research Programme and Intergovernmental Panel on Climate Change in later years.

After completing my Ph.D. with a dissertation on the "Stochastic dynamic analysis of polar sea ice variability", I spent two exciting years at Princeton University, a personally and scientifically rich period, where Kirk Bryan, Suki Manabe, Isaac Held and several others from GFDL provided me with valuable guidance. After my return to the MPI-M in 1983, I continued working towards my habilitation "On the interaction of sea ice with the atmosphere and ocean", again accompanied by Klaus' advice on how to follow the right path.

Following my habilitation in 1988, Ernst Augstein, a former colleague at the MPI-M, asked me to apply for an Associated Professorship at the University of Bremen and the Alfred Wegener Institute for Polar and Marine Research, where Dirk Olbers had taken up a professorship a few years earlier. My application was successful, and my transition to Bremen was planned for November 1989.

However, in June 1989, Rainer Roth, Professor of Meteorology in Hannover, had to decline his planned participation in the upcoming Winter Weddell Gyre Study on board the research icebreaker *Polarstern* which was scheduled for the 6th of September to the 30th of October, and I was asked to replace him as leader of the Hannover Meteorology Group at short notice. When asked Klaus whether I—a climate modeller still employed at the MPI-M—should participate in an Antarctic Winter Expedition, he said: "If I were you, I wouldn't think twice. It's a great opportunity to learn, how observations are made and how they should be interpreted."

This expedition marked a transition point for me, in several ways. It represented a transition from Hamburg to Bremen, from climate modelling to polar climate observations, from giving lectures occasionally to teaching regular courses at the University, from concentrating on my own scientific research to supervising Ph.D. Students. This transition also meant leaving Klaus' sphere of influence. I left the MPI-M with rich memories and a valuable basis of scientific insight provided by Klaus and our colleagues.

Now, while putting these memories to paper, I found myself musing about what would have happened, had I rushed down the stairs of the physics building on that June evening in 1975 just a few minutes earlier or later. What would have been the outcome for me had I not caught the invaluable information which was to determine my scientific career to such a large degree. Was my decision to join Klaus a living model example of a stochastic process? It definitely manifested as a short-term forcing causing a long-term response involving a memory-term.

This decision was not based on pure chance, but it proved to be an excellent opportunity which led to an interesting and fulfilling scientific career.

Klaus was not only the head of our scientific alma mater the MPI-M, but also the Captain of the Institute's soccer team (around 1980, Klaus: top row, 3rd from left; P.L.: bottom row, 2nd from left).[2]

4.6   Martin Heimann

Encountering Klaus

First time I came across Klaus was in the late 1970s when he gave a seminar on stochastic climate models at the physics institute of the University of Bern in Hans Oeschger's department. The department was very much involved in reconstructing climate from paleorecords (radiocarbon, isotopes) and was developing methods to determine past greenhouse gas concentrations from ice cores. The prevailing paradigm within this community at the time was

---

[2] Susanne Hasselmann commented on the photo: "die Fußballmannschaft war ne lustige Angelegenheit, besonders wenn sie gegen die Hamburger Müllabfuhr (harte Burschen) spielten. Am Anfang, erinnere ich noch, dass Klaus nach dem Training mit unserem VW Bus zu Hause vorfuhr, dann passierte erstmal gar nichts, dann ging ganz langsam die Tür auf und ganz langsam kam ein Bein heraus, dann ganz langsam das andere Bein usw. Er hatte so einen Muskelkater".

that variations detected in climate records must be caused by external factors. I am not sure if Klaus' intriguing concept of stochastically driven climate variations were taken very seriously by the departmental scientists. For us students, however, the concept sounded fascinating even though it lacked practical implications, as the department's main research focus was on climate history and budgeting the global carbon cycle but not climate dynamics.

When I was a postdoc in the USA, I came across some fascinating papers by Klaus, which were published as book chapters in the late 1970s early '80s. In these, Klaus envisaged the construction of comprehensive earth models in which biogeochemistry would also play an important role: "…This requires the development of a detailed climate model, which takes account of the oceanic circulation at the global level as well as biological and chemical cycles …" [56]. This visionary research agenda was the critical incentive for me to apply to the Hamburg group later in 1985.

Working at the MPI for meteorology in the late 1980s and early 1990s was a fantastic experience. Klaus' style of running the Institute was much more inspiring, relaxed, and friendly than anything I had experienced before, even in other Max-Planck-Institutes. As a "biogeochemical" outsider I could profit greatly from the frontier Earth System science analysis methods and modelling tools developed by my colleagues and could sometimes apply them successfully to my own work.

Beyond science, life on the 17th and 12th floors of the Geomatikum and later in the "Pavilion" (Klaus: "we don't call this a barrack") was also a lot of fun. Annual highlights included retreats in Salzau, summer excursions, and the Christmas party. And of course, the seminars within the Institute, especially after 1989.

After Perestroika, a seemingly endless stream of eminent Russian turbulence theory scientists visited the MPI for Meteorology to present their research to Klaus. Whilst turbulence theory in the west had already moved to explicit large-scale computer-based numerical modelling, our Russian colleagues still used pencil and paper to calculate smart second, third or higher order turbulence closure schemes. These visits led to very tough seminar experiences. Typically, Klaus would round up the entire Institute department staff in the seminar room at 11am. There we were exposed for up to two hours, well into lunch time, to huge stacks of tightly hand-written transparencies full of equations, often with Cyrillic letters, presented by researchers who were usually not very fluent in English. And all of this related to a topic, that I didn't understand at all. Once, halfway through, a slide had accidentally been copied on paper instead of a transparency. Klaus immediately offered to go to the copying machine to make a transparency.

While we waited, we heard his footsteps from the ceiling of the seminar room in the pavilion where he had his office. Then Klaus came back and gave the transparency to the presenter. But he also brought a thick stack of proposal documents, which he started to read during the second half of the presentation. Nevertheless, in the questions and answering section, Klaus jumped up and posed several sharp Belgrano questions. An amazing feat in multitasking.

What is the legacy of Klaus' scientific work in our field?

Prior to the mid-1980s, the global carbon cycle had been viewed in Earth System science as just a series of passively connected reservoirs. It had been thought that any atmospheric carbon dioxide variation would simply be damped by the redistribution of carbon among these reservoirs. The concept of potentially significant carbon cycle—climate feedbacks was still in its infancy. When the first greenhouse gas concentration records of the last glacial cycle from ice cores became available, this view changed dramatically. One tool used to understand the interplay between climate and biogeochemistry are Earth System models that describe both spheres in spatial and temporal detail in a coupled, physically consistent way. Klaus' research agenda from the late 1970s outlined the way forward for the development of such coupled models. And indeed, Klaus' team in Hamburg took the lead: Ernst Maier-Reimer built the very first dynamic three-dimensional ocean carbon cycle model coupled to a global ocean general circulation model. For the land side Klaus fostered a collaboration with Gerd Esser, a former student of Helmuth Lieth of the University of Osnabrück, yielding the first spatially resolved global terrestrial biosphere model for carbon cycle studies. Unfortunately, however, this model was in many ways too simple to be coupled into a global climate model in a meaningful way. Eventually, despite these pioneering achievements, it took another decade until the very first coupled global carbon cycle climate model was ultimately realised by a group working at the Hadley Centre.

A personal lesson from Klaus that helped me in my career

During my time in Hamburg, I learned two important things from Klaus, which were very helpful when I got into a driving seat later in Jena:

Decide, Klaus was very fast in deciding, mostly on the spot. For example, he offered me a job right after I gave a presentation as an unknown postdoc visiting the institute in 1984. And when I referred back to this offer in a letter in the following year, I received an answer with a prepared contract within

3 working days (which is completely impossible these days, even for a Max-Planck director). Whether right or wrong, Klaus made a decision. In the rare instance when one perhaps had a better alternative, one could challenge the decision and Klaus would be open to revisions. But any alternative really had to be convincing.

The tropical greenhouse. Klaus mentioned once that he envisages the Institute and its inhabitants as a tropical greenhouse full of fast-growing plants, and his own role was the gardener, who simply has to put some fertiliser here, some water there, and perhaps cut a branch or two over there. But the plants are allowed to flourish themselves. This metaphor reflects nicely the inspiring and very free environment we scientists experienced in order to pursue our ideas.

Hamburg, 1990

4.7   Christoph Heinze

A personal memory of Klaus

After completing my diploma in oceanography, I enquired at the Max Planck Institute of Meteorology about potential Ph.D. opportunities. I was invited to an interview with Klaus Hasselmann.

It was a very pleasant and thorough interview at the end of which Klaus told me that I could start if I would like to. I was pleasantly surprised. He also told me that I would be free to knock on the doors of all MPI-researchers to find out more about what it would be like to work at the MPI. Open doors, total trust, and freedom—I found all of that extremely attractive.

What is the legacy of Klaus' scientific work in my field?

In terms of marine biogeochemistry in a climate context, Klaus Hasselmann promoted research on all aspects of ocean carbon cycling of relevance to shaping and changing the Earth's climate. The fruitful collaboration between Klaus and Ernst Maier-Reimer resulted in the first global simulation of the inorganic and organic carbon cycles including a simple prognostic atmospheric reservoir. Ernst and Klaus' respective publication in 1987 [84] includes all key features of modern marine Earth system modelling. It nicely lays out the importance of inorganic carbon chemistry and transport with the ocean currents for uptake of anthropogenic carbon dioxide from the atmosphere. At the same time, the publication documents the fact that an ocean model based solely on the inorganic carbon cycle can never be validated against oceanic measurements because the internal structures of carbon and alkalinity (in contrast to the carbon uptake) are dominated by the organic (i.e., biologically driven) carbon cycle. Therefore, biological processes had also been included in the model to show that it worked. The 1987 paper also covers the relationship between different carbon dioxide emission scenarios, the corresponding uptake by the oceans, and the atmospheric retention over time. Thus the paper anticipates the issue illustrated lateron by the IPCC SRES, RCP, and SSP scenarios and related projections: reducing carbon dioxide emissions effectively helps the ocean to buffer the excess carbon dioxide whilst strong peak emissions of carbon dioxide lead to high atmospheric carbon dioxide concentrations because ocean mixing is kinetically incapable of buffering the emissions to a sufficient degree.

Enabling and furthering the development of the Hamburg ocean carbon cycle model (HAMOCC) still influences modern Earth system models with their inclusion of an interactive carbon cycle to provide quantitatively adequate climate projections. Among other things, the development of the ocean carbon cycle model was possible because of the visionary development of a fast physical prognostic ocean water mass model known as the dynamical Large Scale Geostrophic ocean general circulation model (the "LSG"). In the late 1980s and early 1990s this model was among the very few (if not the only one) dynamical ocean grid point models that could be integrated into full quasi-equilibrium models over at least 2000 years of model time. Combining LSG and HAMOCC was an unbeatable model combination at the time because they allowed drift-free extremely long-term integrations whilst rendering the key features of ocean physics and biogeochemistry in an astonishingly good way. This modelling work was far ahead of its time.

Personal advice from Klaus that helped me in my career

Klaus gave me a lot of advice throughout my time as a Ph.D. student and researcher. For my Ph.D. studies, Klaus had a fantastic idea on how to combine paleoclimatic archives and the results of sensitivity studies with the HAMOCC model on parameter changes and resulting marine tracer changes as well the related shift in atmospheric carbon dioxide concentration to arrive at an estimate of the maximum likelihood for the various hypotheses for what caused the glacial drawdown of atmospheric carbon dioxide. This advice—laid out on a piece of paper in the lobby of a conference that we attended—proved to be a true treasure and a cornerstone for my further Ph.D. work.

How did Klaus' thinking influence my scientific work?

In addition to thinking more in a multivariate and probabilistic way rather than in terms of simple cause-effect relationships, a general attitude towards scientific collaboration comes to mind. I remember the glass cuboid on Klaus' desk that documented an award he received for unselfish collaboration. I thought: "well, this is a really nice award and is characteristic of Klaus' way of handling the scientific process". Creating something together with others without focusing on one's own losses or gains—that helps one to focus on the good, true, and beautiful aspects of scientific work, especially when things may sometimes get difficult in the course of one's daily work.

4.8   Mojib Latif

A personal memory of Klaus

Klaus supported me throughout my career, and I could not be more grateful to him. For example, he thoroughly edited the first draft of my early research papers, and that is how I learned to write scientific papers. Klaus also taught me how to write grant proposals, which also helped me a lot during my later scientific career. Most importantly, however, from the very beginning when I was still a graduate student, he always took great care of me, knowing that I had serious health problems. Klaus even contacted a doctor and made an appointment for me, which demonstrates how much he cared about my health. Suffice it to say that Klaus was my mentor in every respect.

What is the legacy of Klaus' scientific work in your field?

To me, the explanation of climate variability based on the concept of the stochastic climate model is *the* most important scientific achievement Klaus ever made in my field of climate variability. When published in the mid-1970s, the stochastic climate model revolutionised the field of climate variability. The stochastic climate model provides an elegant framework in which climate variability, which is one of the salient features of the climate, can be understood through the interactions between climate subsystems exhibiting vastly different internal timescales. The stochastic climate model concept can be applied to climate variability over a wide range of timescales, from seasonal to multimillennial. Nowadays, complex Earth system models can be integrated for many millennia to investigate such things as the climate-system dynamics during and after the last ice age. These models, which, among other things, include interactive ice sheet dynamics very much support the stochastic nature of climate variability on timescales up to the multimillennial. In comparison to the spectra obtained from standard climate models (simulating the atmosphere–ocean–sea ice system), the spectra obtained from the Earth system models are much "redder", i.e., the variability keeps increasing beyond centennial timescales.

Personal advice from Klaus that helped me in my career

Good work prevails. Colleagues will recognise and acknowledge high-quality research.

How did Klaus' thinking influence your scientific work?

I try to understand climate variability and climate predictability from a stochastic perspective. Klaus also taught me to put things into a wider context.

4.9   Hans Graf

This is a very personal view of what happened two to three decades ago. Since I never kept a diary, some of the details may be less accurate than a historian might hope for.

My first encounter with Klaus was during a seminar talk I gave in the spring of 1987 at the University of Hamburg as a guest speaker from behind the Iron Curtain (that was already beginning to show signs of rust). I had been researching processes that could potentially result in the El Niño phenomenon. Because of the lack of data available to me at Humboldt University in East Berlin, the talk was based on conceptual ideas and hypotheses. Shortly after I began to elaborate on my ideas, Klaus, who was sitting centre front, seemingly started to nod off. "That's it …", I thought "… it's boring." But, to my surprise, after I finished my talk and was basking in some polite desk-knocking, Klaus' hand rose, and he began to bombard me with detailed questions. He understood what I had meant in my doggerel English. And, most importantly, he was supportive and to my very great surprise and satisfaction, invited me to accept a fixed-term five-year position with MPI whenever it suited me. Although I was not able (or allowed) to accept his invitation right away, it strengthened my backbone and let me grow a few centimetres. I finally came back to this offer in 1990 after two extended visits to the Max-Planck-Institute for Meteorology in 1988 and 1989. By that time it was possible for me to move to Hamburg with my family: Germany had finally been re-united. Meanwhile I had incorporated Klaus' PIP and POP concepts into my research and was studying the interaction between tropospheric and stratospheric circulation, a process that would later become very important for the interpretation of continental winter warming following major volcanic eruptions.

When I began my five-year contract in January 1991, Klaus initially suggested that I might be interested in enabling the then active climate model ECHAM2 to be used for paleoclimate studies. This was always typical of his manner—to make suggestions rather than issuing orders.

External events soon put an end to my coding efforts. First, the media reacted to the burning oil wells in Kuwait during the Gulf war with stories of global apocalyptic consequences similar to a nuclear winter. We had long and controversial discussions about these rather extreme visions in the tearoom in which I stated that the relevant effects would only be local during the winter due to the prevailing very strong inversion layer over the Middle East. Klaus joined in and at the end of the day he suggested an effort involving the whole Institute. I received his long leather whip. All work on the climate was put on hold for several weeks, at least for a dozen or so people, and a paper was written and readily accepted and published in *Nature* in which it was proposed that the effects from the soot belching oil wells would only be local. We did not include the summer simulations. This was my first in-depth contact with aerosol science.

When Mt. Pinatubo erupted in June 1991, I asked Klaus if he would be in favour of a study of the effects of the massive volcanic eruption. He suggested writing a proposal to BMBF (the Federal Ministry of Education and Science), which was accepted within two months. I received my first funding for a project of my own, which marked the start of my scientific independence. I also got my own research group.

About a year later, when our research into Pinatubo was well under way, Klaus called me into his office where he talked to me about his wish to include atmospheric chemistry in climate research. It seemed to me that our institute should mark its leading role as the MPI for Chemistry became quite strong in atmospheric research. Finally, Klaus offered me a permanent position on the proviso that I would concentrate on atmospheric chemistry. A PERMANENT POSITION at MPI-M! Since I had no idea of chemistry beyond what I had learned at school, I asked for a day to consider the offer, which he allowed me, saying that: "You can do anything if you are intelligent!". The next day I suggested concentrating on aerosols as a combination of physics and chemistry, which would give me more confidence. I guess that was an intelligent idea. Klaus accepted my proposal, and I began my period of aerosol research, which culminated in the BMBF-funded National Aerosol Research Programme, which eventually resulted in many invaluable contacts including a very close collaboration with the MPI for Chemistry in Mainz.

The last great piece of advice that Klaus gave me in 2002 was to accept the offer to take on a newly installed Chair and Professorship on Environmental Systems Analysis at the University of Cambridge. He dismissed my reservations about missing links and the lack of climate research activity at Cambridge pointing out that: "Once you're there, you'll be able to do whatever you want!".

He was right … again.

4.10 Gabriele Hegerl: Der Alte

I first met Klaus when I came to Hamburg for a job interview as postdoc— I was clearly considered slightly unusual by Hans von Storch and Klaus, and not only because of my very strong Bavarian accent. Klaus found my interest in climate, my weird Ph.D. topic and my language amusing and they hired me to work on a hugely exciting topic, namely the detection of climate change. I followed in Ben Santer's very large footsteps, and worked directly with Klaus and Hans, or rather mostly with Hans at first, as Klaus was too busy and may also have been considered too intimidating. His input also needed translation for a climate science novice like me. It was interesting to see what a huge presence Klaus was within the MPI at that time, and probably even now! The only other time I encountered this level of admiration was when I went out with the Munich philharmonic players after a concert and noticed how they spoke about their then conductor. In both cases, it had to do with people's admiration for someone who had mastered his field and was able to do things we could only dream of. For us, he was the boss, the guru, the man who sets the topics, mentors our work, sees the flaw in our scientific arguments but also has some amazing ideas how to fix them, the decision maker, the person who knows where to go. Presenting results in front of Klaus at the annual retreat in Salzau was an amazing opportunity but also quite terrifying. If there was any flaw in one's work—and surely there was bound to be—then Klaus would be sure to spot it. Everybody had seen him happily dozing through seminars only to wake up and ask THE QUESTION—the one question that really picked on the deep issue somewhere in that problem, the unjustified assumption, the pedestrian approach, or the core of the problem. I know very few scientists who are able to do this—to spot the big issue and latch onto it –so Klaus is a rare and truly outstanding scientist. It was no wonder that the survivors of this opportunity and ordeal would take part in a lot of relaxing activities after the Salzau presentations.

Klaus has very much shaped the field of climate science, and the two pieces of his work that I admire most are his stochastic climate models [38] and signal detection methods [54, 110, 129]. His work is still frequently cited, and his way of thinking about the problem has shaped the field. We no longer search for a deterministic response that can be linked very simply to chains of argument. Instead, we look for the climate system integrating weather phenomena and other noise and resonating in response. This has also worked for me in relation to the role of volcanic activity in the last millennium, where short sharp shocks lead to low frequency variability such as the Little Ice Age. Of course, I am very partial to signal detection—Klaus' 1979 paper [54] was the first to set out a framework for how to achieve this and, whilst the idea has been reshaped by the community, it has survived and his relevant papers still cited frequently. By the way, my Ph.D. student has recently rediscovered principal oscillation patterns [86] and Klaus' arguments against discounting future damage to the climate in integrated assessment modelling has taught us how to think about climate change and the benefit of mitigation. But, I still find stochastic climate models the most beautiful of all Klaus' ideas.

Post meeting relaxation with some beverages (with Ernst Maier-Reimer and Joerg Wolff)

While Klaus tucks into his reward

In addition to teaching me a lot about climate science, Klaus also taught me three essential practical things: the first is that it is worth spending a lot of time polishing papers and getting them just right. I will never forget that long evening when I needed to submit our attribution paper [135] prior to leaving on some trip or other. I thought it was ready. Klaus thought it wasn't. So, throughout the day I got lots of scribbled corrections handed down from the upper floor pavilion to my office. At around 9 pm he ordered pizza and handed over the final corrections. I completed the paper, printed it, and boxed it into a courier box (those were the days). Of course, his corrections were almost illegible but if I didn't manage to decipher them this time, they would come back exactly the same the next time. Very predictable! Even more challenges in reading his handwriting arose when he faxed equations from Sylt. We once had a debate about how to optimise the fingerprint in practice—it's quite hard to decipher faxed scribbled equations and when successful, this is followed by the even harder task of understanding them.

Klaus taught me two other practical things:

To do important things really well. It is fine to focus on the thing one is most dedicated to and to put less effort into less important things. I keep repeating this to myself as a mantra—you want to do research to perfection in midterm project reports or committee reports is not needed.

Research and life are full of opportunities. There is no need to continue doggedly with what one is currently doing. If the topic in question is too busy and the good stuff has already been published then move on and look for new questions. Try something crazy. Don't get stuck on the same thing. I try to do that too and sometimes it works. Following Klaus' advice, at least sometimes, has opened up many really interesting research opportunities. And his optimism also extends to politics: people, he says, will understand that climate change is important and that it needs to be addressed. We will solve this. I certainly hope we do and that the future proves him right!

4.11   Jin-Song von Storch

I belong to the younger generation and got to know Klaus as a generous director interested in science and only in science, although it took me some time to realise this. I started my Ph.D. at the MPI with Hans in 1987, working on predicting ENSO using POPs, and had little idea about Klaus. I remember that at some point I needed to talk to him. I went to see him and was nervous, but found only Elsa Radmann: "Mr. Hasselmann is in a meeting", she snapped. So that was my early impression of Klaus; a director who is not easily reachable.

Like many young scientists, I was keen to present my results. My problem at that time was that I was unsure about how to get Klaus' attention. There were so many great scientists at the Institute such as Mojib who became famous within the TOGA community overnight. And some of them, such as Hans, have always been loud. So, the only way for me to get Klaus' attention was to give a talk. But most talks did not go well, because Klaus usually started asking questions after two minutes and then dominated the entire discussion. One needs to do something about it. I discovered he likes sweets so, when it was my turn again to give a talk, I brought a box of Toffiffee. I was able to keep him busy chewing, but not for too long.

I learned more about Klaus from elder colleagues such as Peter Müller and Dirk Olbers. Actually, I (as a meteorologist) learned physical oceanography from Peter and Dirk (more precisely from their books). I was struck by the rigor and the precision of their theories, both in terms of the fundamental equations and the various approximations derived from these equations. I had a hunch that the way Peter and Dirk work very probably had something to do with their mentor, Klaus. Peter described his experience after he left Klaus' group. He was surprised to learn that Klaus was an exception: when you leave the MPI, you get to know normal people.

The most unforgettable picture I got of Klaus was the one I formed of him at the colloquium held on his 60th birthday shortly after completing my Ph.D. The colloquium was attended by some world-renowned people and Klaus gave his famous talk on his metron theory. Like many others, I didn't understand a word of it, but I do remember him saying: "you can ask me, but you cannot stop me". I witnessed a real scientist talking!

4.12 Hans Von Storch

Encountering Klaus

When I first got into the field of meteorology as a recently graduated mathematician working in the Günter Fischer group, I worked 2 floors below Klaus in the Geomatikum. I was aware that there was a Max-Planck Institute—two friends of mine had done their Ph.D. studies there—but I had no real idea about what they did, and who Klaus was. I didn't meet him until about 1982 or '83, when Klaus organised what we foot soldiers called the "Lütjenseer Wendeparteitag", to which our university group was invited. That was because Klaus had determined that the work being carried out at his Institute had matured to the point that quasi-realistic modelling capacity had to be installed at the MPI, and that our group, in particular Erich Roeckner and Ulrich Schlese, would be useful for this purpose. I had to give a talk, and I decided to talk about the statistical comparison of ensembles of model simulations and went on to discuss non-parametric methods. Klaus did not like it and tried to teach me the significance of red and white noise about which I had no idea. He was probably right, but I didn't want to accept that. It all ended in an unconstructive heated debate. Later Ernst Maier-Reimer comforted me with a good beer. I got my "habilitation" the following year, thanks to Günter Fischer, who then suggested to Klaus that he should hire me. When we met for a chat, I referred to our "discussion" in Lütjensee, but Klaus just waved it away as irrelevant. I found out that he had also had a similar experience, when Reimar Lüst offered him the MPI. Lüst had attended Klaus' now infamous presentation at the Atlantic Hotel, at which he overlayed so many slides on the overhead projector that the screen was just black—an event that Klaus remembers with a certain amount of embarrassment.

I was hired and entered a new scientific world. Eventually I read Jenkins & Watts, learned about red and white noise, about detection and attribution and all that, wrote the book "Statistical Analysis in Climate Science" with Francis Zwiers, and had many more beers with the unforgotten and much missed Ernst Maier-Reimer. What a privilege!

As for Klaus himself, he was always generous, mostly charming and humorous, but strict and impatient when it came to the science. One problem is that his pronunciation is often difficult to understand—but now I know why: his speech is often not a communication to others, but rather the sound that accompanies his thinking. Whenever he mumbles, then he is still doing his intellectual analysis.

What is the legacy of Klaus' scientific work in your field?

Klaus contributed to many scientific challenges, but I only grasped, and perhaps incompletely, his achievements in the field of stochastic framing of the climate system, in ocean wave modelling, and in linking climate and society. Whilst the first two have obviously been enormously successful, I do have certain reservations about the latter.

The really significant part was his statistical thinking, the concept later encapsulated in the concept of "Principal Interaction Patterns", according to which the full phase space of a system is divided into two parts, a small, low dimensional part, where the key dynamics takes place, and the remainder with very high dimensions, which is mostly a slave of the first part and feeds back into the dynamical core through conditional statistical models (commonly named parametrisations). This concept was already encapsulated in his first strike—the stochastic climate model, which predicted that long-term variations would emerge in the climate system, without a forcing acting on these time scales: "smoke without fire". This "noise" was not just a nuisance when it comes to identifying the dynamics and interlinkages but was a generic part of the dynamics. His second strike in 1979 was the discrimination between this unavoidable unprovoked variability, the noise, and any signal reflecting the presence of external forcing. The detection-and-attribution concept developed from this, which justified the assertion that the ongoing emissions of greenhouse gases is changing the climate of Earth. His strike 3a was the formulation of the Principal Interaction Patterns in 1988. He had already developed an early version, which he named Principal Oscillation Patterns, which he asked me to breathe practical life into. I did so—but at the cost of simplification, of vulgarisation. He invented PIPs in response to this. This was his strike 3b. His first two achievements changed climate science and the role it plays in the global economy and policy making. As an abstract concept, the third shaped my thinking.

Klaus attempted to expand these ides to include society as a component within the climate system. However, the basic assumption, namely the persistent existence of a low-dimensional subspace with a dominant dynamic, is questionable in relation to societal dynamics. I do not believe that such a subspace could exist for a sufficiently long time and think that it would be conditioned by a variety of inhomogeneous cultural configurations.

How did Klaus' thinking influence your scientific work?

His thinking guided me—in conceptualising the climate system in the spirit of PIPs, with the detection and attribution being carried out in an appropriate low dimensional subspace of dominant dynamics. He convinced me that noise is ubiquitous in the climate system, on global and regional scales, in the atmosphere and the ocean. My latest research interest was, and continues to be, the emergence of such (hydrodynamical) noise in marginal seas and its scale-dependency.

His thinking also influenced me to ask whatever and whoever: why? Where is the evidence? What are the hidden tacit assumptions?

What piece of personal advice from Klaus has helped you in your career?

The only piece of advice I remember is "don't worry, when time is ripe, a door will open for you. When you are good at something, and it is of interest, an opportunity for an application or for a job will emerge". I now give this same advice to my own Ph D students and co-workers. It works.

Whilst not taking the form of explicit advice, his management approach has informed my own. Never look for finance planning details (Hinzpeter's dogma in this context: "Eine Zahl ist keine Zahl") but base your decision solely on a consideration of the relevance for the work and the issue, whilst keeping the personal implications in mind; decide immediately in most cases. In and of itself, increasing the number of co-workers and of the influx of money is not a legitimate goal when it comes to running a research institute.

4.13   Patrick Heimbach: Interactions with Klaus' Sphere of Influence

It was Spring 1993; I had recently completed my Dipl. Phys. in Bonn and was keen to change my subject of study to climate research. A first application for a Ph.D. position at the University of Hamburg had failed, leaving me deeply disappointed. This was the backdrop for my interview at the MPI, where I first met Klaus and Susanne. At the end of a day's visit, Klaus explained to me that there was no current opening in the climate dynamics division, but that they were looking for a student in the "Seegangsgruppe". I didn't quite know what that was all about, but immediately accepted the offer nonetheless, having been deeply impressed by the person and the interactions I had had just on that day. Two initial personal lessons I learned were: (i) more often than not it is good to follow your instincts; (ii) sometimes an initial rejection opens the door to a much brighter sequel.

Thus, in the summer of 1993 I began work in remote sensing and modelling of ocean surface waves. The context was the recent launch of the first European Remote Sensing satellite ERS-1, which opened up the prospect of being able to observe ocean surface waves on a global, quasi-routine basis, and the ability to perform detailed validation of the third generation Wave Model (WAM). One specific scientific question concerned swell propagation over long distances, and the process of dissipation. The fact that we were picking up a classic field experiment that had been conducted by Walter Munk and colleagues—including Klaus—in the early 1960s, following swell propagation across the Pacific along a great circle [18], but now using remote sensing, offered me a wide range perspective for studying the problem as well as giving me an insight into Klaus' early work.[3]

More happy surprises awaited me soon after starting my work: within days of starting my job, a group of researchers from the USA visited the MPI, and Klaus invited me along to their meetings, despite my almost complete ignorance of the subject. Within months, I took my first trip to Utrecht (KNMI) to celebrate the publication of the now classic book on *Dynamics*

---

[3] There is a beautiful, 30-min documentary about the experiment, which I regard as a must-watch when teaching about ocean surface waves, narrated by Walter Munk, and in which Klaus has several brief appearances: https://www.youtube.com/watch?v=MX5cKoOm6Pk.

and Modelling of Ocean Waves [244]. These are but a few examples of Klaus' trust in people and his ability to develop the deep sense of community that makes research teams successful.

More lessons were learnt along the way, e.g., (iii) that what I had studied in theoretical physics about particles and fields could also be applied in oceanography, as pioneered by Klaus, and (iv) that—arguably—ocean wave research provides the basic training for climate science ("Seegang, die Grundschule der Klimaforschung"); at least this was one (of several) way(s) in which the small "Seegangsgruppe" justified its *raison d'être* within a large climate research institute, sometimes in a slightly tongue-in-cheek manner. Nevertheless, the current renewed interest in the subject provides some vindication (e.g., Villa Bôas et al. 2019).[4]

I was the third "Rheinländer" in the small group, next to Renate Brokopf and Georg Barzel. Renate's cookie box (always filled with "Prinzenrolle") ensured that we'd get regular visits from Klaus. Over the years I would come to represent the group in a variety of project meetings and symposia—earning me the title of "Reisedoktorand" (the travelling doctoral candidate).[5]

The time came for me to produce scientific results. "Schon sehr schön" is what I would get to hear a lot. "Very nice for a start" might be a *precise* translation, but an *accurate* one would emphasise the fact that *lots* of work remained to be done. Those words are telling of Klaus' deeply human approach to mentoring. Always encouraging, setting a positive tone, but just as clearly conveying to the mentee the many ways in which the work he or she presented remained insufficient. Frequently overwhelmed by the deep insights of the mentor, the mentee would walk away from a meeting, wondering how he or she could ever move beyond "Schon sehr schön". What may have saved me was the privilege of being exposed to a rich spectacle of perspectives that Klaus weaved together into a complex story of the climate system, from its physical machinations to its societal interactions.

It is difficult to choose among the many lasting impacts that Klaus has had on the field. Others who have contributed to this volume have provided accounts in the context of surface wave modelling and remote sensing (and see a recent review by Klaus himself [176]), so I will highlight work not done by Klaus himself, but which he had the vision and foresight to support, and which would prove important to my work as a postdoc with Carl Wunsch

---

[4] Bôas, A. B. V., et al. (2019). Integrated Observations of Global Surface Winds, Currents, and Waves: Requirements and Challenges for the Next Decade. *Frontiers in Marine Science, 6*, 2219–2234. https://doi.org/10.3389/fmars.2019.00425.

[5] Among the noteworthy places and people that left an impact on me were encounters with Bertrand Chapron and Harald Krogstad at Ifremer, and David Halpern at the World Expo'98 in Lisbon, Portugal.

at MIT. This was about developing a software tool, initially developed at the MPI by Ralf Giering and Thomas Kaminski and later matured at MIT, that could "differentiate" a model code, i.e., generate code that represents the derivative of some model output with respect to some inputs by means of "automatic differentiation".[6] This tool would prove essential in NASA's ocean data assimilation consortium "Estimating the Circulation and Climate of the Ocean" (ECCO), which now has a 20-year legacy, involving various former and present members of the MPIMet (Marotzke et al. 1999, Stammer et al. 2002, Heimbach et al. 2019).[7]

My personal, deep, and lasting impression is that of an extraordinary individual, not only intellectually, but as a human-being, generous, caring, free of allures, and with a rich sense of humour. The latter shines through in this concluding anecdote: it is the story of an elderly man who appeared regularly in the halls of the Geomatikum with a pamphlet in which he claimed to have proven that Pi is a rational number. At one point, Klaus mused (with a subtle ironical smile) that it might be best for him to join this old man to distribute his own work on the metron model.

4.14   Jörg Wolff: The Shortbread Biscuit

There was a conference in Hawaii, which I really wanted to take part in. To get permission, I grabbed a shortbread biscuit, put it on a small plate, and went straight to Klaus' office. Elsa Radmann allowed me to enter and I told Klaus that this would be an attempt of bribery. He looked longingly at the shortbread. I presented my case, he accepted, and ate the biscuit.

---

[6] Giering, R., & Kaminski, T. (1998). Recipes for adjoint code construction. *ACM Trans Math Softw*, *24*(4), 437–474. https://doi.org/10.1145/293686.293695.

[7] Marotzke, J., Giering, R., Zhang, K. Q., Stammer, D., Hill, C., & Lee, T. (1999). Construction of the adjoint MIT ocean general circulation model and application to Atlantic heat transport sensitivity. *Journal of Geophysical Research*, *104*(29), 529–548. https://doi.org/10.1029/1999jc900236.

Stammer, D., Wunsch, C., Giering, R., Eckert, C., Heimbach, P., Marotzke, J., Adcroft, A., Hill, C. N., & Marshall, J. (2002). Global ocean circulation during 1992–1997, estimated from ocean observations and a general circulation model. *Journal of Geophysical Research*, *107*(C9), 3118–1–27. https://doi.org/10.1029/2001jc000888.

Heimbach, P., Fukumori, I., Hill, C. N., Ponte, R. M., Stammer, D., Wunsch, et al. (2019). Putting It All Together: Adding Value to the Global Ocean and Climate Observing Systems With Complete Self-Consistent Ocean State and Parameter Estimates. *Frontiers in Marine Science*, *6*, 769–10. https://doi.org/10.3389/fmars.2019.0005.

4.15 Ben Santer: A Road Trip with Klaus

One of my favorite personal memories of Klaus was traveling with him to a meeting of the International Detection and Attribution Group (IDAG) in Boulder in the early 1990s. At that time, Klaus was working on the draft of what would later become his seminal 1997 paper on fingerprint detection [129]. When we boarded the international flight to Denver, Klaus informed one of the flight attendants that we were engaged in important scientific research. Were a pair of quiet seats available in business class?

I've never had much luck with polite requests for free upgrades to business class, but Klaus was successful. My "lesson learned" was that it helps to travel with someone who conveys—even to those who do not know him—an impression of quiet authority, of distinction, of being "außergewöhnlich".[8]

And Klaus is "außergewöhnlich". I've never met anyone like him. The essays in this book will surely attest to the extraordinary contributions Klaus has made to many different areas of climate science. Stochastic climate models. PIPs and POPs. Optimal detection of anthropogenic signals. Elucidation of the cold start effect. Development of ocean wave models. Exploration of the economic impacts of climate change. The list of contributions is long and illustrious, each highlighting Klaus's unique ability to see the complex climate system from a novel and interesting perspective.

While such vision and scientific brilliance is "außergewöhnlich", it is the pairing of vision and brilliance with very human qualities—humility, and deep curiosity about the world and people around him—that is truly extraordinary.

Back to our flight to Denver. Klaus worked on the *Climate Dynamics* paper, passed me a draft version, and asked for my comments. I felt that it would be impolite to read a magazine or fall asleep. If your boss is changing the world of anthropogenic signal detection on a flight from Germany to Denver, you don't fall asleep. You pay attention.

During the meeting of the IDAG group at the National Center for Atmospheric Research (NCAR), Klaus provided the scientific direction for the group's efforts to identify a human-caused warming signal. He reminded us of the power of patterns. As he had written back in his famous 1979 paper [54], "*It is necessary to regard the signal and noise fields as multi-dimensional vector quantities and the significance analysis should accordingly be carried out with respect to this multivariate statistical field, rather than in terms of individual grid-point statistics.*" Or put simply: Look at patterns, not at individual grid-points.

---

[8] Langenscheidt's translation of "außergewöhnlich" is "extraordinary, exceptional, outstanding".

Pattern analysis provides you with the power to discriminate between natural internal variability and the forced response to human-caused greenhouse gas increases.

It was a key insight, and it provided a "statistical roadmap for hundreds of climate change detection and attribution studies"—studies which ultimately identified human-caused fingerprints in many different independently monitored climate variables.[9]

After the conclusion of our IDAG meeting in Boulder, Klaus and I had a free afternoon before our return flight to Germany. Why not go for a drive to the Rockies?

What a marvelous experience that was! In Hamburg, given the sheer number of scientists, students, and visitors wanting to see him (and the tight control of his schedule exercised by Frau Radmann), it was difficult to get a few hours of uninterrupted hours of "Hasselmann time." I had that privilege now.

So we drove to Estes Park, the gateway to Rocky Mountain National Park. I recall how good it felt—after hours in airplanes and in a meeting room—to get out and stretch our legs in Estes Park, and to take in the grandeur of the Rockies. And I remember Klaus's humanity. He was genuinely interested in me as a human being, and not just as a scientist. A drive that might have been anxiety-inducing and intimidating for a young post-doc instead became a few truly memorable hours—the opportunity for a fascinating conversation about life and science.

Klaus Hasselmann has accomplished many great things in his scientific career. He published ground-breaking research. He led an institute that became a world-leading research center for climate modeling. He helped the world understand that humans are not merely innocent bystanders in the climate system—human activities are actively changing Earth's climate. But in addition to all of these great achievements, he had a "discernible influence" on the lives of generations of colleagues and students. That contribution will be just as enduring as all of his contributions to climate science.

---

[9] Santer, B.D., C. Bonfils, Q. Fu, J.C. Fyfe, G.C. Hegerl, C. Mears, J.F. Painter, S. Po-Chedley, F.J. Wentz, M.D. Zelinka, and C.-Z. Zou, 2019: Celebrating the anniversary of three key events in climate. *Nature Climate Change*, **9**, 180–182. https://doi.org/10.1038/s41558-019-0424-x.

4.16 Ulrich Cubasch: How a Postdoc Became an IPCC Convening Lead Author

My first encounter with Klaus Hasselmann was at ECMWF in Reading, where I was working on the development of the next generation of the forecasting model. My colleagues mentioned to me that he (they had given him the nickname "The Kaiser") would be coming to Reading for about a fortnight to do research. I was of course curious to meet the scientist with such a nimbus, and I seized the opportunity to have a brief conversation with him. Later I contacted him about the possibility of doing a Ph.D. in Hamburg. It turned out that, unlike the University of Reading, it was possible to do obtain a Ph.D. from the University of Hamburg without being enrolled. Due to the different curriculum structures at UK and German universities, being enrolled would have meant that I would have had to spend a lot of time attending Ph.D.-courses, which merely repeated what I had learned for my German Diploma. Prof. Günter Fischer agreed to supervise the thesis as an official representative of the University of Hamburg jointly with Klaus Hasselmann, and Hans von Storch did the some of the coaching. ECMWF did not mind this set up, as long as it did not interfere with my normal work. At a later stage, its support became stronger, as the thesis dealt with performing extended range predictions using ensemble techniques. It was anticipated that this methodology had the potential to extend weather forecasting for a longer period.

I was later invited by Klaus Hasselmann to join his group as a postdoc. I found the topic of climate science more interesting than the continued attempt to improve weather forecasts, which was the main focus of ECMWF. Some of its member states insisted that that should be its only goal. I was keen to get my teeth into coupling an atmosphere model, something I was familiar with through my work at ECMWF, with a comprehensive ocean model. At that time, only the University of Oregon had accomplished it and published results, but GFDL and NCAR were already performing test runs.

One day, it must have been in late 1988, Klaus Hasselmann came into my office and asked me if I would volunteer to fly to Princeton in his place for a meeting between groups working on coupled ocean–atmosphere modelling. He told me that they were planning a comparison between various examples of this type of model. My job would be to represent the Institute and its research (to fly the flag). I went there, keen to meet all of my colleagues working on this task. It turned out that it was a high-profile international meeting which had been set up in preparation for the first IPCC-report by working group 1. At that time, I (and maybe also Klaus Hasselmann) had not really been aware of the importance of this workshop, so I was a bit surprised by the lion's den atmosphere created by some of my high-profile colleagues. As they had been expecting Klaus Hasselmann, not some little known postdoc, it was a bit of a challenge for me to convince the attendant US- and UK-dominated science community that there was also pertinent research been performed in Germany. As the IPCC strives to achieve an internationally balanced membership, they eventually embraced our effort. Our Institute was selected to compile one of the chapters of the IPCC-report, which dealt with the coupled model comparison. Perhaps feeling a bit snubbed by Klaus Hasselmann's absence, Michael Schlesinger suggested that I should be the author of this chapter. He pointed out that Klaus would probably be too busy to deal with the humble task of comparing models and data. They also assigned Robert Cess, a seasoned scientist with a lot of experience in how to integrate the various scientists' attitudes, as a co-author.

I tried to involve Klaus in the IPCC-activities and discussions when I returned to Hamburg, as considerable rivalries had emerged between the institutes which had been asked to contribute to the comparison. From time to time I approached him for comments or suggestions, particularly when there were conflicts. Knowing the characters of many of the persons involved, he advised me "to keep my head down" and to play an integrative role. During this time, he focused on creating results that would improve the IPCC report. With the MPI being part of the authors team, it was assured that his and the MPI's and University of Hamburg's scientific works would be cited and recognised by the international community.

Due to the IPCC's high international profile, more and more institutions and nations became interested. The IPCC grew larger and larger. To fend off the numerous external attacks by special interest groups, it became increasingly formalistic. Nowadays the author of a chapter is selected in an elaborate procedure, where I as a postdoc would not stand a chance. The IPCCs activities (and all of the people who contributed to its success) were honoured with the Nobel Peace Prize in 2007.

Having been drawn into the IPCC in an early phase of my career, it has influenced my research ever since. I had the fortune to be selected as author and coordinator in all of the following reports. These activities brought me into contract with the international science community, the EU funding agencies, and several German government bodies. I had the opportunity to travel around the world, as the IPCC spreads its meetings around the globe to demonstrate its international character.

In summary, I am grateful that Klaus Hasselmann enabled me to obtain a Ph.D. and that his confidence in delegating tasks to his staff provided me with a once-in-a-lifetime opportunity that shaped my entire career.

4.17 Achim Stössel: From Seaman to Professor Thanks to Klaus Hasselmann

Klaus is the most important person to whom I owe my scientific career, which has extended all the way to a tenured professorship. I never imagined any of that when I was still with the merchant navy some 40 years ago, staring brainlessly out to sea from the navigation bridge of a freighter as a nautical officer on watch. I well remember my first encounter with Klaus in his office at the MPI-M where he justifiably worried about my grades (not a C-candidate, but also not a straight A-candidate), and how he was initially reluctant to accept me as a Ph.D. student. I heard (maybe just a rumour) that Susanne had somehow convinced him of the benefits of having a seaman on board in his institute. I also recall Klaus trying to convince me to work on wind-generated waves rather than sea ice, presumably because of the 4 years of seagoing experience I had by then. At some point much later (I believe it was during one of the Salzau meetings), we even argued about the climate relevance of surface gravity waves versus sea ice.

Anyway, after Klaus and Peter Lemke had decided that I would work on sea ice, I remember coming up with the suggestion to first test Bill Hibler's new viscous-plastic rheology sea-ice model in the Baltic Sea, as this was a region with a dense observational network, which meant that we would readily be able to evaluate the realism of the model simulation. Klaus' response was that the size of the Baltic Sea corresponds to just 2 grid cells of the T21 model, so I was to apply the sea-ice model to the Southern Ocean around Antarctica, for which good forcing and verification data was of course much more difficult to obtain. Peter, Breck Owens, and I nevertheless cranked out a convincing paper, and I eventually defended my dissertation on this topic in December 1990.

By then in my mid-30s, I was confronted with what to do next. With our first child underway, I didn't want to jump from one 3-year project to the next. I recall approaching Klaus one day asking about the possibility of continuing to work at MPI-M as a research scientist. That meeting was rather short: he first asked me about my age, then about the number of publications I had. After hearing my response, he said that he would grant me another 6 months. That was a clear message. I nevertheless stayed on for 3.5 years as a postdoc because of an SFB project for which I obtained funding, but it was clear to me that I would need to look for a more permanent job elsewhere. I therefore submitted some 15 applications for such positions, was interviewed for 3, and was accepted for 1, and that was undoubtedly because of Klaus (recommendation letter) and the fact that I did my doctoral and postdoc research at the MPI-M. Not only that, even when deciding on whether or not to offer me a tenured position, I learned later that my current employers had asked Klaus for a recommendation letter. To sum it all up, I am most grateful and lucky that Klaus accepted me into his Institute back then, and that he continued supporting me all the way to my current position, in spite of my former non-scientific career.

4.18 Robert Sausen: Interactions with Klaus Hasselmann

I first met Klaus Hasselmann when I was a Ph.D. student during a summer school organised by the Studienstiftung des Deutschen Volkes in Alpbach. I was so impressed by his ideas on climate change and the methods he used that I applied for a postdoc position at his Institute and was grateful for the opportunity to start work there in 1982. Once there, I initially found it difficult to understand Klaus's concise way of presenting his ideas as did the other postdocs and Ph.D. students in his group at the MPI for Meteorology. Luckily, we were helped by the "ZKs", the "Zwischenkapazitäten" (the clever minds in between), Jürgen Willebrand and Dirk Olbers. They were already experienced colleagues, both when it came to the science and to understanding Klaus. So, they translated his ideas into a language that a postdoc or a Ph.D. student was able to understand, and, in this way, we learnt a lot.

Following my training phase, Klaus pushed me in the direction of studying averting the initial drift in coupled atmosphere–ocean models. I came up with the idea of "flux correction", whereby a better name would have been "anomaly flux coupling". The method was quite successful, but also controversial. The first time I presented it to an international audience was at the Erice summer school in 1986. The discussion after my short presentation was rather heated, mainly among the lecturers at the summer school, with the Europeans in favour of my ideas and the Americans opposing them. Nevertheless, I, the young scientist, felt fairly safe because I knew that Klaus was protecting me.

I had a similar experience a few years later in 1990 or 1991, when I told a journalist that climate change met with little interest among policy-makers, as the effects of climate change would be felt much later than the legislative period. A high-ranking officer of the German ministry of research complained to Klaus about what I had said. And Klaus simply answered that I told the truth. Klaus taught me to be frank about unpleasant results and news.

I highly appreciated the inspiring and supportive environment that Klaus created at his Institute.

4.19   Dmitry V. Kovalevsky

How did you meet Klaus?

My research collaboration with Klaus began in 2007. Klaus introduced me to socioeconomic modelling related to climate mitigation, and this completely changed my subsequent trajectory in academia. All this began when I was introduced to Klaus at a conference in Berlin in late 2007. Since then, I am indebted to Klaus for all his kind, invaluable, continuous support throughout my career. With the aid of his support with many issues, the socioeconomic research group was established at the Nansen Centre in St. Petersburg (NIERSC)[10] where I was working at that time, and I became the leader of this newly formed group. Klaus provided very active support for the activities of our group and collaborated with us enthusiastically. We developed the models together and published co-authored papers. Thanks to Klaus, we were invited to consortia concerned with a number of major research proposals, and Klaus himself was also a very active contributor to the proposal writing process. As examples of our joint project activities, I would refer to two major EU FP7 projects during the past decade, COMPLEX[11] and EuRuCAS,[12] in the course of which our group at NIERSC collaborated very actively with both Klaus and other project participants on the implementation of the project. Klaus travelled to St. Petersburg several times to present keynote talks at workshops and colloquia organised by our group at NIERSC, and to attend meetings that were important to the group.

During my two research visits to the Max Planck Institute for Meteorology (MPI-M) in 2015, Klaus kindly offered to let me use his desk in the

---

[10] Nansen International Environmental and Remote Sensing Centre (NIERSC), St. Petersburg, Russia.

[11] EU FP7 COMPLEX, Project No. 308601 "Knowledge Based Climate Mitigation Systems for a Low Carbon Economy" (2012–2016).

[12] EU FP7 EuRuCAS, Project No. 295068 "European-Russian Centre for Cooperation in the Arctic and Sub-Arctic Environmental and Climate Research" (2012–2015).

Emeriti office of MPI-M (the other *Emeriti* desk in the office was for Prof. Lennart Bengtsson). Klaus and Susanne kindly invited me, and following my marriage, my family to stay with them and/or visit them in Munich, Glückstadt, on Sylt, and more recently in Hamburg, and we are always welcomed with the warmest hospitality. Another of our unforgettable experiences was when Klaus and Susanne invited us to attend a choir performance in which they were singing. Klaus is always sharing so many interesting stories with us about his life and career, about his family and relatives, and about his travels all over the world.

What is the legacy of Klaus' scientific work in your field?

With what field should I begin? My research career is connected to several areas in which Klaus was very active, including theoretical physics, oceanography and—as mentioned, thanks to Klaus personally—transdisciplinary modelling for climate mitigation. It is the latter area in which his ideas and contributions have shaped my own thinking and research activities to the largest extent.

Has any personal advice from Klaus helped you in your career?

Klaus gave me a lot of invaluable advice on various topics during our lengthy collaboration, and I could gratefully provide many examples here. For instance, he gave me some comprehensive technical advice relating to IT under very non-trivial circumstances, which continues to help me a lot in my research until the present day. That was in 2008 when we had recently begun our collaboration with Klaus on socioeconomic modelling, and I had to master a specialised software programme that Klaus was systematically using for developing his models (to avoid accusations of hidden advertising, I shall refrain from naming this excellent software package here). To help me learn as quickly as possible, Klaus kindly gave me a personal training course in the most wonderful and hospitable environment one could ever imagine: Klaus and Susanne kindly invited me to stay with them on Sylt. For several days I sat with Klaus over a laptop, whilst he used all his pedagogical talent to teach me step-by-step how to use the various features and options of the programme. In a spirit of full disclosure, I should add that after these intense lessons there were wonderful walks with Klaus and Susanne along the seashore and in other beautiful places in Sylt. Having benefited from this personal IT training from a famous scientist, I am still actively using the knowledge and skills I acquired.

How did Klaus' thinking influence your scientific work?

The impact of our nearly 15-year-long collaboration with Klaus on my research activities and, more broadly, on my way of thinking and problem solving, has been enormous. I am very much obliged to Klaus for so many new and inspiring ideas and for the new methods and tools with which he made me familiar. Had it not been for those years of learning from Klaus as well as communicating and collaborating with him, my own mental model of the world would currently have looked completely different.

4.20 Carola Kauhs: A Non-Scientific View on Professor Klaus Hasselmann from the Institute's Librarian

I have known Mr. Hasselmann for 38 years now, and he has accompanied me throughout my entire active professional life. As a librarian who had just completed her degree, I started working at the Max-Planck-Institute in 1983 as the successor to Mrs. Grimminger, whom Mr. Hasselmann had lured away from his own father in 1975 to build up the joint library at the MPI-M and the university institutes of meteorology and geophysics. Mrs. Grimminger had a very special way of dealing with the Managing Director and was able to convince him with her arguments. Lucky me.

Years later, I learned from my library colleagues that the Managing Directors at the MPG Institutes changed regularly. But at our Institute, we had the same Managing Director for years: Mr. Hasselmann. I couldn't understand why my colleagues were so excited about a special event known as the "Scientific Advisory Board". Either we didn't have anything like that at the Institute or it always completely passed me by.

On behalf of the works council, I sometimes had to make Klaus Hasselmann aware of various things. In those days, it was still possible to hold the works meeting in a medium-sized seminar room in the "MPI-Pavilion". Mr. Hasselmann would sit in the front row, face-to-face with the works council members. When a proposal was made to approve educational leave for scientists at the MPI-M to take additional English classes, a firm, non-evasive look was sufficient to give the Director to understand that the proposal was denied on the grounds that scientists at the MPI know enough English and don't need educational leave for that.

I didn't see Mr. Hasselmann that often during his active time at the Institute. One day, a group of architects were strolling through the library discussing expansion plans for the computer centre on the same floor using the library space. No information about these plans had reached me in advance, so I was quite annoyed with this procedure and tried to confront the Director, but all I got from his assistant Ms. Radmann was the information: "He won't be back at the Institute for two days". So, I initially vented my anger at the Director of the computer centre. Two days later, however, I had the opportunity to talk to Mr. Hasselmann. Still full of indignation, I entered his office and was welcomed by a smiling gentleman saying: "I must have been lucky that I wasn't in the office two days ago". This charmingly took the wind out of my sails. We were then able to clarify the matter (almost) peacefully. The library was saved, but for very different reasons.

The first scientific lecture I heard from Klaus Hasselmann was slightly disappointing. As I naturally couldn't understand much of the content, I focused on his presentation style and waited for a gripping performance by a professor. It was still in the days of projectors with acetate slides that had to be changed by hand. Professor Hasselmann replaced his slides so quickly that the audience must have felt dizzy. His flow of speech was similarly rapid and unclear. Did he actually speak English or German? Even some of the scientists probably had difficulties in following the lecture.

Following his retirement and with advances in the electronic supply of literature, I would occasionally receive emails with requests for a given article in PDF format. I was glad to be able to send the requested texts quickly. The full texts were often sent as breakfast reading to Sylt, to Glückstadt or other places. Prior to digitisation however, one request reached me scribbled on a beer mat after he had attended a conference. He was obviously in the "service of science" at all times.

At some point during the first years, he promised me that he would never donate his special print collection to the library. It would be useless for others and is organized in a rather personal manner. That calmed me down considerably, as the days of special print collections seemed to be over. But now in 2021, with its help, I was actually able to verify a few analogue sources for the bibliography of this book. They could not be found using current digital research tools. So, in the end, the collection was very helpful after all.

I have never regretted spending all of my working years at the Institute of which Professor Hasselmann is the founding director, and even now I am always pleased when an email from him is waiting in the mailbox in the morning.

4.21 Gerbrand Komen

Memories

I first encountered Klaus in 1978 in Kiel, where he gave a lecture at a GATE symposium. I was late and had missed the introduction. I expected a typical German accent. So when I heard Klaus my first reaction was: this cannot be him. But it was.

A year later Willem de Voogt and I travelled to Hamburg to meet Klaus to discuss our joining the Sea Wave Modeling Project (SWAMP), an inter-comparison project that Klaus had started. I vividly remember a subsequent meeting with other SWAMP-participants in the periphery of a wave conference in 1981 in Miami in which Klaus and Susanne both took part. They invited us to their hotel room, where we discussed progress whilst eating dinner from fast food boxes.

After I had presented our wave modelling work at the Miami conference Klaus invited me to spend a summer in Hamburg, which I did in 1983. That summer was quite remarkable. It was great working with Klaus and Susanne (see below), it was equally great to experience their wonderful hospitality. It was not so easy to find suitable accommodation for me and my family (wife + 2 kids). But then Klaus and Susanne let us stay with them in Kayhude, for several weeks. And when my family had returned to Holland they let me join them in their choir, the *Altonaer Singakademie,* for the weekly rehearsals and for a special concert trip. A black suit was obligatory, but I didn't have one with me. Fortunately, Klaus had a spare one, his wedding suit, which fitted me nicely. Highlights were performances in Mölln and Lübeck.

I have many precious memories of our frequent interactions during the 15 years or so following that summer. Too many to list, but a few come to the fore.

In 1985 we were at ECMWF with a team to set up the first version of our wave model. One of the staff members invited Klaus to an evening session of his bell ringing group. Klaus took all of us with him. The world of change ringing opened up for us.

The Wave Modelling (WAM) group held annual meetings, in different places. We worked hard, but often the local organiser would arrange a half-day trip, so we could relax and discuss waves in an informal setting. In 1993 we met at Sylt, the very place at which the JONSWAP-experiment was carried out in 1969. I remember our trip by boat to Hallig Hooge as having been most pleasurable.

Klaus was always very busy. Much of our work was done during travel or during leisure time outside official meetings. We would sit together for discussions in places like the Wiener Stadtpark, or during concert breaks (I remember a performance of La Traviata, in Estonian, in Tallinn). Once we met in Copenhagen, in the lobby of his hotel, after he had given a lecture at an important climate meeting. He listened patiently to me and took his time. After we finished discussing ocean waves, I suggested we should relax over dinner, but Klaus declined. He was still full of energy, and wanted to work on his Metron Theory, a unified deterministic theory of fields and particles.

In 1993 I visited Luigi Cavaleri in Venice to work with him on the completion of our monograph on ocean waves. We worked through the weekends in an otherwise empty Palazzo Papadopoli, eating lunch from Luigi's desk in his office on the top floor and listening to Italian opera music in the background. There was a lot we needed to discuss with Klaus, but it was not easy to get hold of him, as he always had many commitments. However, we found out that he didn't mind us calling him on Sundays. So we had lengthy phone calls with him on Sunday mornings whilst looking out over the sunlit roofs of Venice.

Impact on my thinking

Before I met Klaus I had studied his work on the origin of slow climate variations. Klaus had used an analogy with Brownian motion to show that white noise can generate red noise in any system with different time scales. This is an important result because it means that there can be slow variations in the climate system without a cause.

When I actually worked with Klaus I was particularly inspired by his way of writing papers, proposals and minutes and the way in which he led meetings. Also influential was his vision on the development of an integrated wind and wave data assimilation system.

Impact on my career

Jan Sanders of my institute had developed GONO, a numerical model for predicting ocean waves in the North Sea. In 1978 I was charged with the further development. By taking part in SWAMP we were able to connect with the international wave modelling community. This was most stimulating and very fruitful.

My visit in Hamburg in 1983 allowed me to combine the best of my own institute's wave expertise with theoretical and numerical work carried out by Klaus and Susanne. We simulated fetch-limited growth, to see under which conditions a stationary solution can be reached. Comparing the result with observations allowed us to determine an unknown constant in the dissipation source term. The resulting parametrization is still widely used.

In 1984 Klaus established the international wave modelling group (WAM), and he asked me to chair the group. That kept me busy for the next 10 years or so.

Dynamics and Modelling of Ocean Waves, G.J. Komen, L. Cavaleri, M. Donelan, K. Hasselmann, S. Hasselmann and P.A.E.M. Janssen, Plenum Press, New York & London, 532 pp, 1994

with Wave modelling Group, Sintra, Portugal, 1992

Presentation of the WAM book in De Bilt, 1994. Standing, from left: Klaus Hasselmann, Gerbrand Komen, Susanne Hasselmann, Luigi Cavaleri. Kneeling: Peter Janssen und Mark Donelan.

Klaus' legacy in ocean waves

Klaus' fundamental work on the energy balance equation, non-linear wave-wave interaction and wave dissipation in the 1960s helped provide the foundations for modern wave prediction. He then realised his ambitions by mobilising the international wave modelling community in a long sequence of projects: JONSWAP, MARSEN, SWAMP, NORSWAM, WAM, ECAWOM, resulting in the availability of routine wave observations from space, the development of a third-generation ocean wave model and its implementation in the forecasting system of ECMWF. The model and its descendants now run in many centres worldwide. Our knowledge was consolidated in a multi-authored monograph: *Dynamics and Modelling of Ocean Waves* (1994), which is still used as standard reference text for wind driven ocean (surface) waves.

4.22 Luigi Cavaleri: Writing the WAM Book

I was pleasantly surprised and thrilled when Gerbrand Komen approached me inviting to be a co-author of the planned WAM book [244]. I was not even a co-author of the already well known WAM paper; I was dealing mainly with practical problems and wave measurements in the sea, working, in a way, at the opposite end of Klaus et al., who, with the exception of Mark Donelan, were good (actually extremely good) in terms of their thinking and computer expertise, but who had little experience of a real stormy sea. At the end of the adventure (because an adventure it was), this turned out to be a good combination, joining the ones I considered as descending from the sky, and myself climbing my way up with a lot of effort. Of course, Klaus was the master mind behind it all with Gerbrand acting as the front man, working hard to overcome all the practical difficulties, dealing with the bureaucratic and personal aspects of each co-author.

It is amazing how frustrating it can be spending weeks or months writing and assembling a supposedly eloquent chapter only to see it scratched, cancelled, modified, or scribbled over by someone else. On the wings of my enthusiasm, I was running fast, collecting contributions, and seeing what I considered as "my chapter" growing more and more with what were nice pieces of work. At the end (but wait!, it was not the end) I ended up with more than 100 tightly written pages with a lot of figures summarising all the practical problems and successes associated with the direct application of wave modelling. I packaged it all up and sent it proudly to Gerbrand and

Klaus, and left Venice with my young daughter for our house in the moun-
tains to spend a well-deserved Easter vacation there. It was Easter Sunday,
and I had finished cooking our special lunch (I was obviously in very good
mood) and ready to enjoy it with my daughter when the phone rang: it was
Klaus. I was only able to say a single word: "Luigi?", "yes", and then the
storm began. My chapter was a disaster, the worst thing he had ever seen,
a completely useless work, everything had to be done again, it had been a
stupid mistake to involve me in this task, I had spoiled all the efforts by the
other co-authors, the organisation of the chapter was a mess etc. This went
on one-way for 17 min. Klaus must have been extremely angry, and I was
speechless. Without my having said another word, the phone was slammed
down, and I saw the expression of my daughter staring at me with curious
eyes.

I sat at the table looking blankly at the food, my mind running wildly from
the chapter to Klaus. A few minutes later the phone rang again, and it was
Gerbrand to whom Klaus had reported a "perhaps aggressive call" to Luigi.
We talked, two-ways this time, for ten minutes, Gerbrand acting as the wise
man of the group, soothing me, explaining what Klaus really meant, that it
was not a major issue, that with a bit of effort we could achieve a beautiful
(even for Klaus) result. At the end I forced myself to have some food, and
then my young daughter and I went for a relaxing walk in the snow. Indeed,
and in due course (not so long) things did settle, and the rearranged material
ended up as this chapter of the WAM book.

As a matter of fact, apart from the pleasure and satisfaction of contributing
to such a solid piece of scientific literature, the story had also a pleasant and
musical ending. The book was officially presented at KNMI, in De Bilt, in
the Netherlands. Authors, friends, colleagues, and others came together there,
and each of the authors gave a short presentation of his or her contribution.
I decided to do something different. Some weeks before, shuffling CDs in
a music shop, I realised a remarkable fact and I bought a few CDs. To the
amazement of the audience at the presentation at KNMI I did not talk about
the work done or how important are waves in the ocean world. I shaped
my presentation by illustrating the parallel between waves and music, and
how Susanne, a valid pianist, had been the real inspiration to Klaus for the
name WAM: because (and I switched on a tape recorder playing "Eine kleine
Nachtmusik") the true meaning of WAM was Wolfgang Amadeus Mozart,
and with the music playing I distributed the Mozart CDs to my co-authors.

Turning back to the WAM-book and in particular to this chapter , of
course, Klaus's criticism (conveyed in a different form) was correct, and the
final product was much improved. It was also a lesson for life. When one has

a specific target in mind, one should always aim for the best result, without unnecessary compromises. There can be fights, discussions, and clashing opinions. However, if these are done honestly and always in the name of the best science, and if people are open-minded, they will ultimately strengthen mutual friendships and respect and the final result will be better.

Many years later, when Klaus and Susanne were in Venice for one of our meetings, we had dinner at a restaurant together with a group of friends and colleagues. My daughter, now grown up, was there as well, sitting at my side, with Klaus and Susanne in front of us on the other side of the table. She remembered that Easter well and when I explained to her who the person in front of her was, she stared at him, and a smiling Klaus said, "that's the best way to become friends".

Of course, he was right again. I still shuffle through the WAM book with pleasure, remembering the effort, but mainly the, often non-linear, interactions that resulted in that product. Klaus is a great friend, and I hope to host Susanne and Klaus in Venice again and to enjoy an opera or a concert together at our beautiful "La Fenice" theatre, born again, as was science, better than before, from the ashes of a momentary decline.

4.23 Kristina Katsaros

I first met Klaus in 1972, when I was sent to his Institute as a postdoc by Joost Businger, whose group I had joined at the University of Washington after completing my Ph.D. Klaus had invited us to a planning meeting for a JONSWAP 2 experiment. I had brought my two and a half year old daughter along as I was planning to continue on to Sweden to see my mother, and Dieter and Hedi Hasselmann arranged for a babysitter. We were going to borrow an elegant 3-D sonic anemometer from Risø, Denmark and measure the momentum flux from a tripod tower in the North Sea, while others were measuring the wavefield in great detail.

The headquarters for the experiment was on the island of Sylt. Two young German Scientists, Jürgen Müller-Glewe and Eggert Clauss, were measuring similar properties of the air from the same tower. They were. We sailed off on the old Gauss research vessel, to install equipment—it was all very exciting for me and the other young scientists such as a graduate student named Thomas Hauf. The first problem we encountered was that the holes in our mounting plate did not match those of the tower's top plate, which had finally arrived, so we had to go back to the island to make adjustments. Finally, a bit late, we installed our equipment, but all the safety measures, such as buoys to secure distance between tripos and the attending ship had not been arranged as there had been too little time. Just as we were gathering data, a storm came up and caused our attending ship drift, which broke my expensive Danish cable (our technician was on the tower at the time). Others, with their fancy wave devices, had arrays on the bottom of the sea, which were also completely destroyed, which ended the experiment.

Some days later we—i.e., about 20 of us—met in the conference room at University of Hamburg in a rather gloomy mood. Finally, being the Polyanna I am by nature, I raised my voice and said: "Klaus, it may not have been much of an experiment, but it sure was a nice experience". I got a lot of laughs, and jokes are usually not one of my strong points. I think Klaus was glad for the relief—he was already a celebrated theoretician, but the experimental difficulties had been too great this time.

Somehow, I think Klaus had an appreciation for my situation, often being the only woman in a gathering. I had been raised in Sweden where the natural sciences are emphasised in high school, so I hadn't realised that I was something of an oddity elsewhere. Klaus could see that scientific inquiry was important to me, although I think he also knew my limitations. He was definitely supportive over the years—I don't know how often he wrote letters proposing me for a promotion or an award. I was invited to join the Marine Remote Sensing (MARSEN) experiment organised by Klaus the late 1970s and we were quite successful in measuring wind stress and wave field. I co-authored a paper with a graduate student and another one on sea surface temperature (SST) measured by aircraft and ships with several colleagues, notably Armando Fiuza of Portugal and the German aircraft research group. Space-based remote sensing of SST was becoming a matter of routine, but our results of varying SST in the German Bight were new at the time.[13]

Much later, in 1992, I had taken on a new position at IFREMER, France, where we were handling scatterometer and Synthetic Aperture Radar data from the European Earth Research Satellite-1 (ERS-1). Klaus had been involved in the planning and was very anxious to get his hands on some data. So, I was in a good position to get him a tape of Synthetic Aperture Radar data.

About 9 months after my arrival, I wanted to convene a workshop to advance progress on these new data sources. My new colleagues thought it was too soon, but Klaus would be coming to the meeting, and it certainly added some shine to the planning and probably to my prestige and inspired everyone to work extra hard to be ready. Many colleagues from the USA also

---

[13] Katsaros, K.B., A. Fiuza, F. Sousa, and V. Amann: Sea Surface Temperature Patterns and Air-Sea Fluxes in the German Bight during MARSEN 1979, Phase 1. J. Geophys Res. 88, 9871–9882, 1983.

attended, as they were starved for new data of this sort. It was a great and wonderful help to me in this management job and led to some great collaborations and results. I'm sure my bosses in Paris were impressed. It was a great start to my 5 years at IFREMER.

I have hugely enjoyed this long friendship and support by Klaus and Susanne Hasselmann, who both always made me feel at home, even inviting me to lunch with them in the office (Susanne was good at keeping our Klaus well fed and healthy!). Their kindness has been very valuable, even if it often came to me from a great distance, as my main place of work was at University of Washington. Klaus' support was one of the aspects of my career that I really treasure. I consider him to have been an important mentor because he took me seriously and understood me.

4.24 Peter A.E.M. Janssen: Klaus F. Hasselmann—A Giant in Ocean Science

It is well-known that Klaus has had a considerable influence on several developments in various fields of science. I will focus on the field with which I am most familiar, namely that of ocean gravity waves, and I will show that his work has had far-reaching consequences not only for oceanographic applications but also for other fields in which non-linear phenomena play a role.

The history of ocean waves started in the early part of the nineteenth century with the contributions of Poisson and Cauchy who solved the linear initial value problem. This was followed by Stokes who obtained a series expansion for a single finite amplitude gravity wave where the nonlinear dispersion relation was obtained by means of the first application of the renormalisation method. In fact, Stokes renormalised acceleration of gravity to remove secular behaviour which assured convergence of the solution. At the end of the nineteenth century Korteweg and de Vries (KdV) derived solitary wave solutions of permanent shape for shallow water from the famous KdV equation and later, in the 1960's, it was shown by means of the inverse scattering transformation that these solitary waves were in fact stable entities, which were dubbed solitons. For a while, it was fairly quiet at the water wave front until Sverdrup and Munk, stimulated by the practical need for sea state information for landing operations during the second world war, developed the first ocean wave forecasting system.

Then, in the 1950's, a considerable acceleration of the pace of ocean wave research development occurred through the insightful work of Longuet-Higgins (Gaussian statistics), Pierson (introduction of the wave spectrum), Miles (wind input) and Phillips (resonant four wave interaction). Until then, most researchers had viewed ocean waves as essentially linear, but this view began to be challenged in Klaus Hasselmann's seminal work on the statistical theory of four-wave interactions. A number of important insights were put forward: a key role was played by the action density spectrum, the resonant four-wave interactions gave rise to irreversible changes in the spectrum, picking up energy and momentum provided by the wind from the region around twice the peak frequency and transferring it to higher and lower frequencies in such a way that the resonant transfer resulted in a downshift of the wave spectrum whilst the entropy of the wave system increased at the same time. This important work not only stimulated developments in oceanography but also in other fields of physics such as plasma physics where resonant three- and four-wave interactions have played an important role in trying to understand how to contain a plasma sufficiently long to enable the occurrence of nuclear fusion.

Returning to oceanography, all these new insights caused quite a commotion in a field which was notoriously conservative. These researchers not only required a lot of convincing, but there was clearly a need for collecting observations and studying experimental results in the light of the findings from nonlinear resonant interactions. But such tasks required a considerable amount of effort and expertise so it made sense to set up collaborations. One of the first collaborations resulted in the famous JONSWAP campaign which gave a tremendous boost to ocean wave forecasting.

At the same time, a new development emerged that had the potential to improve our knowledge on ocean waves and air-sea interaction: Satellite Oceanography. Klaus has played an important stimulating role in the development of a number of satellite instruments, such as Altimeters, which could determine wind speed and significant wave height, whilst the wave spectrum could be observed using the SAR. A novelty and major asset of the new generation of (European) Remote Sensing Satellites such as ERS-1, ERS-2,… was that it would be able to collect this data on a global scale providing an important stimulus for global wave modelling. This was one of the main reasons for the formation of the WAM group which would set up the software of a new third generation wave model based on the 'correct' physics and Klaus and Susanne played a key role in the development of the WAM model. I really admired the way that Klaus was able to convince two very diverse communities to collaborate. He would tell the Satellite people with a broad smile that it was important to use the products of the wave modellers to improve the satellite observations, and would convince the wave modellers to assimilate the satellite products to improve the wave forecasts. At this early stage of development, both satellite products and ocean wave forecasts were of a fairly low quality so the overall impression this made was of a man lifting himself out of the swamp by his own bootstraps. But this type of interaction worked (!) and over the years it resulted in greatly improved wind and wave products from satellites and ocean wave forecasts. Over the past 25 years one day forecasts, error significant wave height, and wind speed has reduced by a factor of two. Satellite products have improved by a similar amount.

Finally, ocean wave forecasting results are quite sensitive to the quality of the surface wind fields, although ocean wave modellers tend not to emphasise this aspect of the wave forecasting problem. Klaus and Gerbrand Komen, therefore, approached the ECMWF, which was at the forefront of weather forecasting, and there has been a very active group of people responsible for the development of the WAM model and the data assimilation software since the mid-eighties. The wave model became part of the ECMWF's operational suite and a considerable amount of attention was paid to improving the quality of the surface winds. Since 1998 there has been a two-way interaction between wind and waves, which benefited wave height forecast results, surface winds, and geopotential height.

At times, Klaus and I also had some heated debates on esoteric fundamental issues in physics such as the corpuscular nature of light and matter (Klaus has an elegant explanation for this) and the role of the 'arrow-of-time' in statistical mechanics and irreversibility. The concept of an arrow-of-time has been made most popular by Prigogine and it indicates that entropy increases when going forward in time. This asymmetry gives rise to particular problems when one considers only resonant wave-wave interactions, but problems may be removed by introducing non-resonant interactions. This realisation suggested a mechanism for the generation of one-dimensional freak waves.

From this sketch of Klaus's work from my perspective it is clear that Klaus has played a pivotal role in the development of a reliable, high-quality ocean wave forecasting system. His efforts on the nonlinear transfer have also triggered numerous developments in other fields of physics.

I remember the many interactions I have had with Klaus with great pleasure. Most of them were rather formal, mainly because he was always working one way or the other. One exception was a discussion I had with him and Susanne at their home about our favourite football teams, in particular about the Dutch football eleven which in the seventies and eighties were famous for their attractive, elegant play and for their innovative tactics ('total' football) that stimulated football all over the world. If I remember correctly, Klaus favoured Holland to win the 1974 world cup!

4.25 Ola M. Johannessen

I got to know Klaus in 1979 when we both used the JPL SAR flown in NASA CV-990 jet where Klaus used it in the international Maritime Remote Sensing Experiment, MARSEN, in the North Sea in October 1979 and I used it in September in the Norwegian Remote Sensing Experiment, NORSEX 79, in the ice edge region North of Svalbard. After these two major international remote sensing experiments, we began to develop the MIZEX programme as a follow up to the NORSEX 79 project, based on a workshop held at Voss in Norway in October 1980. Klaus was very helpful in getting the German Polar community involved in the MIZEX Programme which was headed by Professor Gotthilf Hempel, Director of the new Alfred Wegener Institute for Polar and Marine Research, which was founded in Bremerhaven in 1980. Professor Hempel agreed that their large new Icebreaker "Polarstern" could take part in the programme. When I held a Chair at the Naval Post Graduate School in Monterey in 1982, I spent a lot of my time drafting the research plan for the Marginal Ice Zone Experiment, which was scheduled the summers of 1983 and 1984, of course with a lot of input from my colleagues based on several earlier workshops. The research plan had to be approved by the MIZEX coordination committee of which Klaus was a member. He was very critical of the draft plan and restructured it, to the great benefit of the programme.

Actually, this MIZEX research plan was later reviewed by a group of eminent US scientists chaired by Professor Richard M. Goody at Harvard University in a meeting in Washington. I was very nervous about presenting it, but fortunately, by chance, Klaus was in Washington that day and took part in the review meeting. He basically told this eminent committee to relax and mentioned the fact that he himself had had some bad experiences with the pilot Wave Experiment, JONSWAP in 1968 (he called it a disaster) before the main experiment had been carried out successfully in 1969. Therefore, he argued, it would be a good idea to do the MIZEX 83 pilot project to sort problems out before the main summer experiment in 1984, which was a huge undertaking which included 7 ships, 8 remote sensing aircraft, 4 helicopters, and over 200 scientists and technicians. Thanks to Klaus' presence and support during this meeting, the review of the MIZEX Programme went well.

After the MIZEX 84 project, I collaborated with Klaus on many occasions. At the Nansen Center we started to become interested in global warming modelling in the 1990s, so I contacted Klaus for help. We were invited to come to his Institute to discuss this and Klaus was very generous as usual. We ended up of going home with a tape containing one of the Max Planck global ocean models for one of our Ph.D. students to implement on our computer. Klaus was also one of the panel members at the student's viva. So, thanks to Klaus, the Nansen Center in Bergen became involved in the field of global modelling.

Both Klaus and I knew Walter Munk very well. Walter had pioneered the field of Acoustic Thermometry of Ocean Climate (ATOC), which was used to measure average temperatures on the basin scale. It has already been shown in 1995 that acoustic transmission across the Arctic Ocean was feasible. This inspired Klaus and me to launch the Acoustic Monitoring of the Ocean Climate in the Arctic modelling project to project what could happen up to 2050 using the Max-Planck ocean model as input for several acoustic models. The result was that acoustic monitoring could be a very useful method to be used in future for monitoring the basin scale temperature in the Arctic Ocean [153]. This is actually now underway in a joint observation programme between the Nansen Center and Scripps Institution of Oceanography.

In 2004, our team, which included Klaus, published the fact in *Tellus* that most of the Ice in the Arctic Ocean would melt during summertime under a doubling of the CO_2, probably before the end of 2100 with much less effect in the winter. This assertion was based on findings made with the Max Planck ECHAM4 global coupled model [158]. This paper led to a lot of subsequent papers about the future of Arctic sea ice.

The European Climate Forum was founded in 2001 with Klaus and Carlo Jaeger as Co-Chairmen. Fortunately, Klaus invited me to be one of the founding members. As a result of the many meetings and workshops, I was able to widen my perspective and knowledge of climate impacts. A group headed by Klaus of which I was a member published the important paper "The Challenge of Long-Term Climate Change" [155] in *Science*, which included projections up to the year 3000. Few papers take in such a long-term perspective.

Klaus and Susanna visited us several times at the Nansen Centers in Bergen and St. Petersburg. In the course of several Nansen Lectures, he introduced us to the important topic of the climate-economy and he launched a programme on this topic with D. Kovalevsky at the Nansen Center in St. Petersburg.

Klaus is my hero in science as well as a very good friend.

4.26 Lennart Bengtsson

I became director of the ECMWF (European Centre for Medium-Range Weather Forecasts) in 1981 after having been involved with the Centre right from its early planning phase. Part of the Centre's remit was to make some of its computer resources available to atmospheric scientists from its member states. Preference was given to projects that were beneficial in terms of the scientific and operational objectives of the ECMWF.

One of my ambitions as director was to try to widen the somewhat limited objectives of the ECMWF to weather predictions in the range 4–10 days. Such ambitions included extending the predictions and to add interesting new products of value for users. Scientifically, I was fascinated by the possibility of using comprehensive models to understand the Earth's climate system and using all possible ways of doing this as an additional task. This was a time when the Global Weather Experiment had just been completed successfully and a serious research effort was launched to gain a better understanding of the climate system. I was initially not particularly interested in climate change issues as I considered this was a bit premature as the models and data were in my view not yet good enough for use in this context. However, there was now a global observation system in operation, methods for assimilating and analysing global data as well as ever more powerful computers that made it possible to undertake realistic climate simulation studies.

Klaus Hasselmann was the head of a European group that regularly visited the ECMWF to develop a forecasting system for wave prediction. To predict the state of the sea and in particular waves was an important task for the meteorological marine services. They had put simple systems into operation that used empirical relations coupled to surface wind speed. The strategy adopted by Klaus' group was to develop a comprehensive approach including the full spectrum of sea waves including the non-linear interaction between the waves. I found this to be a splendid idea that would fit perfectly into a potentially operational task for the ECMWF. This required considerable political efforts as there were certain member states that strongly believed that wave prediction was the task of the individual meteorological services and not a European agency. However, with the support of Klaus and some other leading European scientists, wave prediction based upon Klaus' ideas has now been an important operational task for the ECMWF for many years and some members of Klaus' group later 'went on to join the ECMWF.

Another area in which I had the great pleasure of collaborating with Klaus was at the European Space Agency, ESA, and in the planning of ENVISAT, a major initiative concerning a satellite system devoted to both weather prediction and climate research and monitoring. Klaus and I were members of the planning group for ENVISAT so I had the pleasure of seeing Klaus in action. That was all to my liking. ENVISAT was a complex and very ambitious project, and it took many years before it was finally launched in 2002 almost two decades after the early planning phase. ENVISAT provided very important data for weather prediction and climate monitoring which are crucial in monitoring climate change processes.

After a few years as Director at the ECMWF I proposed a long-term strategy to broaden our objective to include extending weather forecasts beyond the medium range as well as a major extension of model-based experimentation and systematic monitoring of the Earth's climate system. I invited a number of leading scientists including Klaus Hasselmann to provide advice on such a strategy. However, the ECMWF Council was not very pleased, as they had not expected such a wide-ranging initiative. The fact that I had produced a plan in colour was seen as being additionally questionable. I was asked to repeat the exercise with help of the Centre's scientific and technical committees and this time in black and white. I realised that even as a director of an international organisation one's initiatives were still subject to certain limits. Had Klaus been a member of the ECMWF Council I might have succeeded at the first attempt. In the end, and after some years of hard work, I got the strategic plan more or less through. However, I also realised that I would probably fit better in a truly scientific environment. During the whole process I had formed a close friendship with Klaus. A few years later he proposed me as a co-director at the Max Planck Institute in Hamburg, which was later approved by the Max Planck Society. I later accepted the offer, which led to a long and very creative collaboration with Klaus with whom I had discussions almost on a daily basis throughout my time in Hamburg. After many years involvement in operational weather prediction, I was really glad to be back in a true scientific environment once again. I am particularly grateful for the fine scientific collaboration with Klaus over a very long period of time.

4.27 Jürgen Sündermann: Klaus Hasselmann—Colleague and Friend

It was around 1966, when Klaus was thirty-five and I seven years younger, that I first became aware of this rising star in the field of (geo)physics. In a lecture on the large-scale propagation of wind waves in the Pacific given by

Walter Munk in the Auditorium Maximum of the University of Hamburg, he mentioned the significant contribution made by a certain Klaus H. in a recent ocean experiment. That same year I took part in the 2nd Oceanographic World Congress in Moscow and heard an excellent talk given by Klaus as Invited Speaker. A short time later, Klaus, who had relocated to Hamburg by then, was invited to give a general lecture to a public audience at the hotel "Atlantic". This performance wasn't quite as successful, but Klaus was well able to withstand this "baptism of fire" and went on to become the founding director of the new Max Planck Institute for Meteorology (MPI). We became acquainted with each other. As a young assistant in the Institute of Oceanography at the University of Hamburg, I attended his lectures on sea waves and appreciated his inspiring habilitation lecture—a kind of quantum theory in hydrodynamics—among an audience of skeptical classical physicists.

During the following years and decades, he played a significant role in the development and growth of both theoretical and experimental ocean and climate research in Hamburg. This was based on his scientific creativity and on the interdisciplinary collaboration that he inspired and fostered. He was instrumental in overcoming the limitations of the old-fashioned traditional university and in motivating young people from the fields of physics, chemistry, biology, and engineering to work together on new projects. He introduced research structures which integrated the already existing high scientific and logistic potential in Hamburg. He was a key initiator of newly developing model systems for climate simulation and large interdisciplinary field experiments in the North Sea such as JONSWAP (Joint North Sea Wave Project) and FLEX (Fladen Ground Experiment)—to name some highlights.

I was appointed as director of the University of Hamburg's Institute of Oceanography in 1978. Together with Klaus and his MPI colleagues Hans Hinzpeter and Hartmut Graßl—forming what we called the "Gang of Four"—the opportunity arose to greatly strengthen the profile of marine and climate research in Hamburg. Important milestones were the long-term and well-financed Special Research Units (Sonderforschungsbereiche) of the German Research Foundation, the acquisition of the research vessel "Valdivia", the foundation of the German Climate Computing Center (DKRZ), the setting-up of new permanent research units at the university such as "Biogeochemistry" and "Sustainability and Global Change", the foundation of the Climate Service Center Germany, and the establishment of two climate-related Max Planck Research Schools. The next logical step was the formation of a joint research structure: the Center of Marine and Atmospheric Sciences (ZMAW), which included a new common building for the University Institute of Oceanography and the Max Planck Institute for Meteorology. This success was essentially based on the common conception of research priorities and their practical realisation. We exchanged scientists between the institutes, united our libraries, and designed a joint logo and a common email address. To emphasise and accelerate our efforts to concentrate the working groups from both institutes in a new building, Klaus and I even arranged an appointment with the mayor of Hamburg in the Town Hall. We finally got the present joint residence.

North Sea studies, Sylt 2013

Our scientific work together certainly gained from a warm personal understanding and from social events such football competitions and carnival parties. Last, but not least, our private cycle tours and concert visits together with our wives, should be mentioned. Yes, the professional and private friendship with Klaus has certainly enriched my life.

4.28   Klaus Fraedrich: The 1976 Paper on Stochastic Climate Models

"When the institute (MPI-M) was created, I had two goals. One was understanding the origin of the natural variability of climate. This was not understood at all, but was clearly a key issue if we wished to distinguish between natural climate variability and human made climate change. I had just developed my stochastic model of climate variability, so I could build on that work as a starting point " (see Interview in Section IIa):

... while Joseph Egger (Munich) employed Hasselmann's Brownian motion analogue to a low-order large scale atmospheric circulation model (1981), which is based on the Jule Charney multiple equilibrium theory of blocking (1979),

... and I (Berlin) incorporated this stochastic noise Ansatz to a low-order climate model (1979) to introduce catastrophes (now: tipping points) and resilience. Both, Jule Charney and I, after participating in the 1975 IIASA-workshop 'Analysis and Computation of Equilibria and Regions of Stability' (H.R. Grümm, Editor), have applied this workshop's new ideas on dynamics in chemistry, climatology, ecology, and economics to their own fields of interest at that time.

This archive photo (below) may also document the fast spread of Klaus Hasselmann's novel approach within the German meteorological community outside of Hamburg. And here it is acknowledged by a private toast on his inspiring idea at the 1978 Berlin international conference on 'Man's Impact on Climate'.

As to the second goal it appears that Klaus has come closer to achieving it: "We needed a good coupled atmosphere–ocean model, but we had no global ocean circulation model of comparable quality to the available global atmospheric circulation models." He introduced this goal in Berlin, setting it out as a conceptual sketch, which has since entered lecture notes introducing the climate prediction and the predictability problem in classes. A seamless prediction which, in those days, has not yet occurred on the horizon, is included here simply by broadening the band width spanning the prognostic-deterministic climate models. Those are just two of the so many highlights of Klaus Hasselmann's achievements, which have stimulated not only us but also a large number of our colleagues to follow various initiatives; and they motivated us to continue on the arduous and cumbersome pursuit of our own goals. These, after all, are open ended.

A toast with Klaus Hasselmann on his landmark 1976-paper

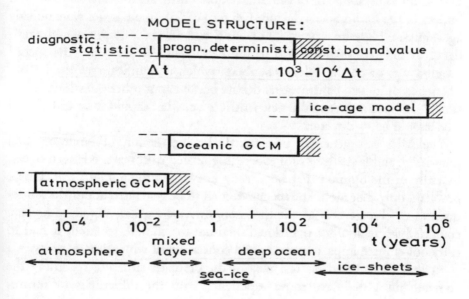

Natural time scales of the climate system and the time-scale band width associated with prognostic-deterministic models.

4.29   Udo Simonis: Klaus, the PIK and Me

The reunification of Germany led to a number of significant scientific innovations accompanied by active collaboration and growing friendship among the scientists involved. In early 1991, the Federal Ministry of Research decided to establish an institute for climate research in Brandenburg under the auspices of the Leibniz Society. The concept for this had been developed by environmentally conscious ministry officials. It was then reviewed by the Science Council in July and—with a significantly reduced scope—recommended for implementation.

The first meeting of a ten-member founding committee (the later Board of Trustees) was held in October 1991, and included Klaus Hasselmann, Director of the Max Planck Institute for Meteorology Hamburg (MPI) and Udo E. Simonis, Director of the International Institute for Environment and Society of the Berlin Social Science Center (WZB). Some issues were quickly agreed upon: Hasselmann was chosen as chairman of the committee, the residence of the new institute was to be the city of Potsdam, and the special location was to be the Telegraphenberg, which is significant in the history of science. A longer, controversial debate began about other questions: What should the special task of the new institute be, what should it be called, and who should be its director?

The MPI was generally considered the incarnation of environmental knowledge and first address for everything related to climate. Klaus, too, was basically of the opinion that he already knew everything about the climate problem, only that more knowledge needed to be generated about the consequences of climate change for the economy, society, and nature. So, there couldn't just be another traditional institute for climate research; it had to carry out climate impact research, and concern itself with climate policy.

At that time, I had no real idea of the dramatic situation regarding the climate, but I did have some experience with the difficulties of formulating and substantiating consistent international environmental policy: I had coined the term "Weltumweltpolitik" (for world environmental policy) at the WZB. Everybody knows that a person less experienced in the field in question can really annoy the expert in the field, but only a few are aware that he can also animate the expert to learn to think differently. I only had to adjust to the relatively precise natural sciences, but Klaus had to get involved with the diverse, occasionally diffuse social sciences. It became a mutual learning process characterised by an increasing respect for one another and a growing genuine friendship.

This learning process had been facilitated when a ten-member "International Scientific Advisory Board (SAB)" was appointed in February 1994, and I became—and remained for eight years—its chairman. The SAB met frequently and usually drafted very detailed minutes, which the PIK Board of Trustees then had to discuss. A recurring dictum appeared in many of these advisory board minutes: the call for a good balance between the natural and social sciences and interactions between the respective practitioners carried out in good faith.

This permanent demand was based on the insight of the American geographer Gilbert F. White, who had formulated it in anticipation of the Anthropocene era as follows: "The future of the globe's interlocking natural and social systems might depend more on human behaviour than on the further investigation of natural processes." Another postulate was also repeatedly called for by the SAB on suitable occasions: "Your work should be theoretically demanding, empirically relevant, and done at the right time".

Whilst the realisation of the second postulate can be considered to have been accomplished well at the PIK, the first one is a task that remains outstanding. However, much work has been and is being done to address this issue. In addition to important natural scientists, significant social scientists were invited to the institute; in addition to the training of young natural scientists, young social scientists were actively promoted; in addition to men, a particularly large number of women were recruited and, what may well be the most important thing, everyone learned to collaborate constructively and to communicate effectively.

In 1992, the year of its founding, the PIK had just 39 employees, 8 of them in the administration; in 2012, twenty years later, the ratio was 340 to 11—a significant indicator of the institute's successful development, but also of efficiency of the institute's administration.

After these 20 years, however, our relationship was by no means at an end. When one's years of membership in the board of trustees and in the advisory board were over, there was first a proper farewell party with the appointment as "Honorary member of the PIK". This immediately gave rise to a new idea: when it was time for Klaus and Udo to leave, we needed more external supporters, because a successful institute not only has internal friends, but also external enviers and opponents.

According to German law, seven members are needed to establish an association; they were quickly at hand and so the "Association of Friends and Supporters of the PIK" was founded in 2002. I was elected chairman and Klaus was elected vice-chairman of the association. In the following years, we regularly held annual meetings, organised numerous award ceremonies for institute staff and ran events to increase public empathy for the institute. We transferred the chairmanship of the association into other hands in 2016.

For both Klaus and me, the following years were years of reflection and relaxation, but also and especially of joy at the birth and development of a "common child" that had become known worldwide in a relatively short period of time—the "Potsdam Institute for Climate Impact Research (PIK)".

4.30   Hartmut Graßl: Klaus Hasselmann as Creator of Science Infrastructure

Scientists should not only create new knowledge leading to the well-being of humankind, but should also help to improve the conditions to create new knowledge. In order to achieve that we need to convince politicians to invest in science infrastructure. Here I report about only two out of many of Klaus Hasselmann's initiatives to enhance the infrastructure for science, because I could directly observe it. Finally, I mention a science highlight to which the entire Max Planck Institute for Meteorology contributed.

The German Climate Computing Centre

When the Federal Research Ministry's expert panel on "Basic Climatological Research" (Sachverständigenkreis Klimatologische Grundlagenforschung) met in Bonn in 1987, the head of unit for climate, Dr. Irmhild Tannhäuser, approached Klaus Hasselmann prior to the start of the meeting with a surprising message: "Herr Hasselmann, I have found 18 Million Deutschmarks in the Marine Technology budget line earmarked for this year, which cannot be spent this year. Now you could start your long-desired climate computing centre in Hamburg". As chairman of this expert panel I proposed to Klaus to add an item to our agenda called "Discussion about a German Climate Computing Centre". When discussing this agenda item two out of eight members of the panel, from southern Germany, argued for the installation at their research centres. After a long and partly controversial debate I asked for a voting (having realized that a majority for Hamburg is probable), which then went positive for Hamburg. Hence, a small expert panel at the Ministry for Research and Technology has decided to bring this major infrastructure element as close as possible to the MPI for Meteorology institute. Klaus has determined the fate of both institutions at least until the end of 1999, when he had to retire. The expert panel on Basic Climatological Research became—after an initiative of the Bavarian Prime Minister

Franz Josef Strauß in the second chamber of our parliament (Bundesrat) in November 1987—in 1988 the Scientific Advisory Council on Climate of the German Federal Government. We became also members of this council. Klaus' next initiative for new research infrastructure followed soon.

Potsdam Institute for Climate Impact Research (PIK)

This Advisory Council to the Federal German Government recommended, following Klaus' repeated wish in its meetings in 1988 and 1989: Establishment of an Institute for Climate Impact Research in Germany. Klaus' main argument has been: Natural science based climate research has reached a high international level in Germany, but we need an internationally competitive climate impact research institution as well, which also should answer socio-economic research questions related to anthropogenic climate change. The Federal Government accepted the council's recommendation benevolently. When the Berlin wall fell, it became obvious that an institute for climate impact research had to be established in east Germany. In January 1992 the PIK started with its founding director Hans Joachim Schellnhuber.

At the end of my report on joint activities with Klaus Hasselmann I remember a scientific activity of all groups at the MPI for Meteorology, which had as its basis the coupled climate model development in Klaus Hasselmann's group.

Climate change as a consequence of a war

During a meeting in Bonn in early 1991 the federal research minister Riesenhuber approached me and proposed: The Max Planck Institute for Meteorology in Hamburg should assess the climate consequences of the burning oil wells in Kuwait after the attack by Iraq. Our institute, equipped with the only coupled atmosphere/ocean-model in Europe, at that time, would be worldwide the only one to perform these calculations, because it is forbidden to publish such model results for our colleagues in the USA during the Gulf war. Back in Hamburg I learned that Klaus had also been asked and in three weeks the coupled model results were ready to be submitted to Nature [97]. Klaus Hasselmann's coworker Mojib Latif had the task to ask daily all participants in all three departments of the institute about progress and problems. For example, in my group we had a simple but serious calculation error in the amount of solar radiation absorbed by soot (black carbon). The key model result: There is only a small regional cooling around Kuwait and the soot is mostly wet deposited in Asian mountain ranges and East Africa. After all the "doomsday" messages by other groups in the media the echo to our results in the public was modest. Later we learned that the model results had even overestimated the observed cooling at the surface. However, the exaggerated climate change estimates caused by the fires published in the media by other scientists fell silent.

5

Statistics

See also https://pure.mpg.de/cone/persons/resource/persons37172?lang=en.

5.1 Curriculum Vita

25 October 1931: Born in Hamburg.
1934: Emigrated to England with family.
1936–1949: Elementary and Grammar School (High School) in Welwyn Garden City, Herts., England.
July 1949: Final High School Exam (Cambridge Higher School Certificate).
August 1949: Return to Hamburg with family.
September 1949–April 1950: Practical course in Mechanical Engineering, Menck und Hambrock, Hamburg.
May 1950–July 1955: Study of Physics and Mathematics at the University of Hamburg.
November 1952: Pre-Diplom Exam.
July 1955: Diplom Exam (Diplom thesis on Turbulence, advisor: Professor K. Wieghardt).
November 1955–July 1957: Study of Physics and Fluid Dynamics at the University of Göttingen and the Max-Planck-Institute of Fluid Dynamics.
July 1957: Ph.D., University of Göttingen (Professor W. Tollmien).
August 1957: Marriage to Susanne Barthe.
August 1957–October 1961: Research Assistant to Professor K. Wieghardt at the Institute of Naval Architecture at the University of Hamburg.

© The Author(s) 2022
H. von Storch, *From Decoding Turbulence to Unveiling the Fingerprint of Climate Change*,
https://doi.org/10.1007/978-3-030-91716-6_5

October 1961–October 1964: Assistant, then Associate Professor at the Institute for Geophysics and Planetary Physics and Scripps Institution of Oceanography, University of California, La Jolla, USA.

February 1963: Habilitation in Hamburg.

November 1964–November 1966: Lecturer at the University of Hamburg.

November 1966–February 1969: Professor at the University of Hamburg (leave of absence September 1967–February 1968).

September 1967–February 1968: Visiting Fellow, University College, Cambridge University.

February 1969–September 1972: Department Director and Professor at the University of Hamburg (leave of absence, July 1970–July 1972).

July 1970–July 1972: Doherty Professor, Woods Hole Oceanographic Institution, Woods Hole, Mass., USA.

September 1972–January 1975: Full Professor for Theoretical Geophysics, Managing Director, Institute of Geophysics at the University of Hamburg.

February 1975–November 1999: Director of the Max-Planck-Institute of Meteorology, Hamburg.

January 1988–November 1999: Scientific Director at the German Climate Computer Centre, Hamburg.

November 1999: Emeritus.

5.2   Publication List[1,2]

Most of the full texts are available through the publication repository of the Max-Planck-Society (MPG.PuRe) via: https://pure.mpg.de/cone/persons/resource/persons37172.

Papers in Refereed Journals or Equivalent Publications

1. Hasselmann, K. (1958). Zur Deutung der dreifachen Geschwindigkeitskorrelationen der isotropen Turbulenz. *Deutsche Hydrographische Zeitschrift, 11*, 207–217. doi:https://doi.org/10.1007/BF02020016.
2. Hasselmann, K. (1958). Die Totalreflexion von kugelförmigen Kompressionsfronten in elastischen Medien; v. Schmidtsche Kopfwellen. *Zeitschrift für angewandte Mathematik und Mechanik, 38*, 310–312. doi:https://doi.org/10.1002/zamm.19580380734.

---

[1] By Lennart Bengtsson.
[2] Prepared by Carola Kauhs.

3. Hasselmann, K. (1960). Grundgleichungen der Seegangsvoraussage. *Schiffstechnik, 7*, 191–195.

4. Hasselmann, K. (1960). Die Totalreflexion einer kugelförmigen Kompressionsfront an der Trennungsebene zweier elastischer Medien. *Zeitschrift für angewandte Mathematik und Mechanik, 40*, 464–472. doi:https://doi.org/10.1002/zamm.19600401005.

5. Hasselmann, K. (1961). Über den nichtlinearen Energieaustausch innerhalb eines Seegangsspektrums. *Zeitschrift für angewandte Mathematik und Mechanik, 41* (S1), T137-T138. doi:https://doi.org/10.1002/zamm.19610411372.

6. Hasselmann, K. (1962). On the non-linear energy transfer in a gravity-wave spectrum: Part 1. General theory. *Journal of Fluid Mechanics, 12*, 481–500. doi:https://doi.org/10.1017/S0022112062000373

7. Hasselmann, K. (1962). Über zufallserregte Schwingungssysteme. *Zeitschrift für angewandte Mathematik und Mechanik, 42*, 465–476. doi:https://doi.org/10.1002/zamm.19620421005.

8. Hasselmann, K. (1963). On the non-linear energy transfer in a gravity wave spectrum: Part 2. Conservation theorems; wave-particle analogy; irreversibility. *Journal of Fluid Mechanics, 15*, 273–281. doi:https://doi.org/10.1017/S0022112063000239.

9. Hasselmann, K. (1963). On the non-linear energy transfer in a gravity-wave spectrum: Part 3. Evaluation of the energy flux and swell-sea interaction for a Neumann spectrum. *Journal of Fluid Mechanics, 15*, 385–398. doi:https://doi.org/10.1017/S002211206300032X.

10. Hasselmann, K. (1963). On the nonlinear energy transfer in a wave spectrum. In *Ocean wave spectra: Proceedings of a conference* (pp. 191–200). Englewood Cliffs: Prentice-Hall.

11. Hasselmann, K., Munk, W., & MacDonald, G. (1963). Bispectra of ocean waves. In M. Rosenblatt (Ed.), *Proceedings of the Symposium on time series analysis* (pp. 125–139). Wiley: New York.

12. Hasselmann, K. (1963). A statistical analysis of the generation of microseisms. *Reviews of Geophysics, 1*, 177–210. doi:https://doi.org/10.1029/RG001i002p00177.

13. Munk, W., & Hasselmann, K. (1964). Super-resolution of tides. In K. Yoshida (Ed.), *Studies on Oceanography* (pp. 339–344). University of Tokyo Press.

14. Hasselmann, K. (1965). Über Streuprozesse in nichtlinear gekoppelten Wellenfeldern. *Zeitschrift für angewandte Mathematik und Mechanik, 45*(S1), T114–T115. doi:https://doi.org/10.1002/zamm.19650459058.

15. Hasselmann, K. (1966). On nonlinear ship motions in irregular waves. *Journal of Ship Research, 10*, 64–68.

16. Hasselmann, K. (1966). Feynman diagrams and interaction rules of wave-wave scattering processes. *Reviews of Geophysics, 4*, 1–32. doi:https://doi.org/10.1029/RG004i001p00001.

17. Hasselmann, K. F. (1966). Generations of waves by turbulent wind. In R. D. Cooper (Ed.), *Sixth Symposium Naval Hydrodynamics* (pp. 585–592). Washington: Office of Naval Research.

18. Snodgrass, F. E., Groves, G. W., Hasselmann, K. F., Miller, G. R., Munk, W. H., & Powers, W. H. (1966). Propagation of ocean swell across the Pacific. *Philosophical Transactions of the Royal Society of London, Series A: Mathematical and Physical Sciences, 259*(1103), 431–497. doi:https://doi.org/10.1098/rsta.1966.0022.

19. Hasselmann, K. (1967). A criterion for nonlinear wave stability. *Journal of Fluid Mechanics, 30*, 737–739. doi:https://doi.org/10.1017/S00221 12067001739.

20. Hasselmann, K. (1967). Nonlinear interactions treated by methods of theoretical physics (with application to generation of waves by wind). *Proceedings of the Royal Society A: Mathematical, Physical and Engineering Sciences, 299*(1456), 77–103. doi:https://doi.org/10.1098/rspa. 1967.0124.

21. Hasselmann, K. (1968). Weak-interaction theory of ocean waves. In M. Holt (Ed.), *Basic developments in fluid dynamics* (pp. 117–182). New York: Academic Press. doi:https://doi.org/10.1016/B978-0-12-395520-3.50008-6.

22. Hasselmann, K., & Collins, J. (1968). Spectral dissipation of finite-depth gravity waves due to turbulent bottom friction. *Journal of Marine Research, 26*, 1–12.

23. Hasselmann, K., & Wibberenz, G. (1968). Scattering of charged particles by random electromagnetic fields. *Zeitschrift für Geophysik, 34*, 353–388.

24. Hasselmann, K. (1969). The sea surface. In Morning review lectures of the Second International Oceanographic Congress (pp. 49–54). Paris: UNESCO.

25. Essen, H.-H., & Hasselmann, K. (1970). Scattering low-frequency sound in the ocean. *Zeitschrift für Geophysik, 36*, 655–678.

26. Hasselmann, K., & Schieler, M. (1970). Radar backscatter from the sea surface. In *Eighth Symposium Naval Hydrodynamics* (pp. 361–388). Arlington: Office of Naval Research.

27. Hasselmann, K. (1970). Wave-driven inertial oscillations. *Geophysical Fluid Dynamics, 1,* 463–502. doi:https://doi.org/10.1080/030919270 09365783.

28. Hasselmann, K., & Wibberenz, G. (1970). A note on the parallel diffusion coefficient. *The Astrophysical Journal, 162,* 1049–1051. doi:https://doi.org/10.1086/150736.

29. Wibberenz, G., Hasselmann, K., & Hasselmann, D. (1970). Comparison of particle-field interaction theory with solar proton diffusion coefficients. *Acta Physica Academiae Scientiarum Hungaricae, 29*(Suppl. 2), 37–46.

30. Hasselmann, K. (1971). Determination of ocean wave spectra from Doppler radio return from the sea surface. *Nature—Physical Science, 229,* 16–17. doi:https://doi.org/10.1038/physci229016a0.

31. Hasselmann, K. (1971). On the mass and momentum transfer between short gravity waves and larger-scale motions. *The Journal of Fluid Mechanics, 50,* 189–205. doi:https://doi.org/10.1017/S00221120710 02520.

32. Hasselmann, K. (1972). Die Vorhersage in der Meeresforschung. *Meerestechnik—Marine Technology, 3,* 96–99.

33. Hasselmann, K., Barnett, T., Bouws, E., Carlson, H., Cartwright, D., Enke, K., Ewing, J., Gienapp, A., Hasselmann, D., Kruseman, P., Meerburg, A., Müller, P., Olbers, D., Richter, K., Sell, W., & Walden, H. (1973). Measurements of wind-wave growth and swell decay during the joint North Sea wave project (JONSWAP). *Ergänzungsheft zur Deutschen Hydrographischen Zeitschrift, Reihe A, Nr. 12.*

34. Hasselmann, K. (1973). On the characterisation of the wave field in the problem of ship response. *Schiffstechnik, 20,* 56–60.

35. Hasselmann, K. (1974). On the spectral dissipation of ocean waves due to white capping. *Boundary-Layer Meteorology, 6,* 107–127. doi:https://doi.org/10.1007/BF00232479.

36. Alpers, W., Hasselmann, K., & Schieler, M. (1975). Fernerkundung der Meeresoberfläche von Satelliten aus. *Raumfahrtforschung, 19,* 1–7.

37. Hasselmann, K., Ross, D. B., Müller, P., & Sell, W. (1976). A parametric wave prediction model. *Journal of Physical Oceanography, 6,* 200–228. doi: https://doi.org/10.1175/1520-0485(1976)006 <0200:APWPM>2,0.CO;2.

38. Hasselmann, K. (1976). Stochastic climate models—1. Theory. *Tellus, 28,* 473–485. doi:https://doi.org/10.3402/tellusa.v28i6.11316.

39. Frankignoul, C., & Hasselmann, K. (1977). Stochastic climate models—2. Application to sea-surface temperature anomalies and thermocline variability. *Tellus, 29*, 289–305. doi:https://doi.org/10.3402/tellusa.v29i4.11362.

40. Hasselmann, K., & Herterich, K. (1977). Klima und Klimavorhersage. *Annalen der Meteorologie, 12*, 42–46.

41. Hasselmann, K. (1977). Application of 2-timing methods in statistical geophysics. *Journal of Geophysics—Zeitschrift für Geophysik, 43*, 351–358.

42. Hasselmann, K., Ross, D., Müller, P., & Sell, W. (1977). A parametric wave prediction model—a reply. *Journal of Physical Oceanography, 7*, 134–137. doi:https://doi.org/10.1175/1520-0485(1977)007<0134:R>2.0.CO;2.

43. Leipold, G., & Hasselmann, K. (1977). Lösung von Bewegungsgleichungen durch Projektion auf Parametergleichungen, dargestellt an der ozeanischen Deckschicht. *Annalen der Meteorologie, 12*, 50–51.

44. Alpers, W., Hasselmann, K., & Kunstmann, J. (1978). Validity of weak particle-field interaction theory for description of cosmic-ray particle diffusion in random magnetic-fields. *Astrophysics and Space Science, 58*, 259–271. doi:https://doi.org/10.1007/BF00644516.

45. Alpers, W., & Hasselmann, K. (1978). The two-frequency microwave technique for measuring ocean-wave spectra from an airplane or satellite. *Boundary-Layer Meteorology, 13*, 215–230. doi:https://doi.org/10.1007/BF00913873.

46. Crombie, D., Hasselmann, K., & Sell, W. (1978). High-frequency radar observations of sea waves travelling in opposition to the wind. *Boundary-Layer Meteorology, 13*, 45–54. doi:https://doi.org/10.1007/BF00913861.

47. Hasselmann, K., Alpers, W., Barick, D., Crombie, D., Flachi, C., Fung, A., van Hutten, H., Jones, W., De Loor, G., Lipa, B., Long, R., Ross, D., Rufenach, C., Sandham, W., Shemdin, O., Teague, C., Trizna, D., Valenzuela, G., Walsh, E., Wentz, F., & Wright, J. (1978). Radar measurements of wind and waves. *Boundary-Layer Meteorology, 13*, 405–412. doi:https://doi.org/10.1007/BF00913885.

48. Hasselmann, K. (1978). On the spectral energy balance and numerical prediction of ocean waves. In A. Favre, & K. Hasselmann (Eds.), *Proceedings of the NATO Symposium on Turbulent Fluxes Through the Sea Surface, Wave Dynamics, and Prediction* (pp. 531–545). Plenum Publ. Corp. doi:https://doi.org/10.1007/978-1-4612-9806-9_35.

49. Shemdin, O., Hasselmann, K., Hsiao, S. V., & Herterich, K. (1978). Nonlinear and linear bottom interaction effects in shallow water. In A. Favre, & K. Hasselmann (Eds.), *Proceedings of the NATO Symposium on Turbulent Fluxes Through the Sea Surface, Wave Dynamics, and Prediction* (pp. 347–372). Plenum Publ. Corp. doi:https://doi.org/10.1007/978-1-4612-9806-9_23.

50. Barnett, T. P., & Hasselmann, K. (1979). Techniques of linear prediction, with application to oceanic and atmospheric fields in the tropical Pacific. *Reviews of Geophysics, 17*, 949–968. doi:https://doi.org/10.1029/RG017i005p00949.

51. Gunther, H., Rosenthal, W., Weare, T. J., Worthington, B. A., Hasselmann, K., & Ewing, J. A. (1979). A hybrid parametrical wave prediction model. *Journal of Geophysical Research: Oceans, 84*, 5727–5738. doi:https://doi.org/10.1029/JC084iC09p05727.

52. Hasselmann, K. (1979). Linear statistical models. *Dynamics of Atmospheres and Oceans, 3*, 501–521. doi:https://doi.org/10.1016/0377-0265(79)90029-0.

53. Hasselmann, K. (1979). On the problem of multiple time scales in climate modelling. In W. Bach (Ed.), *Man's impact on climate: Proc. of an Int. Conference, Berlin, 1978* (pp. 43–55). Amsterdam u.a.: Elsevier. doi:https://doi.org/10.1016/B978-0-444-41766-4.50011-4.

54. Hasselmann, K. (1979). On the signal-to-noise problem in atmospheric response studies. In D. B. Shaw (Ed.), *Meteorology over the tropical oceans* (pp. 251–259). Bracknell: Royal Meteorological Society.

55. Long, R. B., & Hasselmann, K. (1979). Variational technique for extracting directional spectra from multicomponent wave data. *Journal of Physical Oceanography, 9*, 373–381. doi:https://doi.org/10.1175/1520-0485(1979)009<0373:AVTFED>2.0.CO;2.

56. Hasselmann, K. (1980). Ein stochastisches Modell der natürlichen Klimavariabilität. In H. Oeschger (Ed.), *Das Klima: Analysen und Modelle, Geschichte und Zukunft* (pp. 259–260). Berlin, Heidelberg: Springer. doi:https://doi.org/10.1007/978-3-642-67813-4_17.

57. Hasselmann, K. (1980). A simple algorithm for the direct extraction of the two-dimensional surface image spectrum from the return signal of a synthetic aperture radar. *International Journal of Remote Sensing, 1*, 219 240. doi:https://doi.org/10.1080/01,431,168,008,948,234.

58. Herterich, K., & Hasselmann, K. (1980). A similarity relation for the non-linear energy-transfer in a finite-depth gravity-wave spectrum. *Journal of Fluid Mechanics, 97*, 215–224. doi:https://doi.org/10.1017/S0022112080002522.

59. Lemke, P., Trinkl, E. W., & Hasselmann, K. (1980). Stochastic dynamic analysis of polar sea ice variability. *Journal of Physical Oceanography, 10,* 2100–2120. doi:https://doi.org/10.1175/1520-0485(1980)011<2100: SDAOPS>2.0.CO;2.

60. Shemdin, O. H., Hsiao, S. V., Carlson, H. E., Hasselmann, K., & Schulze, K. (1980). Mechanisms of wave transformation in finite-depth water. *Journal of Geophysical Research: Oceans, 85,* 5012–5018. doi:https://doi.org/10.1029/JC085iC09p05012.

61. Barnett, T. P., Preisendorfer, R. W., Goldstein, L. M., & Hasselmann, K. (1981). Significance tests for regression model hierarchies. *Journal of Physical Oceanography, 11,* 1150–1154. doi:https://doi.org/10.1175/ 1520-0485(1981)011<1150:STFRMH>2.0.CO;2.

62. Cardone, V., Carlson, H., Ewing, J. A., Hasselmann, K., Lazanoff, S., McLeish, W., & Ross, D. (1981). The surface wave environment in the GATE B/C Scale—Phase III. *Journal of Physical Oceanography, 11,* 1280–1293. doi:https://doi.org/10.1175/1520-0485(1981)011<1280: TSWEIT>2.0.CO;2.

63. Hasselmann, K. (1981). Construction and verification of stochastic climate models. In A. Berger (Ed.), *Climatic Variations and Variability: Facts and Theories* (pp. 481–497). Dordrecht: D. Reidel Publ. Comp. doi:https://doi.org/10.1007/978-94-009-8514-8_28.

64. Hasselmann, K., & Barnett, T. P. (1981). Techniques of linear prediction for systems with periodic statistics. *Journal of the Atmospheric Sciences, 38,* 2275–2283. doi:https://doi.org/10.1175/1520-0469(198 1)038<2275:TOLPFS>2.0.CO;2.

65. Hasselmann, K. (1981). Modeling the global oceanic circulation for climatic space and time scales. In E. B. Kraus, & M. Fieux (Eds.), *NATO Advanced Research Institute on 'Large Scale Transport of Heat and Matter in the Oceans'* (pp. 112–122). Paris: Laboratoire d'Océanographie Physique.

66. Alpers, W., & Hasselmann, K. (1982). Spectral signal to clutter and thermal noise properties of ocean wave imaging synthetic aperture radars. *International Journal of Remote Sensing, 3,* 423–446. doi:https:// doi.org/10.1080/01431168208948413.

67. Hasselmann, K., & Shemdin, O. H. (1982). Remote sensing experiment in MARSEN (Foreword). *International Journal of Remote Sensing, 3,* 359–361.

68. Hasselmann, K. (1982). An ocean model for climate variability studies. *Progress in Oceanography, 11,* 69–92. doi:https://doi.org/10.1016/0079- 6611(82)90004-0.

69. Herterich, K., & Hasselmann, K. (1982). The horizontal diffusion of tracers by surface waves. *Journal of Physical Oceanography, 12,* 704–711. doi:https://doi.org/10.1175/1520-0485(1982)012 <0704:THDOTB>2.0.CO;2.

70. Hasselmann, K., & Herterich, K. (1983). Application of inverse modelling techniques to paleoclimatic data. In A. Ghazi (Ed.), *Paleoclimatic Research and Models (PRaM): Report and Proceedings of the Workshop* (pp. 52–68). Dordrecht: D. Reidel Publ. Comp.

71. Barnett, T. P., Heinz, H.-D., & Hasselmann, K. (1984). Statistical prediction of seasonal air temperature over Eurasia. *Tellus Series A-Dynamic Meteorology and Oceanography, 36,* 132–146. doi:https://doi.org/10.3402/tellusa.v36i2.11476.

72. Komen, G. J., Hasselmann, S., & Hasselmann, K. (1984). On the existence of a fully developed wind-sea spectrum. *Journal of Physical Oceanography, 14,* 1271–1285. doi:https://doi.org/10.1175/1520-048 5(1984)014<1271:OTEOAF>2.0.CO;2.

73. Attema, E., Bengtsson, L., Bertotti, L., Cavaleri, L., Cavanie, A., Frassetto, R., Guymer, T., Hasselmann, K., Kaneshige, T., Komen, G., Offiler, D., Larsen, S., Louet, J., Pierdicca, N., Powell, J., Rapley, C., Rosenthal, W., Schwenzfeger, K., Thomas, J., Trivero, P., & de Voogt, W. (1985). Report on the Working Group on Wind and Wave Data. In *The use of satellite data in climate models: Proc. of a conference held in Alpach, Austria, 10–12 June 1985* (pp. XIII–XVI). Noordwijk: ESA Scientific and Technical Publications Branch.

74. Hasselmann, K. (1985). Assimilation of microwave data in atmospheric and wave models. In *The use of satellite data in climate models: Proc. of a conference held in Alpach, Austria, 10–12 June 1985* (pp. 47–52). Noordwijk: ESA Scientific and Technical Publications Branch.

75. Hasselmann, K., Raney, R. K., Plant, W. J., Alpers, W., Shuchman, R. A., Lyzenga, D. R., Rufenach, C. L., & Tucker, M. J. (1985). Theory of synthetic aperture radar ocean imaging: A MARSEN view. *Journal of Geophysical Research: Oceans, 90,* 4659–4686. doi:https://doi.org/10. 1029/JC090iC03p04659.

76. Hasselmann, S., & Hasselmann, K. (1985). The Wave Model EXACT-NL. In *Ocean wave modeling* (pp. 249–251). Heidelberg u.a.: Springer. doi:https://doi.org/10.1007/978-1-4757-6055-2_24.

77. Hasselmann, S., & Hasselmann, K. (1985). Computations and parameterizations of the nonlinear energy transfer in a gravity-wave spectrum. Part I: A new method for efficient computations of the exact nonlinear transfer integral. *Journal of Physical Oceanography, 15,*

1369–1377. doi:https://doi.org/10.1175/1520-0485(1985)015<1369: CAPOTN>2.0.CO;2.

78. Hasselmann, S., Hasselmann, K., Allender, J. H., & Barnett, T. P. (1985). Computations and parameterizations of the nonlinear energy transfer in a gravity-wave spectrum. Part II: Parameterizations of the nonlinear energy transfer for application in wave models. *Journal of Physical Oceanography*, *15*, 1378–1391. doi:https://doi.org/10.1175/ 1520-0485(1985)015<1378:CAPOTN>2.0.CO;2.

79. Hasselmann, K. (1986). Wave modelling activities of the WAM Group relevant to ERS-1. In *Proceedings of an ESA Workhop on ERS-1 Wind and Wave Calibration, Schliersee 1986* (pp. 173–175). Noordwijk: ESA Scientific and Technical Publications Branch.

80. Hasselmann, K., Guymer, T., Johnson, D., Kaneshige, T., Lefebvre, M., Rapley, C., Mollo-Christensen, E., Lecomte, P., Conde, J., Svendson, E., & Liferman, A. (1986). The feasibility of an ERS-1 oriented, but scientifically autonomous, international experiment campaign. Report of Working Group 6. In *Proceedings of an ESA Workhop on ERS-1 Wind and Wave Calibration, Schliersee 1986* (pp. 223–227). Noordwijk: ESA Scientific and Technical Publications Branch.

81. Hasselmann, K., & Alpers, W. (1986). The response of Synthetic Aperture Radar to ocean surface waves. In O. M. Phillips, & K. Hasselmann (Eds.), *Wave dynamics and radio probing of the ocean surface: Proc. IUCRM Symposium* (pp. 393–402). Plenum Publ. Corp. doi:https:// doi.org/10.1007/978-1-4684-8980-4_27.

82. Kruse, H. A., & Hasselmann, K. (1986). Investigation of processes governing the large-scale variability of the atmosphere using low-order barotropic spectral models as a statistical tool. *Tellus Series A—Dynamic Meteorology and Oceanography*, *38*, 12–24. doi:https://doi.org/10.3402/ tellusa.v38i1.11694.

83. Herterich, K., & Hasselmann, K. (1987). Extraction of mixed layer advection velocities, diffusion coefficients, feedback factors and atmospheric forcing parameters from the statistical analysis of North Pacific SST anomaly fields. *Journal of Physical Oceanography*, *17*, 2145–2156. doi:https://doi.org/10.1175/1520-0485(1987)017<2145: EOMLAV>2.0.CO;2.

84. Maier-Reimer, E., & Hasselmann, K. (1987). Transport and storage of CO_2 in the ocean—an inorganic ocean-circulation carbon cycle model. *Climate Dynamics*, *2*, 63–90. doi:https://doi.org/10.1007/BF0 1054491.

85. Young, I. R., Hasselmann, S., & Hasselmann, K. (1987). Computations of the response of a wave spectrum to a sudden change in wind direction. *Journal of Physical Oceanography, 17*, 1317–1338. doi:https://doi.org/10.1175/1520-0485(1987)017<1317:COTROA>2.0.CO;2.

86. Hasselmann, K. (1988). PIPs and POPs: The reduction of complex dynamical systems using principal interaction and oscillation patterns. *Journal of Geophysical Research: Atmospheres, 93*, 11015–11021. doi:https://doi.org/10.1029/JD093iD09p11015.

87. Hasselmann, K. (1988). Some problems in the numerical simulation of climate variability using high-resolution coupled models. In M. E. Schlesinger (Ed.), *Physically-based modelling and simulation of climate and climatic change: Part 1* (pp. 583–614). Dordrecht: Kluwer Academic Publ. doi:https://doi.org/10.1007/978-94-009-3041-4_14.

88. Sausen, R., Barthel, K., & Hasselmann, K. (1988). Coupled ocean–atmosphere models with flux correction. *Climate Dynamics, 2*, 145–163. doi:https://doi.org/10.1007/BF01053472.

89. von Storch, H., Bruns, T., Fischer-Bruns, I., & Hasselmann, K. (1988). Principal oscillation pattern analysis of the 30- to 60-day oscillation in general circulation model equatorial troposphere. *Journal of Geophysical Research: Atmospheres, 93*, 11022–11036. doi:https://doi.org/10.1029/JD093iD09p11022.

90. WAM Development and Implementation Group (1988). The WAM Model—A third generation ocean wave prediction model. *Journal of Physical Oceanography, 18*, 1775–1810. doi:https://doi.org/10.1175/1520-0485(1988)018<1775:TWMTGO>2.0.CO;2.

91. Winebrenner, D. P., & Hasselmann, K. (1988). Specular point scattering contribution to the mean Synthetic Aperture Radar image of the ocean surface. *Journal of Geophysical Research: Oceans, 93*, 9281–9294. doi:https://doi.org/10.1029/JC093iC08p09281.

92. Hasselmann, K. (1989). Das Klimaproblem—eine Herausforderung an die Forschung. In R. Gerwin (Ed.), *Wie die Zukunft Wurzeln schlug: Aus der Forschung der Bundesrepublik Deutschland* (pp. 145–159). Berlin u.a.: Springer-Verlag.

93. Bruening, C., Alpers, W., & Hasselmann, K. (1990). Monte-Carlo simulation studies of the nonlinear imaging of a two dimensional surface wave field by a synthetic aperture radar. *International Journal of Remote Sensing, 11*, 1695–1727. doi:https://doi.org/10.1080/01,431,169,008,955,125.

94. Hasselmann, K. (1990). Climate and development: scientific efforts and assessment—The state of the art. In H.-J. Karpe, D. Otten, & S. C. Trinidade (Eds.), *Climate and development: climatic change and variability and the resulting social, economic and technological implications* (pp. 67–122). Berlin, Heidelberg: Springer. doi:https://doi.org/10.1007/978-3-642-45670-1_11.

95. Hasselmann, K. (1990). Waves, dreams, and visions. *Johns Hopkins APL Technical Digest, 11*, 366–369.

96. Hasselmann, K. (1990). Equation punctuation argumentation. *Physics Today, 43*, 15.

97. Bakan, S., Chlond, A., Cubasch, U., Feichter, J., Graf, H. F., Graßl, H., Hasselmann, K., Kirchner, I., Latif, M., Roeckner, E., Sausen, R., Schlese, U., Schriever, D., Schult, I., Schumann, U., Sielmann, F., & Welke, W. (1991). Climate response to smoke from the burning oil-wells in Kuwait. *Nature, 351*, 367–371. doi:https://doi.org/10.1038/351367a0.

98. Donelan, M., Ezraty, R., Banner, M., Hasselmann, K., Janssen, P., Phillips, O., & Dobson, F. (1991). Research needs for better wave forecasting: LEWEX Panel Discussion. In R. C. Beal (Ed.), *Directional ocean wave spectra: measuring. modeling. predicting, and applying* (pp. 196–204). Baltimore: Johns Hopkins University Press.

99. Hasselmann, K. (1991). How well can we predict the climate crisis? In H. Siebert, & Institut für Weltwirtschaft an der Universität Kiel (Eds.), *Environmental Scarcity: The International Dimension* (pp. 165–183). Tübingen: J.C.B. Mohr (Paul Siebeck).

100. Hasselmann, K. (1991). Waves, dreams, and visions (Epilogue). In R. C. Beal (Ed.), *Directional ocean wave spectra: measuring, modeling, predicting, and applying* (pp. 205–208). Baltimore: Johns Hopkins University Press.

101. Hasselmann, K., Hasselmann, S., Brüning, C., & Speidel, A. (1991). Interpretation and application of SAR wave image spectra in wave models. In R. C. Beal (Ed.), *Directional ocean wave spectra: measuring, modeling, predicting, and applying* (pp. 117–124). Baltimore: Johns Hopkins University Press.

102. Hasselmann, K., & Hasselmann, S. (1991). On the nonlinear mapping of an ocean wave spectrum into a synthetic aperture radar image spectrum and its inversion. *Journal of Geophysical Research: Oceans, 96*, 10,713–10,729. doi:https://doi.org/10.1029/91JC00302.

103. Hasselmann, K. (1991). Ocean circulation and climate change. *Tellus Series B—Chemical and Physical Meteorology, 43*, 82–103. https://doi.org/10.3402/tellusb.v43i4.15399

104. Bauer, E., Hasselmann, K., & Young, I. (1992). Satellite data assimilation in the wave model 3G-WAM. In *Proceedings of the Central Symposium of the "International Space Year" Conference, Munich, Germany, 30. March 4. April 1992* (pp. 377–380). Noordwijk: ESA Publishing Division.

105. Bauer, E., Hasselmann, S., Hasselmann, K., & Graber, H. C. (1992). Validation and assimilation of Seasat altimeter wave heights using the WAM wave model. *Journal of Geophysical Research: Oceans, 97*, 12,671–12,682. doi:https://doi.org/10.1029/92JC01056.

106. Cubasch, U., Hasselmann, K., Höck, H., Maier-Reimer, E., Mikolajewicz, U., Santer, B. D., & Sausen, R. (1992). Time-dependent greenhouse warming computations with a coupled ocean–atmosphere model. *Climate Dynamics, 8*, 55–69. doi:https://doi.org/10.1007/BF00209163.

107. Hasselmann, K., Sausen, R., Maier-Reimer, E., & Voss, R. (1992). Das Kaltstartproblem bei Klimasimulationen mit gekoppelten Atmosphäre-Ozean-Modellen. *Annalen der Meteorologie, 27*, 153–154.

108. Brüning, C., Hasselmann, S., Hasselmann, K., Lehner, S., & Gerling, T. (1993). On the extraction of ocean wave spectra from ERS-1 SAR wave mode image spectra. In *Proceedings of the first ERS-1 Symposium: Space at the Service of our Environment, 4–6 November 1992, Cannes, France* (pp. 747–752). Noordwijk: ESA Publishing Division.

109. Hasselmann, K., Sausen, R., Maier-Reimer, E., & Voss, R. (1993). On the cold start problem in transient simulations with coupled atmosphere–ocean models. *Climate Dynamics, 9*, 53–61. doi:https://doi.org/10.1007/BF00210008.

110. Hasselmann, K. (1993). Optimal fingerprints for the detection of time-dependent climate change. *Journal of Climate, 6*, 1957–1971. doi:https://doi.org/10.1175/15200442 (1993)006<1957:OFFTDO>2.0.CO;2.

111. Heinze, C., & Hasselmann, K. (1993). Inverse multiparameter modeling of paleoclimate Carbon cycle indices. *Quaternary Research, 40*, 281 296. doi:https://doi.org/10.1006/qres.1993.1082.

112. Maier-Reimer, E., Mikolajewicz, U., & Hasselmann, K. (1993). Mean circulation of the Hamburg LSG OGCM and its sensitivity to the thermohaline surface forcing. *Journal of Physical Oceanography, 23*, 731–757. doi:https://doi.org/10.1175/1520-0485(1993)023 <0731: MCOTHL>2.0.CO;2.

113. Pennell, W. T., Bamett, T. P., Hasselmann, K., Holland, W. R., Karl, T. R., North, G. R., MacCracken, M. C., Moss, M. E., Pearman, G., Rasmusson, E. M., Santer, B. D., Smith, W. K., von Storch, H., Switzer, P., & Zwiers, F. (1993). The detection of anthropogenic climate change. *Fourth Symposium on Global Change Studies* (pp. 21–28). American Meteorological Society.

114. Snyder, R. L., Thacker, W. C., Hasselmann, K., Hasselmann, S., & Barzel, G. (1993). Implementation of an efficient scheme for calculating nonlinear transfer from wave-wave interactions. *Journal of Geophysical Research: Oceans, 98*, 14,507–14,525. doi:https://doi.org/ 10.1029/93JC00657.

115. Brüning, C., Hasselmann, K., Hasselmann, S., Lehner, S., & Gerling, T. (1994). A first evaluation of ERS-1 Synthetic Aperture Radar wave mode data. *The Global Atmosphere and Ocean System, 2*, 61 -98.

116. Santer, B. D., Brüggemann, W., Cubasch, U., Hasselmann, K., Höck, H., Maier-Reimer, E., & Mikolajewicz, U. (1994). Signal-to-noise analysis of time-dependent greenhouse warming experiments. Part 1: Pattern analysis. *Climate Dynamics, 9*, 267–285. doi:https://doi.org/10. 1007/BF00204743.

117. Santer, B. D., Mikolajewicz, U., Brüggemann, W., Cubasch, U., Hasselmann, K., Höck, H., Maier-Reimer, E., & Wigley, T. M. L. (1995). Ocean variability and its influence on the detectability of greenhouse warming signals. *Journal of Geophysical Research: Oceans, 100*, 10,693–10,725. doi:https://doi.org/10.1029/95JC00683.

118. von Storch, H., & Hasselmann, K. (1995). Climate variability and change. In G. Hempel (Ed.), *The ocean and the poles: Grand challenges for European cooperation* (pp. 33–58). Jena u.a.: Gustav Fischer Verl.

119. Barzel, G., Long, R. B., Hasselmann, S., & Hasselmann, K. (1996). Wave model fitting using the adjoint technique. In M. A. Donelan, W. H. Hui, & W. J. Plant (Eds.), *The Air-Sea Interface: Radio and Acoustic Sensing, Turbulence and Wave Dynamics* (pp. 347–354). Miami, Florida: Rosenstiel School of Marine and Atmospheric Science, Univ. Miami.

120. Bauer, E., Hasselmann, K., Young, I., & Hasselmann, S. (1996). Assimilation of wave data into the wave model WAM using an impulse response function method. *Journal of Geophysical Research: Oceans, 101,* 3801–3816. doi:https://doi.org/10.1029/95JC03306.

121. Hasselmann, K. (1996). The metron model: Towards a unified deterministic theory of fields and particles, Part 1: The Metron concept. *Physics Essays, 9,* 311–325.

122. Hasselmann, K. (1996). The metron model: Towards a unified deterministic theory of fields and particles, Part 2: The Maxwell-Dirac-Einstein system. *Physics Essays, 9,* 460–475.

123. Hasselmann, S., Bruning, C., Hasselmann, K., & Heimbach, P. (1996). An improved algorithm for the retrieval of ocean wave spectra from synthetic aperture radar image spectra. *Journal of Geophysical Research: Oceans, 101,* 16,615–16,629. doi:https://doi.org/10.1029/96JC00798.

124. Hasselmann, S., Hasselmann, K., & Brüning, C. (1996). Extraction of wave data from ERS-1 SAR wave mode image spectra. In M. A. Donelan, W. H. Hui, & W. J. Plant (Eds.), *The Air-Sea Interface: Radio and Acoustic Sensing, Turbulence and Wave Dynamics* (pp. 773–780). Miami, Florida: Rosenstiel School of Marine and Atmospheric Science, Univ. Miami.

125. Hegerl, G. C., von Storch, H., Hasselmann, K., Santer, B. D., Cubasch, U., & Jones, P. D. (1996). Detecting greenhouse-gas-induced climate change with an optimal fingerprint method. *Journal of Climate, 9,* 2281–2306. doi:https://doi.org/10.1175/1520-0442(1996)009<2281:DGGICC>2.0.CO;2.

126. Lehner, S., Bruns, T., & Hasselmann, K. (1996). Test of a new onboard shiprouting system. In *Proceedings of the Second ERS Applications workshop* (pp. 297–301). Noordwijk: ESA/ESTAC.

127. Lionello, P., Hasselmann, K., & Mellor, G. I. (1996). On the coupling between a surface wave model and a model of the mixed layer in the ocean. In M. A. Donelan, W. H. Hui, & W. J. Plant (Eds.), *The Air-Sea Interface: Radio and Acoustic Sensing, Turbulence and Wave Dynamics* (pp. 195–201). Miami, Florida: Rosenstiel School of Marine and Atmospheric Science, Univ. Miami.

128. Bauer, E., Hasselmann, S., Lionello, P., & Hasselmann, K. (1997). Comparison of assimilation results from an optimal interpolation and the Green's function method using ERS-1 SAR wave mode spectra. In *Third ERS Symposium on Space at the Service of our Environment* (pp. 1131–1136).

129. Hasselmann, K. (1997). Multi-pattern fingerprint method for detection and attribution of climate change. *Climate Dynamics, 13*, 601–611. doi:https://doi.org/10.1007/s003820050185.

130. Hasselmann, K. (1997). Climate-change research after Kyoto. *Nature, 390*(6657), 225–226. doi:https://doi.org/10.1038/36719.

131. Hasselmann, K. (1997). The metron model: Towards a unified deterministic theory of fields and particles, Part 3: Quantum phenomena. *Physics Essays, 10*, 64–86.

132. Hasselmann, K. (1997). The metron model: Towards a unified deterministic theory of fields and particles, Part 4: The standard model. *Physics Essays, 10*, 269–286.

133. Hasselmann, K. (1997). Climate change—Are we seeing global warming? *Science, 276*(5314), 914–915. doi:https://doi.org/10.1126/science.276.5314.914.

134. Hasselmann, K., Hasselmann, S., Giering, R., Ocaña, V., & von Storch, H. (1997). Sensitivity study of optimal CO_2 emission paths using a simplified Structural Integrated Assessment Model (SIAM). *Climatic Change, 37*, 345–386. doi:https://doi.org/10.1023/A:1005339625015.

135. Hegerl, G. C., Hasselmann, K., Cubasch, U., Mitchell, J. F. B., Roeckner, E., Voss, R., & Waszkewitz, J. (1997). Multi-fingerprint detection and attribution analysis of greenhouse gas, greenhouse gas-plus-aerosol and solar forced climate change. *Climate Dynamics, 13*, 613–634. doi:https://doi.org/10.1007/s003820050186.

136. Heimbach, P., Hasselmann, S., & Hasselmann, K. (1997). Three year global intercomparison of ERS-1 SAR wave mode spectral retrievals with WAM model data. In *Third ERS Symposium on Space at the Service of our Environment* (pp. 1143–1149).

137. Hasselmann, K., & Hasselmann, S. (1998). Multi-actor optimization of greenhouse gas emission paths using coupled integral climate response and economic models. In H.-J. Schellnhuber, & V. Wenzel (Eds.), *Earth systems analysis: integrating science for sustainability—Complemented results of a symposium* (pp. 381–415). Springer. doi:https://doi.org/10.1007/978-3-642-52354-0_20.

138. Hasselmann, K. (1998). Conventional and Bayesian approach to climate-change detection and attribution. *Quarterly Journal of the Royal Meteorological Society, 124*, 2541–2565. doi:https://doi.org/10.1002/qj.49712455202.

139. Hasselmann, K. (1998). The metron model: Towards a unified deterministic theory of fields and particles. In A. K. Richter (Ed.), *Understanding Physics* (pp. 155–186). Katlenburg-Lindau: Copernicus Gesellschaft.

140. Heimbach, P., Hasselmann, S., & Hasselmann, K. (1998). Statistical analysis and intercomparison of WAM model data with global ERS-1 SAR wave mode spectral retrievals over 3 years. *Journal of Geophysical Research: Oceans, 103*, 7931–7977. doi:https://doi.org/10.1029/97JC03203.

141. Barnett, T. P., Hasselmann, K., Chelliah, M., Delworth, T., Hegerl, G., Jones, P., Rasmusson, E., Roeckner, E., Ropelewski, C., Santer, B., & Tett, S. (1999). Detection and attribution of recent climate change: A status report. *Bulletin of the American Meteorological Society, 80*, 2631–2659. doi:https://doi.org/10.1175/1520-0477(1999)080<2631:DAAORC>2.0.CO;2.

142. Hasselmann, K. (1999). Cooperative and non-cooperative multiactor strategies of optimizing greenhouse gas emissions. In H. von Storch (Ed.), *Anthropogenic climate change* (pp. 209–256). Berlin u.a.: Springer-Verlag. doi:https://doi.org/10.1007/978-3-642-59992-7_7.

143. Hasselmann, K. (1999). Intertemporal accounting of climate change—Harmonizing economic efficiency and climate stewardship. *Climatic Change, 41*, 333–350. doi:https://doi.org/10.1023/A:1005441119269.

144. Hasselmann, K. (1999). Climate change—Linear and nonlinear signatures. *Nature, 398*(6730), 755–756. doi:https://doi.org/10.1038/19635.

145. Hasselmann, K. (1999). Climate prediction is heavy weather. *Physics World, 12*, 24.

146. Hasselmann, K. (1999). Modellierung natürlicher und anthropogener Klima-änderungen. *Physikalische Blätter, 55*, 27–30. doi:https://doi.org/10.1002/phbl.19990550109.

147. Petschel-Held, G., Schellnhuber, H. J., Bruckner, T., Toth, F. L., & Hasselmann, K. (1999). The tolerable windows approach: Theoretical and methodological foundations. *Climatic Change, 41*, 303–331. doi:https://doi.org/10.1023/A:1005487123751.

148. Hasselmann, K. (2000). The outlook for climate change. In H. Siebert, & Institut für Weltwirtschaft an der Universität Kiel (Eds.), *The Economics of International Environmental Problems* (pp. 27–49). Tübingen: Mohr Siebeck.

149. Hasselmann, K. (2001). Optimizing long-term climate management. In E.-D. Schulze, & M. Heimann (Eds.), *Global biogeochemical cycles in the climate system* (pp. 333–343). San Diego: Academic Press. doi:https://doi.org/10.1016/B978-012631260-7/50029-7.

150. Hooss, G., Voss, R., Hasselmann, K., Maier-Reimer, E., & Joos, F. (2001). A nonlinear impulse response model of the coupled carbon cycle climate system (NICCS). *Climate Dynamics, 18*(3–4), 189–202. doi:https://doi.org/10.1007/s003820100170.

151. Joos, F., Prentice, I. C., Sitch, S., Meyer, R., Hooss, G., Plattner, G.-K., Gerber, S., & Hasselmann, K. (2001). Global warming feedbacks on terrestrial carbon uptake under the Intergovernmental Panel on Climate Change (IPCC) emission scenarios. *Global Biogeochemical Cycles, 15*, 891–907. doi:https://doi.org/10.1029/2000GB001375.

152. Hasselmann, K. (2002). Is climate predictable. In A. Bunde, J. Kropp, & J. Schellnhuber (Eds.), *The science of disasters: climate disruption, heart attacks, and market crashes* (pp. 141–169). Berlin: Springer. doi:https://doi.org/10.1007/978-3-642-56257-0_4

153. Johannessen, O., Sagen, H., Hamre, T., Hobaek, H., Hasselmann, K., Maier-Reimer, E., Mikolajewicz, U., Wadhams, P., Kaletzky, A., Bobylev, L., Evert, E., Troyan, V., Naugolnykh, K., & Esipov, I. (2002). Acoustic monitoring of ocean climate in the Arctic (AMOC). In N. C. Flemming, & S. Vallerga et al. (Eds.), *Operational Oceanography— Implementation at the European and regional Scales* (pp. 371–378). Amsterdam: Elsevier Science BV. doi:https://doi.org/10.1016/S0422-9894(02)80043-5.

154. Bruckner, T., Hooss, G., Füssel, H.-M., & Hasselmann, K. (2003). Climate system modeling in the framework of the tolerable windows approach: The ICLIPS climate model. *Climatic Change, 56*, 119–137. doi:https://doi.org/10.1023/A:1021300924356.

155. Hasselmann, K., Latif, M., Hooss, G., Azar, C., Edenhofer, O., Jaeger, C. C., Johannessen, O. M., Kemfert, C., Welp, M., & Wokaun, A. (2003). The challenge of long-term climate change. *Science, 302*(5652), 1923–1925. doi:https://doi.org/10.1126/science.1090858.

156. Hasselmann, K., Schellnhuber, H. J., & Edenhofer, O. (2004). Climate change: complexity in action. *Physics World, 17*, 31–35. doi:https://doi.org/10.1088/2058-7058/17/6/34.

157. Hasselmann, K., & Hasselmann, S. (2004). The metron model: a unified deterministic theory of fields and particles—a progress report. In *Proceedings of Institute of Mathematics of NAS of Ukraine* (pp. 788–795). Kyiv: Institute of Mathematics of NAS of Ukraine.

158. Johannssen, O. M., Bengtsson, L., Miles, M. W., Kuzmina, S. I., Semenov, V. A., Alekseev, G. V., Nagurnyi, A. P., Zakharov, V. F., Bobylev, L. P., Pettersson, L. H., Hasselmann, K., & Cattle, A. P. (2004). Arctic climate change: observed and modelled temperature and sea-ice variability. *Tellus Series A-Dynamic Meteorology and Oceanography, 56*(4), 328–341. doi:https://doi.org/10.1111/j.1600-0870.2004.00060.x.

159. Barth, V., & Hasselmann, K. (2005). Analysis of climate damage abatement costs using a dynamic economic model. *Vierteljahreshefte zur Wirtschaftsforschung (DIW), 74*, 148–163.

160. Schnur, R., & Hasselmann, K. (2005). Optimal filtering for Bayesian detection and attribution of climate change. *Climate Dynamics, 24*(1), 45–55. doi:https://doi.org/10.1007/s00382-004-0456-3.

161. The International Ad Hoc Detection and Attribution Group (2005). Detecting and attributing external influences on the climate system: a review of recent advances. *Journal of Climate, 18*, 1291–1314. doi:https://doi.org/10.1175/JCLI3329.1.

162. Weber, M., Barth, V., & Hasselmann, K. (2005). A multi-actor dynamic integrated assessment model (MADIAM) of induced technological change and sustainable economic growth. *Ecological Economics, 54*(2–3), 306–327. doi:https://doi.org/10.1016/j.ecolecon.2004.12.035.

163. von Laer, D., Hasselmann, S., & Hasselmann, K. (2006). Gene therapy for HIV infection: what does it need to make it work? *Journal of Gene Medicine, 8*, 658–667. doi:https://doi.org/10.1002/jgm.908.

164. von Laer, D., Hasselmann, S., & Hasselmann, K. (2006). Impact of gene-modified T cells on HIV infection dynamics. *Journal of Theoretical Biology, 238*, 60–77. doi:https://doi.org/10.1016/j.jtbi.2005.05.005.

165. Hasselmann, K., & Barker, T. (2008). The Stern Review and the IPCC fourth assessment report: implications for interaction between policy-makers and climate experts. An editorial essay. *Climatic Change, 89*, 219–229. doi:https://doi.org/10.1007/s10584-008-9435-8.

166. Jaeger, C. C., Krause, J., Haas, A., Klein, R., & Hasselmann, K. (2008). A method for computing the fraction of attributable risk related to climate damages. *Risk Analysis, 28*, 815–823. doi:https://doi.org/10.1111/j.1539-6924.2008.01070.x.

167. Hasselmann, K. (2009). Simulating human behavior in macroeconomic models applied to climate change. Dahlem Conference "Is there a mathematics of social entities", Berlin, 14.-19. December 2008. *ECF Working Paper, 2/2009.*

168. Hasselmann, K. (2009). What to do? Does science have a role? *European Physical Journal-Special Topics, 176*, 37–51. doi:https://doi.org/10.1140/epjst/e2009-01147-x.

169. Hasselmann, K. (2010). Application of system dynamics to climate policy assessment. In A. Fitt, J. Norbury, H. Ockendon, & E. Wilson (Eds.), *Progress in Industrial Mathematics at ECMI 2008* (pp. 203–208). Berlin: Springer. doi:https://doi.org/10.1007/978-3-642-12110-4_27.

170. Hasselmann, K. (2010). The climate change game. *Nature Geosciences, 3*, 511–512. doi:https://doi.org/10.1038/ngeo919.

171. Hasselmann, K., & Voinow, A. (2012). The actor-driven dynamics of decarbonization. In C. C. Jaeger, & et al. (Eds.), *Reframing the problem of climate change* (pp. 131–159). Milton Park: Earthscan. doi:https://doi.org/10.4324/9780203154724.

172. Jaeger, C. C., Hasselmann, K., Leipold, G., Mangalagiu, D., & Tabara, J. D. (2012). Conclusion—Action for climate. In C. C. Jaeger, & et al. (Eds.), *Reframing the problem of climate change* (pp. 237–244). Milton Park: Earthscan. doi:https://doi.org/10.4324/9780203154724.

173. Jaeger, C. C., Hasselmann, K., Leipold, G., Mangalagiu, D., & Tabara, J. D. (2012). Introduction: Beyond the zero sum game: from shirking burdens to sharing benefits. In C. C. Jaeger, & et al. (Eds.), *Reframing the problem of climate change* (pp. 1–14). Milton Park: Earthscan. doi:https://doi.org/10.4324/9780203154724.

174. Giupponi, C., Borsuk, M., de Vries, B., & Hasselmann, K. (2013). Innovative approaches to integrated global change modelling. *Environmental Modelling and Software, 44*, 1–9. doi:https://doi.org/10.1016/j.envsoft.2013.01.013.

175. Hasselmann, K., & Kovalevsky, D. V. (2013). Simulating animal spirits in actor-based environmental models. *Environmental Modelling and Software, 44*, 10–24. doi:https://doi.org/10.1016/j.envsoft.2012.04.007.

176. Hasselmann, K., Chapron, B., Aouf, L., Ardhuin, F., Collard, F., Engen, G., Hasselmann, S., Heimbach, P., Janssen, P., Johnsen, H., Krogstad, H., Lehner, S., Li, J.-G., Li, X.-M., Rosenthal, W., & Schulz-Stellenfleth, J. (2013). The ERS SAR wave mode: A breakthrough in global ocean wave observations. In Y. Desnos (Ed.), *ERS Missions: 20 Years of Observing the Earth* (pp. 167–197). Noordwijk: ESA/ESTEC.

177. Hasselmann, K. (2013). A classical path to unification. *Journal of Physics Conference Series, 437*: 012,023. doi:https://doi.org/10.1088/1742-6596/437/1/012023.

178. Hasselmann, K. (2013). Ernst Maier-Reimer: The discovery of silence. *Nature Geoscience, 8*(10), 809–809. doi:https://doi.org/10.1038/nge o1953.

179. Hasselmann, K. (2013). Detecting and responding to climate change. *Tellus, Series B—Chemical and Physical Meteorology, 65*: 20,088. doi:https://doi.org/10.3402/tellusb.v65i0.20088.

180. Kovalevsky, D., & Hasselmann, K. (2014). A hierarchy of out-of-equilibrium actor-based system-dynamic nonlinear economic models. *Discontinuity, Nonlinearity, and Complexity, 3*, 303–318. doi:https://doi.org/10.5890/DNC.2014.09.007

181. Kovalevsky, D., & Hasselmann, K. (2014). Assessing the transition to a low-carbon economy using actor-based system-dynamic models. In *Proceedings—7th International Congress on Environmental Modelling and Software, iEMSs 2014* (pp. 1865–1872).

182. Kovalevsky, D. V., & Hasselmann, K. (2014). Modelling the impacts of a national carbon tax in a country with inhomogeneous regional development: an actor-based system-dynamic approach. In *ERSA 54th Congress "Regional development & globalisation: Best practices", 26–29 August 2014, St. Petersburg, Russia.* Louvain-la-Neuve: European Regional Science Association (ERSA).

183. Hasselmann, K., Cremades, R., Filatova, T., Hewitt, R., Jaeger, C., Kovalevsky, D., Voinov, A., & Winder, N. (2015). Free-riders to forerunners. *Nature Geoscience, 8*, 895-898. doi:https://doi.org/10.1038/ngeo2593.

184. Kovalevsky, D. V., & Hasselmann, K. (2016). Actor-based system dynamics modelling of win–win climate mitigation options. In *The 8th International Congress on Environmental Modelling and Software (iEMSs 2016), 10–14 July 2016, Toulouse, France.*

185. Kovalevsky, D., Hewitt, R., de Boer, C., & Hasselmann, K. (2017). A dynamic systems approach to the representation of policy implementation processes in a multi-actor world. *Discontinuity, Nonlinearity, and Complexity, 6*, 219–245. doi:https://doi.org/10.5890/DNC.2017.09.001.

186. Heinze, C., & Hasselmann, K. (2019). Preface: Ernst Maier-Reimer and his way of modelling the ocean. *Biogeosciences, 16* (Spec. Iss.: Progress in quantifying ocean biogeochemistry – in honour of Ernst Maier-Reimer), 751–753. doi:https://doi.org/10.5194/bg-16-751-2019.

187. Pettersson, L. H., Kjelaas, A. G., Kovalevsky, D. V., & Hasselmann, K. (2020). Climate change impact on the Arctic economy. In O. M. Johannessen, L. P. Bobylev, E. V. Shalina, & S. Sandven (Eds.), *Sea Ice in the Arctic: Past, Present and Future* (pp. 465–506). Cham: Springer International Publishing. doi:https://doi.org/10.1007/978-3-030-21301-5_11.

188. Hewitt, R., Cremades, R., Kovalevsky, D., & Hasselmann, K. (2021). Beyond shared socioeconomic pathways (SSPs) and representative concentration pathways (RCPs): climate policy implementation scenarios for Europe, the US and China. *Climate Policy, 21*, 434–454. doi:https://doi.org/10.1080/14693062.2020.1852068.

Other Publications and Grey Literature

189. Hasselmann, K. (1955). Über die Trägheitskräfte der isotropen Turbulenz. Diploma Thesis, Technische Universität Hamburg. doi:https://doi.org/10.15480/882.516.

190. Hasselmann, K. (1955). Potentialtheoretische Druckverteilung an einigen drehsymmetrischen Halbkörpern. *Schriftenreihe Schiffbau, 29*. doi:https://doi.org/10.15480/882.526.

191. Hasselmann, K. (1957). Über eine Methode zur Bestimmung der Reflexion und Brechung von Stoßfronten und von beliebigen Wellen kleiner Wellenlängen an der Trennungsfläche zweier Medien. PhD Thesis, Universität Göttingen, Göttingen.

192. Hasselmann, K. (1958). Zur Deutung der dreifachen Geschwindigkeitskorrelationen der isotropen Turbulenz. *Schriftenreihe Schiffbau, 84*. doi:https://doi.org/10.15480/882.558.

193. Hasselmann, K. (1960). Decay of wave-induced velocity fluctuations in the small HSVA Tank. *Schriftenreihe Schiffbau, 66*. doi:https://doi.org/10.15480/882.554.

194. Hasselmann, K. (1960). Über den Einfluß nichtlinearer Wechselwirkungen auf die Energieverteilung in einem Seegangsspektrum. *Schriftenreihe Schiffbau, 81*. doi:https://doi.org/10.15480/882.779.

195. Hasselmann, K. (1961). Interpretation of Phillips' wave growth mechanism. [In *Ocean wave spectra: Proceedings of a conference*] (pp. nicht im Konferenzband 1963 enthalten).

196. Hasselmann, K. (1970). Der Sonnenwind. Jahrbuch der Akademie der Wissenschaften in Göttingen, 22–25.

197. Sell, W., & Hasselmann, K. (1972). *Computations of nonlinear energy transfer for JONSWAP and empirical wind wave spectra.* Hamburg: Institut für Geophysik, Universität Hamburg.

198. Hasselmann, K. (1977). The dynamical coupling between the atmosphere and the ocean. In *The influence of the ocean on climate* (pp. 31–44). Genf: WMO. (Reports on marine science affairs; 11)

199. Hasselmann, K. (1980). Klimamodelle. *Annalen der Meteorologie, N.F. 15*, 81–82

200. Hasselmann, S., & Hasselmann, K. (1981). A symmetrical method of computing the nonlinear transfer in a gravity wave spectrum. *Hamburger Geophysikalische Einzelschriften: Reihe A, Wissenschaftliche Abhandlungen, 52.*

201. Maier-Reimer, E., Müller, D., Olbers, D., Willebrand, J., & Hasselmann, K.(1982). *An ocean circulation model for climate variability studies.* Hamburg: Max-Planck-Institut für Meteorologie.

202. Maier-Reimer, E., Müller, D., Olbers, D., Willebrand, J., & Hasselmann, K.(1982). *Ein Modell der ozeanischen Zirkulation zur Untersuchung von Klimaschwankungen.* Hamburg: Max-Planck-Institut für Meteorologie.

203. Young, I. R., Hasselmann, S., & Hasselmann, K. (1985). Calculation of the nonlinear wave-wave interactions in cross seas. *Hamburger Geophysikalische Einzelschriften—Reihe A: Wissenschaftliche Abhandlungen, 74.*

204. Hasselmann, K., Hasselmann, S., Bauer, E., Brüning, C., Lehner, S., Graber, H., & Lionello, P. (1988). Development of a Satellite SAR Image Spectra and Altimeter Wave Height Data Assimilation System for ERS-1. *Report/Max-Planck-Institute for Meteorology, 019.* doi:https://doi.org/10.17617/2.2578369.

205. Oberhuber, J., & Hasselmann, K. (1988). Ozeanmodelle. Promet, 18 (Nos. 1–3—Das Max-Planck-Institut für Meteorologie), 14–21

206. Hasselmann, K., Hasselmann, S., & Barthel, K. (1990). Europan Space Agency Contract Report use of a wave model as a validation tool for ERS-1 AMI Wave products and as an input for the ERS-1 Wind Retrieval Algorithms. *Report/Max-Planck-Institut für Meteorologie, 055.* doi:https://doi.org/10.17617/2.2556396.

207. Latif, M. (Ed.). (1991). Strategies for future climate research: A collection of papers presented at the birthday colloquium in honour of Klaus Hasselmann's 60th anniversary. Hamburg: Max-Planck-Institut für Meteorologie.

208. Hasselmann, K. (1993). Das Klimamodell: zu den Grundlagen des Klimasystems. In Ruprecht-Karls-Universität Heidelberg (Ed.), *Klima: Vorträge im Wintersemester 1992/93 [Sammelband der Vorträge des Studium Generale]* (pp. 9–29). Heidelberg: Heidelberger Verl.-Anst.

209. Hasselmann, K., Sell, W., Blum, W., & Thierbach, D. (1994). *Deutsches Klimarechenzentrum.* Hamburg: DKRZ.

210. Hasselmann, K., Bengtsson, L., Cubasch, U., Hegerl, G. C., Rodhe, H., Roeckner, E., von Storch, H., Voss, R., & Waszkewitz, J. (1995). Detection of anthropogenic climate change using a fingerprint method. *Report/Max-Planck-Institut für Meteorologie, 168.* doi:https://doi.org/10.17617/2.2534307.

211. Santer, B. D., Cubasch, U., Hasselmann, K., Brüggemann, W., Höck, H., Maier-Reimer, E., & Mikolajewicz, U. (1995). Selecting components of a greenhouse-gas fingerprint. In *Global change: Proceedings of the first Demetra meeting held at Chianciano Terme, Italy from 28 to 31 October 1991* (pp. 164–183). Luxemburg: Office for Official Publications of the European Community.

212. Hasselmann, K. (1996). Optimierte Klimaschutzstrategien. In *Klima— Umwelt—Gesellschaft: ein interdisziplinäres Seminar der Universität Hamburg am 16./17. November 1995 im Haus Rissen* (pp. 9–23). Hamburg: Universität Hamburg.

213. Heimbach, P., Hasselmann, S., Brüning, C., & Hasselmann, K. (1996). Application of wave spectral retrievals from ERS-1 wave mode data for improved wind and wave field analyses. In *Proceedings of the Second ERS Applications workshop* (pp. 303–308). Noordwijk: ESA/ESTAC.

214. Czakainski, M., & Hasselmann, K. (1997). Klimaforschung im Kreuzfeuer der Interessen: Interview mit Prof. Dr. Klaus Hasselmann. *Energiewirtschaftliche Tagesfragen: ET, 47,* 568–574.

215. Hasselmann, K. (1997). Die Launen der Medien: eine Antwort auf die Kritik an der Klimaforschung. *Die ZEIT,* (32/1997)

216. Hasselmann, S., Bennefeld, C., Hasselmann, K., Graber, H., Jackson, F. C., Hauser, D., Vachon, P. W., Walsh, E. J., & Long, R. B. (1998). Intercomparison of two-dimensional wave spectra obtained from microwave instruments, buoys and WAModel simulations during the surface wave dynamics experiment. *Report/Max-Planck-Institut für Meteorologie, 258.* doi:https://doi.org/10.17617/2.3185922.

217. Tett, S., Mitchell, J., Hasselmann, K., & Komen, G. (1998). Attribution beyond discernible—Workshop aims. In S. Tett, & et al. (Eds.), *Attribution: Beyond discernible. Euroclivar Workshop on Climate Change Detection and Attribution (Report Eucliv; 10)* (pp. 31–41)

218. Hasselmann, K. (2000). (Über)Leben auf dem Raumschiff Erde. In H. Adamski (Ed.), *Der Gott der Fakultäten* (pp. 181–202). Münster: Lit.

219. Hasselmann, K., Lehner, S., & Schulz-Stellenfleth, J. (2000). *FEME ESA Report: ERS SAR Observations of ocean waves in the marginal ice zone.*

220. Johannessen, O. M., Sandven, S., Sagen, H., Hamre, T., Haugen, V. J., Wadhams, P., Kaletzky, A., Davis, N. R., Hasselmann, K., Maier-Reimer, E., Mikolajewicz, U., Soldatov, V., Bobylev, L., Esipov, I. B., Evert, E., & Naugolnykh, K. A. (2001). Acoustic Monitoring of the Ocean Climate in the Arctic Ocean (AMOC): Final Report. *NERSC Technical Report, 198.*

221. Hasselmann, K. (2002). Der Kyoto-Prozess zum Klimaschutz: Hintergründe und Entwicklungsoptionen aus Sicht der Klimaforschung. In *Kraft-Wärme-Kopplung als Beitrag zu Klimaschutz und Energieeinsparung* (pp. 7–16). Braunschweig: Cramer.

222. Welp, M., Hasselmann, K., & Jaeger, C. C. (2003). Climate change and paths to sustainability: the role of science based stakeholder dialogues. *Reference Magazine, 19*, 8–13.

223. Kovalevsky, D., & Hasselmann, K. (2013). Out-of-equilibrium actor-based system-dynamic modeling of the economics of climate change. In *GSS Preparatory Workshop for the 3rd Open Global Systems Science Conference (2014)*. Beijing, China.

224. Arto, I., Capellán-Pérez, I., Filatova, T., González-Eguinob, M., Hasselmann, K., Kovalevsky, D. V., Markandya, A., Moghayer, S. M., & Tariku, M. B. (2014). Review of existing literature on methodologies to model non- linearity, thresholds and irreversibility in high-impact climate change events in the presence of environmental tipping points. *EU FP7 COMPLEX Report, D5.2.*

225. Filatova, T., Moghayer, S., Arto, I., Belete, G. F., Dhavala, K., Hasselmann, K., Kovalevsky, D. V., Niamir, L., Bulavskaya, T., & Voinov, A. (2014). Dynamics of climate-energy-economy systems: development of a methodological framework for an integrated system of models. *EU FP7 COMPLEX Report, D5.3.*

226. Moghayer, S., Capellán-Pérez, I., Arto, I., Markandya, A., González-Eguino, M., Flatova, T., Pinouche, F., Chahim, M., Kovalevsky, D., & Hasselmann, K. (2013). State of the art review of climate-energy-economic modeling approaches. *EU FP7 COMPLEX Report, D5.1.*

227. von Storch, H., Barkhordarian, A., Hasselmann, K., & Zorita, E. (2013). Can climate models explain the recent stagnation in global warming? http://academia.edu/4210419/Can_climate_models_explain_the_recent_stagnation_in_global_warming

228. Kovalevsky, D., Arto, I., Dhavala, K., Filatova, T., Hasselmann, K., Moghayer, S. M., Niamir, L., & Voinov, A. (2015). Report on integration of climate scenarios in the modeling system. *EU FP7 COMPLEX Report, D5.4.*

229. Arto, I., Boonman, H., Capellán-Pérez, I., Husby, T. G., Filatova, T., González-Eguinob, M., Hasselmann, K., Kovalevsky, D., Markandya, A., Moghayer, S. M., Niamir, L., Tariku, M. B., & Voinov, A. (2016). Coupled environment-ecology models. In N. Winder, & H. Liljenström (Eds.), *EU FP7 COMPLEX Final Scientific Report, Vol. 2: Non-linearities and System-Flips* (pp. 81–108). Sigtuna, Sweden: Sigtunastiftelsen.

230. Arto, I., Boonman, H., Capellán-Pérez, I., Husby, T. G., Filatova, T., González-Eguinob, M., Hasselmann, K., Kovalevsky, D., Markandya, A., Moghayer, S. M., Niamir, L., Tariku, M. B., & Voinov, A. (2016). Climate mitigation policies. In N. Winder, & H. Liljenström (Eds.), *EU FP7 COMPLEX Final Scientific Report, Vol. 2: Non-linearities and System-Flips* (pp. 67–80). Sigtuna, Sweden: Sigtunastiftelsen.

231. Arto, I., Boonman, H., Capellán-Pérez, I., Husby, T. G., Filatova, T., González-Eguinob, M., Hasselmann, K., Kovalevsky, D., Markandya, A., Moghayer, S. M., Niamir, L., Tariku, M. B., & Voinov, A. (2016). Lake system. In N. Winder, & H. Liljenström (Eds.), *EU FP7 COMPLEX Final Scientific Report, Vol. 2: Non-linearities and System-Flips* (pp. 55–66). Sigtuna, Sweden: Sigtunastiftelsen.

232. Arto, I., Capellán-Pérez, I., Filatova, T., Gonzá-lez-Eguinob, M., Hasselmann, K., Kovalevsky, D., Markandya, A., Moghayer, S. M., & Tariku, M. B. (2016). Socio-ecological system. In N. Winder, & H. Liljenström (Eds.), *EU FP7 COMPLEX Final Scientific Report, Vol. 2: Non-linearities and System-Flips* (pp. 49–54). Sigtuna, Sweden: Sigtunastiftelsen.

233. Arto, I., Capellán-Pérez, I., Filatova, T., Gonzá-lez-Eguinob, M., Hasselmann, K., Kovalevsky, D., Markandya, A., Moghayer, S. M., & Tariku, M. B. (2016). The climate system. In N. Winder, & H. Liljenström (Eds.), *EU FP7 COMPLEX Final Scientific Report, Vol. 2: Non-linearities and System-Flips* (pp. 43–48). Sigtuna, Sweden: Sigtunastiftelsen.

234. Arto, I., Capellán-Pérez, I., Filatova, T., Gonzá-lez-Eguinob, M., Hasselmann, K., Kovalevsky, D., Markandya, A., Moghayer, S. M., & Tariku, M. B. (2016). Definitions. In N. Winder, & H. Liljenström (Eds.), *EU FP7 COMPLEX Final Scientific Report, Vol. 2: Non-linearities and System-Flips* (pp. 39–42). Sigtuna, Sweden: Sigtunastiftelsen.

235. Hasselmann, K., & Kovalevsky, D. (2016). A hierarchy of out-of-equilibrium actor-based system-dynamic nonlinear economic models. In N. Winder, & H. Liljenström (Eds.), *EU FP7 COMPLEX Final Scientific Report, Vol. 2: Non-linearities and System-Flips* (pp. 109–117). Sigtuna, Sweden: Sigtunastiftelsen.

236. Kovalevsky, D., & Hasselmann, K. (2016). Actor-based system dynamics modelling of abrupt climate change scenarios. In N. Winder, & H. Liljenström (Eds.), *EU FP7 COMPLEX Final Scientific Report, Vol. 2: Non-linearities and system-flips* (pp. 118–127). Sigtuna, Sweden: Sigtunastiftelsen.

237. Kovalevskiy, D., Shchiptsova, A., Rovenskaya, E., & Hasselmann, K. (2016). Narrowing uncertainty of projections of the global economy-climate system dynamics via mutually compatible integration within multi-model ensembles. *IIASA Working Paper, WP-16–015*

238. Hasselmann, K. (2017). 12 Fragen An. 12 Questions to. *Gaia-Ecological Perspectives for Science and Society, 26,* 4–5. doi:https://doi.org/10.14512/gaia.26.1.2

239. Hewitt, R., Hasselmann, K., Kovalevsky, D. V., & Cremades, R. (2019). The transformative role of actor interactions: new approaches to the climate policy narrative. In *The 11th International Social Innovation Research Conference (ISIRC 2019)—ISIRC Abstract Booklet*. Glasgow, UK.

Books

240. Favre, A., & Hasselmann, K. (Eds.). (1978). Turbulent fluxes through the sea surface, wave dynamics, and prediction. Berlin u.a.: Springer-Verlag. doi:https://doi.org/10.1007/978-1-4612-9806-9.

241. Hunt, J. J., Bengtsson, L., Bolle, H.-J., Gudmandsen, P., Hasselmann, K., Houghton, J., & Morel, P. (Eds.). (1985). The use of satellite data in climate models: Proceedings of a conference held in Alpach, Austria, 10–12 June 1985. Noordwijk: ESA Scientific & Technical Publications Branch.

242. The SWAMP Group (1985). Ocean wave modeling. New York: Plenum Publ. Corp. doi:https://doi.org/10.1007/978-1-4757-6055-2.

243. Phillips, O. M., & Hasselmann, K. (Eds.). (1986). Wave dynamics and radio probing of the ocean surface. New York: Plenum Press. doi:https://doi.org/10.1007/978-1-4684-8980-4.

244. Komen, G., Cavaleri, L., Donelan, M., Hasselmann, K., Hasselmann, S., & Janssen, P. (Eds.). (1996). Dynamics and modelling of ocean waves. Cambridge: Cambridge Univ. Press. doi:https://doi.org/10.1017/CBO9780511628955.

245. von Storch, H., & Hasselmann, K. (2010). Seventy years of exploration in oceanography: A prolonged weekend discussion with Walter Munk. Berlin u.a.: Springer. doi:https://doi.org/10.1007/978-3-642-12087-9.

246. Jaeger, C., Hasselmann, K., Leipold, G., Mangalagiu, D., & Tàbara, J. (Eds.). (2012). Reframing the problem of climate change: From zero sum game to win–win solutions. Milton Park: Earthscan. doi:https://doi.org/10.4324/9780203154724.

5.3 Awards

January 1963: Carl Christiansen Commemorative Award.
April 1964: James B. Macelwane Award of the American Geophysical Union.
November 1970: Academic Award for Physics from the Academy of Sciences in Göttingen.
January 1971: Sverdrup Medal of the American Meteorological Union.
December 1981: Belfotop-Eurosense Award of the Remote Sensing Society.
April 1990: Robertson Memorial Lecture Award of the US National Academy of Sciences.
September 1990: Förderpreis für die Europäische Wissenschaft of the.
Körber-Stiftung, Hamburg.
June 1993: Nansen Polar Bear Award, Bergen, Norway.
December 1994: Oceanography Award sponsored by the Society for Underwater Technology, Portland, UK.
March 1996: Oceanology International Lifetime Achievement Award.
October 1996: Premio Italgas per la Ricerca e L'Innovazione 1996.

May 1997: Symons Memorial Medal of the Royal Meteorological Society.
November 1998: Umweltpreis 1998 der Deutschen Bundesstiftung Umwelt.
May 1999: Karl-Küpfmüller-Ring der Technischen Universität Darmstadt.
July 2000: Dr. honoris causa, University of East Anglia.
April 2002: Vilhelm Bjerknes Medal of the European Geophysical Society.
November 2005: Gold medal of the University of Alcala, Spain. https://www.nobelprize.org/prizes/physics/2021/summary.
2007: achievement award of the International Meeting of Statistical Climatology
2009: BBVA Frontiers of Science Award
2021: Nobel Prize in Physics

Printed in the United States
by Baker & Taylor Publisher Services